THE END OF FOOD

BOOKS BY PAUL ROBERTS

THE END OF OIL

THE END OF FOOD

The
END
of FOOD

 Paul Roberts

MARINER BOOKS
HOUGHTON MIFFLIN HARCOURT
BOSTON • NEW YORK

First Mariner Books edition 2009

Copyright © 2008 by Paul Roberts

www.hmhbooks.com

Library of Congress Cataloging-in-Publication Data

Roberts, Paul, date.
The end of food / Paul Roberts.
p. cm.
ISBN 978-0-618-60623-8
1. Food supply. 2. Nutrition. 3. Food industry and trade —
Environmental aspects. 4. Food industry and trade —
Social aspects. 5. Nutrition policy. I. Title.
HD9000.5.R578 2007
363.8 — dc22 2007036031

ISBN 978-0-547-08597-5 (pbk.)

Printed in the United States of America

DOC 10 9 8 7 6 5 4 3 2 1

For Hannah and Isaac

Contents

Prologue

IN LATE OCTOBER of 2006, seven weeks after the first reports that *E. coli* O157:H7 had been found in bags of fresh spinach, investigators working the farms in California's Salinas Valley got a break. On a cattle ranch near Highway 101, authorities shot a wild boar and found in its guts the same strain of *E. coli* that had killed three people and sickened some two hundred others. The ranch was nearly a mile from the suspect spinach farm, but the evidence of a connection was compelling; that same *E. coli* strain had also been found in cattle manure and in streams on the ranch. The fence around the spinach farm had been trampled down by wild boars, and boar tracks had been spied among the spinach plants. While refusing to label the pigs the sole culprits, Dr. Kevin Reilly, with the California Department of Health Services, told reporters that the animals presented a "real clear vehicle" by which the *E. coli* bacteria could have moved from ranch to farm. Noting that no one had been sickened by tainted spinach since late September, when the Food and Drug Administration had lifted its ban on spinach, Reilly told reporters that "all evidence points to this outbreak having concluded."[1]

As reassuring as Reilly's words may have seemed, anyone with even a passing knowledge of the modern food system knew the story was far from over. As Reilly himself cautioned, there were plenty of other plausible sources for the contamination — irrigation water, farm runoff, and fertilizer, to name a few — and thus plenty of reasons to worry that other outbreaks were simply waiting to happen. And in fact, just as the spinach crisis was winding down, another *E. coli* outbreak, this one among customers of Taco Bell, was traced back to lettuce from the Salinas Valley. Nor would these outbreaks be the last bad news for the food industry. Over the next twelve months, consumers would lurch from one food-safety debacle to the next, among them, a massive salmonella outbreak from tainted peanut butter, the unfolding Chinese import scandal, and, in October 2007, an outbreak of *E. coli* in hamburger so widespread that the recall effort bankrupted the nation's largest seller of ground beef.

Granted, food-borne illness isn't exactly a novelty in the United States today; *E. coli* outbreaks in particular have almost become an annual autumn ritual. What is new, however, is that even the food industry itself has begun to acknowledge, if only tacitly, that the problem of food safety is getting out of control. Soon after the 2006 spinach outbreaks, executives at Kroger, Safeway, Costco, and other grocery chains sent a revealing letter to U.S. produce companies demanding major improvements in safety procedures — for spinach and lettuce, of course, but also for items not yet under public scrutiny. Leafy greens might have "the most immediate priority," the retailers wrote, but "we expect that the [produce] associations share our urgency to have standardized food safety requirements and commensurate auditing criteria for additional crops . . . including: melons, tomatoes, and green onions." More telling, the letter stated that because there was no means of totally eradicating bacteria from fresh produce — no "kill step," in the parlance — an "outreach effort" would be needed to make clear to consumers that any "intervention steps" the industry undertook were intended to only "minimize risk to the extent possible."[2] In other words, there would no longer be even the implicit guarantees that fresh produce was safe from contamination.

Corporate candor of this kind is generally either accidental or purely tactical, and clearly, food retailers have been trying to push as much responsibility for food-borne illnesses and contamination onto their suppliers as they possibly can — just as retailers have always done with costs they do not wish to pay. Yet the letter also underlines the powerlessness felt by both the food companies and the agencies that regulate them; the safety problem was moving beyond the industry's capacity to manage, or even fully understand. It took Taco Bell weeks before it realized the pathogen had entered the food chain on lettuce, not green onions, and even after six months of study, federal and state investigators admitted that "no definitive determination could be made" as to how *E. coli* O157:H7 had gotten into the spinach.[3] Public confidence has fallen so low that when FDA commissioner Andrew von Eschenbach tried to argue that the melamine pet-food scandal "has demonstrated our effectiveness at detecting and containing a problem,"[4] he was roundly ridiculed by critics who noted, among other things, that of the three hundred million tons of food imported into United States each year (roughly a seventh of the nation's food supply), government inspectors are able to examine less than 2 percent of it.[5]

After decades of hearing that our food system is the best, it almost seemed as if a curtain had been drawn back and we'd been allowed a glimpse of the shadowy structures behind the food system — the huge networks of production and distribution and retailing that convey millions of tons of food to the hundreds of millions of consumers — only to find those structures broken or derailed. As successive food-safety stories began to blur into a single narrative of incompetence and uncertainty, as admissions and equivocations poured forth from investigators, policymakers, and industry executives, the impression grew stronger that these enormous food companies and powerful agencies were conceding not only that the modern food system was in trouble but also that there was little they or anyone else could do about it. For the first time, we were acknowledging the widening gap between the modern food economy and the billions of people it was ostensibly built to serve.

This is not the food narrative that most of us grew up with. Until late in the twentieth century, the modern food system was celebrated as a monument to humanity's greatest triumph. We were producing more food — more grain, more meat, more fruits and vegetables — than ever before, more cheaply than ever before, and with a degree of variety, safety, quality, and convenience that preceding generations would have found bewildering. Critics and malcontents might complain about farm chemicals or exploited migrant workers or certainly the blandness of much of our processed food. But for the rest of a grateful world, these were trivial costs to pay for a superabundance that had liberated humankind from a long night of hunger and drudgery.

Today, it's becoming ever more obvious that our triumph was never complete. The same supply chains that undergird our global supermarket, making fresh produce and meat available in every hemisphere and every season, have also created perfect opportunities for both familiar foodborne pathogens, such as *E. coli* and salmonella, as well as emerging varieties, such as avian flu, the rapidly mutating virus that may well be the basis of the next global pandemic. And for all our miraculous productivity, nearly a billion people — one in seven — remain "food insecure," to use Washington's sanitized term, and their ranks swell by about 7.5 million a year. Where hunger has been banished, populations now struggle with the less desirable consequences of the modern diet, such as obesity, heart disease, and diabetes. Worse, many of the same methods that unleashed such abundance, such as large-scale livestock operations and chemically intensive farming, have so degraded the productive capacities of our natural systems that it's not clear how we'll feed the nearly ten billion people expected by midcentury, or even how long current food production levels can be maintained.

In the meantime, the very act of eating, the basis of many of our social, family, and spiritual traditions — not to mention the one cheap pleasure that could ever rival sex — has for many devolved into an exercise in irritation, confusion, and guilt. In North America, Europe, and even emerging Asia, hundreds of millions of anxious consumers flit from one diet to the next, obsessing over bad carbs and good fats, additives and allergies, worrying over food as if we were hunter-gatherers on some primeval veldt instead of citizens in the wealthiest, most sophisticated cultures in human history. The very meaning of food is being transformed: food cultures that once treated cooking and eating as central elements in maintaining social structure and tradition are slowly being usurped by a global food culture, where cost and convenience are dominant, the social meal is obsolete, and the art of cooking is fetishized in coffee-table cookbooks and on television shows.

On nearly every level, we are reaching the end of what may one day be called the "golden age" of food, a brief, near-miraculous period during which the things we ate seemed to grow only more plentiful, more secure, more nutritious, and simply *better* with each passing year. Thus, even as we struggle to understand why the safety of our food is becoming so much harder to assure, it's clear that safety is only one of a cluster of concerns and that we need to be asking a much broader set of questions: What is happening to our food? How could our immensely successful food system have become so overextended? How close are we to the point of breakdown? And what practical solutions might help restore balance to the system before that point is reached?

When consumers in rich Western nations turn their attention to the failings in the food system, blame is generally divided between food companies, which, we're certain, put profits before all else, and government regulators, who are outmaneuvered or corrupted by that profit-driven power. Yet although this conventional economic narrative does explain some of our food problems, it misses and often obscures those problems' true origins. Today's food crisis is fundamentally economic, but not in the familiar sense that food companies operate for financial gain or that consumers shop for the best price. Rather, the crisis is economic in the sense that our food system can only truly be understood as an *economic* system, one that, like all economic systems, has winners and losers, suffers periodic and occasionally profound instability, and is plagued by the same inherent and irreducible gap between what we demand and what is actually supplied.

This is not, on its face, a radical proposition. Food was our first form of wealth, and its production was our first economic enterprise, generating not only most of our employment and prosperity but many of the tools with which the larger economy would eventually be built. Agriculture gave rise to rudimentary economic organization and specialization, to accounting and management, to trade and speculation, and, ultimately, to an explicit economic paradigm — capitalism — which was probably invented on sixteenth-century sugar plantations. And later, when surging population growth in eighteenth-century Europe exceeded existing production methods and threatened famine, the things that saved us — the move to labor-saving technologies, the shift to larger production scales, and the creation of a global system of food trade — were precisely the sparks that ignited the subsequent Industrial Revolution.

And this relationship worked both ways. Even as food production influenced the way we made everything else (Henry Ford invented his assembly line after watching a line of butchers methodically disassembling cattle in a meatpacking plant), the way we made everything else began to influence the making of food. Farms came to be run like integrated factories, turning "inputs" of seed, feed, and chemicals into steady "outputs" of grain and meat. Individual shopkeepers such as the butcher, the baker, and the greengrocer were consolidated into huge, efficient, one-stop supermarkets, which were then merged into sprawling retail chains, which in turn used their enormous volume and market share to squeeze discounts from food companies, just as other big-box retailers were squeezing manufacturers of clothing, cosmetics, and other consumer goods. Even the process of cooking and eating took on a businesslike efficiency, with new time-saving appliances in our kitchens and a plethora of premade packaged foods.

At every level, in fact, the modern food sector has become a miniature version of the industrial economy it once inspired. Raw materials such as No. 2 yellow corn or BSCB (boneless, skinless chicken breasts) are now handled like any other commodity: produced wherever costs are lowest, shipped to wherever demand is highest, and managed via the same contracts, futures, and other instruments used for timber, or tin, or iron ore. Food-processing companies employ the same technologies and business models of other high-volume manufacturers: The continuous advances in technology and the ever larger scales of production that drive down costs in cars and home electronics are now also standard in the food business, as is the relentless product innovation one finds in clothing or cosmetics. Each year, thousands of new food items, from prepackaged salad kits to microwavable bacon, arrive on grocery shelves or restaurant buffets — often via the same distribution channels that bring DVDs, disposable razors, deodorant, cosmetics, toys, and other consumer products. To an important degree, the success of the modern food sector has been its ability to make food behave like any other consumer product.

But this is the paradox of the food economy and, in my view, the source of most of its current problems: for all that the food system has evolved like other economic sectors, food itself is fundamentally *not* an economic phenomenon. Food production may follow general economic principles of supply and demand; it may indeed create employment, earn trade revenues, and generate profits, sometimes considerable profits; but the underlying product — the thing we eat — has never quite conformed to rigors of the modern industrial model.

Physically, food is so unsuited to mass production that we've had to re-engineer our plants and livestock to make them more readily harvested and processed (and even these updated materials remain so fragile they must be amended with preservatives, flavorings, and other additives). Our farming and manufacturing methods incur such enormous "external" costs — from farm-chemical runoff to the inequities of cheap labor to a choking surplus of calories — that the longevity of the system is now in serious doubt. Even the shift in cooking from the home to the factory, though it has left us free to engage in other pursuits, has also left us with far less knowledge of, and control over, what we eat.

This isn't to argue that markets don't work, or that ever greater "productivity" — that is, making more for less, the cornerstone of modern economic progress — hasn't been essential in saving the world from starvation and drudgery. It is, however, to suggest that in the case of food, there are incompatibilities between system and product, and, more precisely, to propose that the attributes of food that our economic system tends to value and to encourage — mass producibility, for example, or cheapness, or uniformity, or heavy processing — aren't necessarily the attributes that work best for the people eating the food, or the culture in which that food is consumed, or the environment in which it is produced. It is to suggest that much of what is standard practice in business today — the drive toward lower costs, higher production scales, and larger markets — brings perverse consequences in the food sector. It is to argue, in short, that there is a widening discrepancy between what is demanded and what is actually supplied, and to propose that it is within this gap, between food as an economic proposition and food as a biological phenomenon, that today's biggest challenges arise.

Ironically, the problems with the modern food system begin with its very success. For all the many benefits of high-volume, low-cost production, the capacity to generate enormous streams of food at ever lower prices has also effectively locked producers into a vicious cycle: the more food they produce, the more food they must continue to produce. To grow wheat at a competitive price, for example, a farmer must continually lower his costs of production. Generally, he does this by adopting better technology, like a new combine, which lets him harvest each acre more quickly. The new machine will be expensive (a big harvest combine runs about $400,000), but because it lets the farmer produce more wheat, he can spread that cost over more bushels. Of course, our farmer isn't alone; all his neighbors, and all those farmers in nearby counties, and farmers everywhere else in the industrialized world are also trying to spread their costs over more bushels, which has the effect of increasing the total supply of wheat. And if supply rises faster than demand, as has been the reality for much of the last half century, prices will fall, which means farmers must invest in more technology to produce still more bushels, and so on.

Fifty years ago, a University of Minnesota economist named Willard Cochrane identified this "technology treadmill" as an emerging problem for farmers. But in fact, the treadmill applies to anyone in the supply chain that now runs from the maker of fertilizers to the seller of groceries. For these players, success is now determined increasingly by their capacity to continually cut their "per unit" costs, which typically means making or moving ever greater numbers of units. This volume imperative, in turn, explains a great many things about the modern food landscape, from the aggressive way the United States and other surplus producers push those surpluses into foreign markets to the ubiquity of value-size meals to the relentless expansion of restaurants, vending machines, and other food opportunities. It is probably not entirely coincidental that the 1980s, when the high-volume, low-cost food model emerged in its mature form, also marked the point when obesity rates tilted sharply upward. Obesity is, in fact, one of the more visible illustrations of food's poor fit as an economic phenomenon; although consumers can consume as many DVDs or sneakers as their credit card companies will allow, the same cannot be said for food, no matter how cheap it becomes.

But bulging waistlines are only one unintended effect of our low-cost, high-volume model. We should hardly be surprised that the same relentless focus on cost that has led to massive recalls of imported Chinese toys would have similar consequences for imported Chinese food. Even American food companies are finding it harder and harder to operate robust food-safety programs under a business model fixated on cost and volume. And when our safeguards fail and pathogens and other contaminants do enter the supply chain (whose very size and "just in time" efficiency all but ensure that the outbreak spreads rapidly), our ability to counterattack is fading. Because many livestock farmers still routinely dose their animals with antibiotics, pathogens such as salmonella are becoming more resistant to antibiotics, thus harder to kill. Or as Michael Blum, a medical officer in the U.S. Food and Drug Administration, put it recently, "We've come to a point, for certain infections, where we don't have antibiotic agents available."

Most egregiously, for all its abundance, the modern food system has not come close to eliminating hunger. Despite the extraordinary breakthroughs in crop science that saved Asia from outright starvation in the 1960s, efforts to make this so-called Green Revolution permanent have failed; relief efforts are all but continuous in sub-Saharan Africa, as well as parts of Asia and South America, with impacts that defy the imagination. Beyond the thirty-six million hunger-related deaths each year,[6] chronic malnutrition freezes entire populations in a medieval nightmare of constant exhaustion, stunted bodies and minds, and ravaged potential. By one estimate, the lack of a single micronutrient, vitamin A, has left more than three million children under the age of five in sub-Saharan Africa permanently blind.[7]

We can hardly blame the modern food industry for all the natural (and political) disasters that have left places like sub-Saharan Africa in such wretched poverty. But Africa's problems do expose yet another weakness in the modern food model. A food economy based on large-scale, technology-intensive production methods and rising consumer demand doesn't work in countries where consumers can barely afford raw commodities. More and more, food companies that depend on the sale of high-value packaged foods are bypassing countries like Ethiopia and Bangladesh altogether in favor of wealthy markets in Europe and North America, or booming regions like Asia, where wealthy consumers aspire to steak, ice cream, energy drinks, and other elements of the high-value, high-calorie Western-style diet. Thus, despite the fact that food costs are half what they were fifty years ago and despite a global food supply that now exceeds per capita caloric needs by about 20 percent, the world has nearly as many malnourished citizens as it does overnourished ones, a perverse symmetry that is emblematic of the system's larger failings.

Hunger, food-borne pathogens, and an epidemic of nutrition-related health problems may be the most visible and discussable signs of that failure, but there are others that run as deeply, if not quite so quantifiably. Like all forms of industrialism, the rise of the modern food system has been accompanied by extraordinary, and not altogether beneficial, social change. The great transformation of farming and food processing that began in the United States after World War II brought greater productivity and lower prices, but it also nearly destroyed the rural culture upon which this country still claims to base many of its values. Thirty years later, the rise of big-box grocery retailers like Wal-Mart pushed down prices further still while crushing a generation of small producers and neighborhood grocers, and leaving the food sector in the hands of a few very large, very powerful global players.

Similarly, if our liberation from the time and expense of home processing and preparation freed middle-class consumers, it also upended a critical piece of our social structure. Many of the things we claim to cherish — family relationships, cultural identity, ethnic diversity — were all intimately linked to the making and eating of food and are now changing as we outsource more and more of our food preparation to restaurants and industrial kitchens. Not only do we cook less than we used to, but more of us eat alone — at our desks, in our cars, standing at our kitchen counters. In America, the average family shares a meal fewer than five times a week, and similar trends are seen wherever incomes are rising and free time is declining — Saudi Arabia, Mexico, even France, a bastion of traditional food culture, where nearly a quarter of all meals are now taken outside the home. This is not a plea for us all to become producers again; it is, however, a suggestion that there will be something seriously wrong when no one is a producer, when no one is a cook, and when the closest that anyone gets to making a meal is at a restaurant salad bar.

It can certainly be argued that problems like overabundance, a confusing food culture, and even declining food safety are acceptable tradeoffs for the many advantages of the modern food system, and that in a world that is becoming richer, smaller, and more crowded, the high-volume, low-cost model probably represents the best we can hope for. But even such a pragmatic view no longer holds. For all that the current system is built to overproduce, it's now clear that this great food machine is already encountering limits to what it can generate.

Over the next forty years, demand for food will rise precipitously — both because global population will continue climbing and because the developing world, where most of that growth will occur, will continue to catch up with Western dietary patterns, particularly the love of meat. And while eating more meat can bring health improvements in many poor nations, meatier diets also geometrically increase overall food demand because meat is one of the least efficient ways to obtain calories. On average, it takes eight pounds of grain to make a pound of meat, which is why so much land must now be devoted to feed crops and why as largely vegetarian populations in South Asia and Africa begin to approach the dietary practices of Europe and North America, demand for feed crops will more than double — a worrying event, given that most of the world's readily available arable land is already being farmed, and much of what remains is forested or has soils poorly suited to intensive agriculture. Factor in the new demand for grain from the emerging biofuels industry — which now claims nearly a third of the entire U.S. corn supply — and suddenly, the massive surpluses that choked global markets for decades have all but vanished.

Granted, humanity has been here before. When land grew scarce in industrializing Europe, and a dour eighteenth-century economist named Thomas Malthus declared that famine was inevitable, farmers learned to grow more grain per acre with a steady progression of labor-saving machines, better fertilizers, and higher-yielding crops. But this technology-driven progress is now struggling. In much of the developing world, the famously high-yield growth of the Green Revolution is tapering off and in some cases even declining — partly for lack of fertilizers and other chemicals, but partly because the overuse of those same chemicals has exhausted the soil's productive capacity. Even where yields can be maintained, as in the American Midwest, the price has been the contamination of water supplies and rivers by potent chemicals that can turn bays and offshore waters into massive dead zones devoid of fish or much else.

Researchers at the big input companies, such as Monsanto and Dow, insist that a new generation of farm technologies based on genetic manipulation will usher in breakthroughs that can meet these new challenges. But even if these claims bear out — and there is considerable skepticism about both the safety and the efficacy of such technologies — tomorrow's high-tech farmers will face a very different world. Not only will they be trying to feed more people, but they will be doing so without the benefit of three critical advantages their predecessors took for granted — cheap energy, abundant water, and a stable climate.

Even by conservative estimates, the combination of rising temperatures and shifting patterns in rainfall and storm frequency will push down total global food output, and this while demand is rising. By the year 2070, Africa, a continent already on the brink of a food-system collapse, may be entirely unable to produce certain crops, such as wheat.

And well before then, the food economy will have crossed another threshold that is just as critical: petroleum, perhaps the single most important input in modern food production (it serves both as a fuel for tractors and food transportation and as the chemical base for fertilizers and pesticides), is gradually becoming so scarce and expensive that many of the assumptions underlying a global industrial food system are now in question. Nearly everything about the way our food system has developed over the last half century — from our ability to manufacture fertility to our capacity to move food to import-dependent nations — could not have occurred without cheap energy, and the degree to which that system can continue in a world of high energy prices is a frightening unknown.

Perhaps more worrying, farmers are rapidly running out of water.

Agriculture consumes more water than any other sector, and rising demand is depleting water resources in nearly every part of the world, from North Africa and China, where water tables are falling so fast farmers must drill their wells thousands of feet deep, to the United States, where the massive Ogallala Aquifer, the world's largest underground lake and the key resource for an area that runs from South Dakota to Texas and from Colorado to Missouri, may run dry within three decades.

Given that much of the huge gains in food output were possible only because farmers found ways to use more water, declining water supplies raise chilling questions about where and how farmers will produce the food the world will need twenty or even ten years from now. Already, water depletion is leading countries to curb their own farm production and instead to import their water indirectly, in the form of grain purchased from the United States, Europe, Brazil, Argentina, and other big grain exporters. Such solutions are, of course, only temporary; eventually, these big exporters will deplete their own water supplies. Already, the surging demand for water-as-grain is fostering a subtle competition among the big importers that is strikingly similar to the current rivalry among big oil importers — and which might someday lead to similar international tensions. In the meantime, this shifting balance of world food supplies is producing a new axis of global food power between big importers, most notably China and India, and ambitious new exporters, like Brazil — while sidelining traditional food powerhouses, such as the United States, which is slowly losing influence over global food markets much as it lost control over oil markets thirty years ago.

The food economy is hardly the only system to have encountered its limits. All sectors — from energy to housing to automotive — are now coming to grips with various constraints and external costs, and many of the risks the food system now faces, such as declining supplies of energy and the problems of cheap labor, are simply extensions of risks now at play within the larger economic system. But for several reasons, the crisis in the food economy is likely to be more problematic.

First, because food production now occurs in an entirely global context, with different commodities produced wherever costs are lowest and then simply moved to areas of demand, the system is extraordinarily vulnerable. Any disruption in the means of transportation (say, because of rising energy costs) or in the ability of exporters to export (perhaps due to massive climate-related crop failures) poses the very real risk of isolating entire regions that now specialize in only single crops or, worse, produce very little of their own food and depend mainly on imports.

Second, the food economy is hugely resistant to sudden change. Not only is this system carried along by enormous economic momentum — nearly every piece of the modern supply chain, from the individual farm to the largest retailer, is designed for, and dependent on, continuous increases in output — but it is extraordinarily adept at either deflecting its critics or absorbing them. The organic movement, which emerged in the 1940s as a direct critique of large-scale industrial food production, had by the 1990s been co-opted by that same system, and many organic products are now produced with the same high-volume, low-cost methods and sold by the same big-box retailers. And while some consumers are waking up to the dangers of the current food system and are demanding healthier, more sustainable products, this revolt is hugely constrained, in that it occurs within the relatively narrow boundaries of that industry's fiscal, structural, and technical limits. Thus, although food companies will often replace "bad" products with "good" products to satisfy the latest consumer anxiety, these "good" products are almost always created within existing systems of production, processing, distribution, marketing, and financing, and still represent a carefully calculated compromise between consumer desires and the manufacturers' strategic imperatives and financial and technical capabilities. This is a system designed to perpetuate itself; it is not a system from which one should expect radical change.

For these reasons, many critics have argued that change must be driven from outside the system, and that the objectives must be to replace some or all of that system with something entirely new. But this is hardly a scenario that will be supported by companies and other players that have already made substantial investments in the large-scale, low-cost model and that exert considerable influence on the political process. Even reform-minded policymakers must battle a system of government regulations and programs so large, so entrenched, and so interwoven with the food industry itself that it has become a veritable force of nature. Indeed, despite widespread agreement that America's use of farm subsidies is helping drive the food-production system into the ground, no one has been able to muster the votes to change it. Despite growing evidence that the global food trade, for all its benefits, is increasing the risk of food-safety problems, Washington and other governments continue to press for more access to foreign markets. And even though it's increasingly clear that our food is cheap because we ignore many of the real costs of production, mainstream consumers show little inclination to start paying more for their food.

More problematic, there is no alternative food system waiting in the wings. Although researchers, farmers, and others have come up with thousands of new ideas for producing food differently, the public research money that is used to help bring those ideas to market has been rapidly dwindling, and what funds remain are channeled into conventional agricultural practices. And even if money were suddenly made available for a truly alternative food system, there is no coherent vision of what that system should look like. For some, the answer lies in simply making current methods more efficient and productive, much as our predecessors did in past crises. But an equally vocal camp is arguing that these past technological successes were simply short-term gains based on unsustainable practices, and what is needed is a fundamental reimagining of the way we make and think about food. Yet even advocates of a total rebuild cannot agree on the new design. Should we all be eating organic? Is it even possible to grow enough food without synthetic pesticides? Should all food be produced locally? Or on small farms? Is there room for big-scale agribusiness, transgenic food, or other controversial technologies? Will we have to give up meat?

That we are now asking such questions is undeniably an encouraging sign. As a culture, we probably haven't paid this much attention to the food system and its flaws since the turn of the last century, when muckraking books like *The Jungle* revealed an emerging food industry that sought to cut costs by selling rotten meat, diseased milk, and canned foods adulterated with toxic chemicals. Food is again in the news — because of the disasters and accidents, of course, but also because of some encouraging developments as well. Alternative food systems are beginning to emerge, or re-emerge, depending on your perspective. Small farms are the fastest-growing sector in U.S. agriculture, and the farmers' market is now so ubiquitous it is almost cliché. And although public funding for alternative agriculture is waning, some of the slack is being picked up by foundations, nongovernmental organizations, and even some food companies[8] that are supporting research and new models of food production. Community groups and nonprofits are bringing farming to urban environments, real food to school cafeterias, and cooking skills to the classroom. Slowly, parts of society have once again begun to look at food as if it mattered.

More fundamentally, there is a growing awareness among some consumers, policymakers, and even industry executives that our food and food systems are flawed and that these flaws show every sign of devolving into serious and destabilizing crises. The recent outbreaks of *E. coli* and salmonella and the widening Chinese import scandal have forced the food industry and the government to acknowledge real problems with the status quo — and to begin to ask how these problems might be addressed. Even the surge in grain prices caused by the ethanol boom (itself spawned by high oil prices) has demonstrated convincingly just how tight global food supplies have become — and offered a glimpse of what could happen to food prices when demand in Asia *really* kicks in.

Will such awareness be sufficient? Three decades ago, the Organization of Petroleum Exporting Countries (OPEC) pushed up the price of oil and inadvertently forced the United States and other big importers into an era of conservation, efficiency, and massive innovation. Over the next several decades, growing pressures on the food system will create a similar impetus for change and will induce a new generation of innovators to look more aggressively at alternative food systems. But these food pioneers will be operating with far less certainty than did their energy predecessors. Food, unlike nearly all other commodities, cannot be replaced. Markets are adept at substituting a scarce product with another, more readily available one, but there is no substitute for food. Once we deplete the resources on which food production depends — primarily soil, water, and the natural stock of plants and animals, all of which are threatened by industrial agriculture — we have no synthetic version of food. Which means that once yields begin to fall, as they have in regions like Asia, the decline in output will be extraordinarily difficult to reverse. Thus, even as we are daunted by the complexity of remaking this system, we grow more aware of the risks of failure.

The food economy and its many problems are hardly virgin territory, and in recent years several writers have done much to illuminate this large and changing landscape. Eric Schlosser, the author of *Fast Food Nation,* introduced a generation of consumers to the often-unappetizing realities underlying the modern cuisine (he remains Public Enemy Number One for many in the meat industry), while Marion Nestle's *Food Politics* detailed the food industry's growing political influence. In *The Omnivore's Dilemma*, Michael Pollan deftly explored the hidden costs of industrial-strength meat (his passages on the links between *E. coli* and large-scale beef production in particular should be required reading for all carnivores) as well as the challenges of keeping alternatives, such as organic agriculture, truly alternative. And since the 1990s, in books such as *Plan B 2.0* and *Who Will Feed China*, Lester Brown has called attention to the issues of sustainability and the looming mismatch between supply and demand.

What is still needed, in my view, is a way to consider such critical questions and concerns in a larger, more global context. Thus, in *The End of Food*, I've worked to extend the current and often familiar narratives about food into a broader exploration of the food economy as a whole, in the hopes that readers might see how many of the seemingly disparate problems we're now confronting — obesity, the prevalence of food-borne disease, the persistence of hunger, even the transformation of Third World wilderness into enormous, export-oriented farms — are in fact related to, and even driven by, the same basic economic mechanisms that gave rise to the modern food system in the first place.

To provide this comprehensive perspective, I've approached the food economy from multiple angles. I've studied the different links in the global food chain — the farmers and livestock operators, of course, but also input companies such as Monsanto, commodity traders like Cargill, processors such as Nestlé and Kraft, and retailers and food-service companies like Wal-Mart and McDonald's. Similarly, I've explored the various national actors that control, or are controlled by, the global food economy. I studied food systems in Europe, birthplace of the modern food industry, which is now struggling to balance older food customs with the new need to industrialize. I crisscrossed the United States, whose massive production capacities have given it an OPEC-like influence over world food markets, and also China, whose rising population and enormous demand are usurping some of that power while reshaping the entire food system. Last, I traveled to Africa, most of whose inhabitants are not participants in, or in many cases even aware of, the modern food economy.

Because the modern food system is the result of a long and complex evolution, I've also addressed the significant periods in that history — the "meat revolution" that made us human; the agricultural revolutions that made us farmers; the postwar confluence of science and business that gave us superabundance; and, finally, beginning in the early 1980s, the great unraveling that would culminate in the current crises. I've interviewed the scientists and technicians, the farmers and CEOs, that control the existing food economy, the advocates and visionaries who are trying to build the next one, and the relief workers and others who are trying to bridge the gap. But I've also spent time with the individual players — the family farmers, the shopkeepers, and the consumers, many of whom are simply trying to make sense of a transformation they cannot control and often do not even fully understand. Some have spoken willingly and openly; others, only with the protection of anonymity.

This book is organized somewhat like the economic system it seeks to reveal. In the first of three parts, I explore the origins and operations of the food system. Chapter 1 gives the backstory of the food economy, from our carnivorous origins to the first agricultural revolution, and from our near extinction in the eighteenth century to our rescue by the rise of industrialized food production. Chapter 2 explores the industrializing of food; it draws on the operations of the world's largest food company, Nestlé, to show how raw materials are disassembled and reassembled in ways that give consumers greater control of their time but that have given the food industry ever greater control over what we eat. In Chapter 3, we look at the retail revolution, when large grocery companies leveraged their new size and market share to seize control of the supply chain and launch a low-cost, high-volume model that would change not just food production but food itself. Chapter 4 examines the fallout from these transformations — from the declining nutritional quality of processed food to the emergence of obesity, diabetes, and other health consequences of a food system geared toward overproduction.

Part 2 looks broadly at the impacts of food production. Chapter 5 outlines the rise of the global food trade, which, like industrialism, has brought incredible benefits but new dangers as well: the easier transmission of diseases, a vulnerability to rising energy costs, and a growing competition among a handful of food superpowers — the United States, Europe, Brazil, and China — for access to agricultural production. Chapter 6 lays out the main paradox of plentitude — persistent global hunger — and examines how, in an age of superabundance, one billion people, many of them in sub-Saharan Africa, can be excluded from the global food economy. Chapter 7 looks into the shifting battle against food-borne disease, particularly the way industrialization, even as it defeated many of humanity's traditional bacterial enemies, has inadvertently created the conditions for a new generation of pandemics. Chapter 8 wraps up the section with a survey of the many factors — from soil contamination and land scarcity to declining supplies of energy and water — that will require us to completely reengineer the current system of food production — and perhaps to reconsider a diet centered so heavily on meat.

The third and final part of the book explores the challenges of fixing our food system. Chapter 9 compares the two primary contenders in the battle over the next food system — transgenic and organic food — and finds them both wanting. Chapter 10 extends this critique into other food alternatives, showing how each offers important potential but also faces resistance from the economic, political, and cultural status quo. The final chapter lays out scenarios, both hopeful and less so, of how the needed transformation could take place.

The modern food economy is far too vast, and the debate over its problems and future far too contentious, to be condensed into a single book, and there are many elements of this crucial topic that I simply could not include. Nor do I expect that all readers will agree with the priority I've assigned the various food crises, or with the space I've devoted to certain proposals; few fields are evolving as rapidly as that of food technology and agronomy — a reality that is both maddening and encouraging. Rather, my aim has been to map out broad contours of the modern food economy in a way that lets those most affected by that economy — consumers — understand how it works, why it is failing, and, above all, what options exist to make real and lasting change.

1

Starving for Progress

I N THE LATE 1940s, anglers who fished the waters of the Hudson River near Orangetown, New York, noticed something odd about the trout they were reeling in: every year, the fish were getting larger. Fishermen rarely complain about big fish, but because the creatures in question were being hooked downstream from Lederle Laboratories, a pharmaceutical company, some may have wondered whether the phenomenon was entirely natural. Eventually, the fish stories reached the upper management at Lederle, where they piqued the curiosity of Thomas Jukes, a brilliant biophysicist and expert in the new field of vitamin nutrition, who decided to investigate. Jukes knew that Lederle discharged its factory wastes in great piles near the river. He also knew that one such waste product was a residual mash left over from the fermentation process that Lederle used to make its hot new antibiotic tetracycline. Jukes surmised that the mash was leaching into the river and being eaten by the fish, and that something in the mash — Jukes dubbed it a "new growth factor" — was making them larger.

Initially, Jukes suspected the factor might be vitamin B_{12}, a newly identified nutrient that was known to boost growth in laboratory animals. The vitamin was a byproduct of fermentation, so it was very likely to be in the mash. But when Jukes and a colleague, Robert Stokstad, tested the mash, they found something quite unexpected, and even world-changing: although B_{12} was indeed present, the new growth factor wasn't that vitamin but the tetracycline itself. When mixed with cornmeal and fed to baby chickens, even tiny doses of the amber-colored antibiotic boosted growth rates by an unprecedented 25 percent.[1]

Jukes wasn't sure why this was happening. He speculated (correctly, as it turned out) that the tetracycline was treating the intestinal infections that are routine in closely confined farm animals, and that calories that normally would have been consumed by the chicks' immune system were going instead to make bigger muscles and bones. In any case, the phenomenon wasn't limited to baby chickens. Other researchers soon confirmed that low, subtherapeutic doses of tetracycline increased growth in turkeys, calves, and pigs by as much as 50 percent, and later studies showed that antibiotics made cows give more milk and caused pigs to have more litters, more piglets per litter, and piglets with larger birth weights. When the discovery was announced to the world in 1950, Jukes's new growth factor was the closest thing anyone had ever seen to free meat and a welcome development amid rising concerns over food supplies in war-torn Europe and burgeoning Asia. As the *New York Times* put it, tetracycline's "hitherto unsuspected nutritional powers" would have "enormous long-range significance for the survival of the human race in a world of dwindling resources and expanding populations."[2]

Jukes's discovery would indeed have enormous long-range significance, although not quite in the ways the *Times* envisioned. By the middle of the twentieth century, the global food system was in the throes of a massive transformation. In even the poorest of nations, thousand-year-old methods of farming and processing were being replaced by a new industrial model of production that could generate far more calories than had been possible even a generation earlier — and which seemed poised to end the cycle of boom and bust that had plagued humanity for eons. But the great revolution was incomplete. For all our great success in industrializing grains and other plants, the more complex biology of our cattle, hogs, chickens, and other livestock defied the mandates of mass production. By the early twentieth century, meat — the food that humans were built for and certainly the food we crave — was still so scarce that populations in Asia, Europe, and even parts of the United States suffered physical and mental stunting, and by the end of World War II, experts were predicting global famine.

Then, abruptly, the story changed. In the aftermath of the war, a string of discoveries by researchers like Thomas Jukes in the new fields of nutrition, microbiology, and genetics rendered it possible to make meat almost as effortlessly as we produced corn or canned goods. We learned to breed animals for greater size and more rapid maturation. We moved our animals from pastures and barnyards and into far more efficient sheds and feed yards. And we boosted their growth with vitamins and amino acids, hormones and antibiotics (it would be years before anyone thought to ask what else these additives might do). This livestock revolution, as it came to be known, unleashed a surge in meat production so powerful that it transformed the entire food sector and, for a brief time, allowed many of us to return to the period of dietary history that had largely defined us as a species — and where the story of the modern food economy properly begins.

By most accounts, that narrative started about three million years ago, with *Australopithecus*, a diminutive ancestor who lived in the prehistoric African forest and ate mainly what could be found there — fruits, leaves, larvae, and bugs. *Australopithecus* surely ate some meat (probably scavenged from carcasses, as he was too small to do much hunting), but most of his calories came from plants, and this herbivorous strategy was reflected in every element of *Australopithecus*'s being. His brain and sensory organs were likely optimized to home in on the colors and shapes of edible (and poisonous) plants. His large teeth, powerful jaws, and oversize gut were all adapted to coarse, fibrous plant matter, which is hard to chew and even harder to digest. Even his small size — he stood barely four feet tall and weighed forty pounds — was ideal for harvesting fruit among the branches.

So perfectly did *Australopithecus* match his herbaceous diet that our story might well have ended there. Instead, between 3 million and 2.4 million years ago, *Australopithecus* got a shove: the climate began to cool and dry out, and the primeval jungle fragmented into a mosaic of forest and grasslands, which forced our ancestors out of the trees and into a radically new food strategy. In this more open environment, early humans would have found far less in the way of fruits and vegetables but far more in the way of animals, some of which ate our ancestors, and some of which our ancestors began to eat. This still wasn't really hunting, but scavenging carcasses left by other predators[3] — yet now with an important difference: our ancestors were using stone tools to crack open the leg bones or skulls, which other predators typically left intact, to get at the calorie-rich, highly nutritious marrow and brains.[4] Gradually, their feeding strategies improved. By around 500,000 years ago, the larger, more upright *Homo erectus* was using crude weapons to hunt rodents, reptiles, even small deer. *Erectus* was still an omnivore and ate wild fruit, tubers, eggs, bugs, and anything else he could find.[5] But animal food — muscle, fat, and the soft tissues like brains and organs — now made up as much as 65 percent of his total calories, almost the dietary mirror image of *Australopithecus*.

On one level, this shift away from plants and toward animal food was simple adaptation. All creatures choose feeding strategies that yield the most calories for the least effort (anthropologists call this optimal foraging behavior), and with fewer plant calories available, our ancestors naturally turned to animal foods as the simplest way to replace those calories. But what is significant is this: even if the move toward meat began out of necessity, the consequences went far beyond replacing lost calories. In the economics of digestion, animal foods give a far greater caloric return on investment than plants do. It might take more calories to chase down a frisky antelope on the veldt than to pluck fruit in the forest. But for that extra investment, *Homo erectus* earned more calories — far more. Fat and muscle are more calorie dense than plants are and thus offer more energy per mouthful. Animal foods are also easier to digest, so their calories can be extracted faster. In all, meat provided more calories, and thus more energy, that could then be used for hunting, fighting, territorial defense, and certainly mating. Meat was also a more reliable food source; by shifting to meat, prehistoric man could migrate from Africa to Europe, where colder winters and lack of year-round edible vegetation would have made an herbivorous diet impossible.[6]

But meat's real significance to human evolution was probably not the quantity of calories it contained but the *quality* of these new calories. Because animal and human tissues have the same sixteen amino acids (whereas most plant-based proteins contain just eight), animal converts readily into human: meat is the ideal building block for meat. That's why bodybuilders eat a lot of meat; it also helps explain why, as our ancestors ate more animal foods, their bodies grew larger. Whereas *Australopithecus* stood four feet tall, *Homo erectus* was a strapping six feet in height, and much stronger, which made him better at eluding predators and hunting.*
As important, *Homo erectus*'s skull was a third larger than that of *Austra-*

---

* The point isn't that meat made us big but that by eating more meat, our ancestors could then adapt more readily to an environment where greater size and strength were advantageous. But once attained, our new stature had to be maintained, which is one reason our ancestors sought out larger prey animals; not only did these big beasts supply a lot of calories, they also supplied more fat per pound than did smaller animals.

lopithecus, and the brain inside vastly more developed — an adaptation known as encephalization that was also related to the meatier diet. Just as muscle grows best on a diet of meat, brains thrive on the fatty acids, and especially on two long-chain fatty acids, the omega-3 fat *docosahexaenoic acid* (DHA) and the omega-6 fat *arachidonic acid* (AA), which are abundant in animal fats and soft tissues.[7] Plants have omega-3 and omega-6 fatty acids, too, but these are shorter forms and can't provide the same nutritional benefits.

Fatty acids were just the start. The brain is what's known as expensive tissue — not only does it need lots of DHA to grow so large, it also needs lots of calories to create all the chemical neurotransmitters upon which mental activity depends. The bigger the brain, the more calories it requires, which is why across the zoological spectrum, bigger brains tend to be found with bigger bodies. A sperm whale, for example, can support a twenty-pound brain mainly because it also has a massive stomach and heart. But humans defied this brain-body pattern. In the millions of years between *Australopithecus* and *Homo erectus,* brain size nearly tripled, yet body size barely doubled. Somehow, the human body was fueling a very large brain with a relatively small set of body organs. How? Again, the likely answer was meat. Recall that meat is more calorie dense and easier to digest than plants. According to paleoanthropologist Leslie Aiello, coauthor of the expensive-tissue theory, as our ancestors ate more meat and fewer plants, they no longer needed the large primate gut to digest all the plant matter. Over time, the gut shrank to about 60 percent of the size of other primates' — a critical development, as digestive systems themselves consume lots of energy and having a smaller gut meant more available calories for larger brains. (In a similar development, because we weren't required to grind up so much plant matter, our jaws and teeth became smaller.) This is not a claim for dietary determinism: meat didn't "make" monkeys human. Many factors interacting in complex ways spurred the changes in our ancestors' physiology that ultimately produced modern humans. But it's also clear that without more animal foods, their bodies and brains couldn't have gotten larger. And without those bigger bodies and brains, they couldn't have become the intelligent, tool-using, highly effective hunters who were able to spread so quickly from Africa to the Middle East, Asia, and finally Europe. It's probably not entirely coincidental that the several offshoots of *Australopithecus* that remained herbivorous became extinct.

In any case, by around 180,000 years ago, as the first of the four ice ages began, animal foods dominated and defined the human food strategy. Neanderthals and, later, the Cro-Magnons, the first anatomically modern humans, were primarily hunters. Each had its own strategies, but both relied heavily on the mastodon, bison, woolly rhinoceros, and other arctic megafauna that had been driven southward into human territory by expanding glaciers. To prehistoric hunters, these big animals were walking meat markets — risky to hunt, but offering a huge payoff. By some estimates, Cro-Magnon hunters were earning as much as fifteen thousand calories per hour — far more than their predecessors. In fact, although Cro-Magnon foraged for plants and tubers, eggs, insects, fruits, and honey, two-thirds of their calories came from animal foods, making their diet, as Mike Richards at Oxford University has shown, nearly identical to that of bears, wolves, and other "top-level carnivores."

By the start of the last Ice Age, about ninety thousand years ago, big-game hunting had become the brutally efficient practice that was celebrated in cave paintings — and that elevated humans to a kind of dietary elitism. Daily life must still have been solitary, poor, nasty, brutish, and short: infant mortality was high, work was dangerous, and treatment for injury or infection was nonexistent, which helps explain why the average life expectancy may have been eighteen years. That, coupled with low birthrates (in part because infants couldn't digest meat and unprocessed plants and so had to be breastfed longer, which delayed subsequent pregnancies), kept population growth nearly flat. By some estimates, world population cycled at around one million for tens of thousands of years. Still, from the strict standpoint of food economics — that is, the quantity and quality of available calories — Cro-Magnon was fabulously wealthy. Indeed, so ideally matched were these ancestors to their diets that those who did survive childhood trauma and hunting accidents were probably healthier than most of their modern decedents: according to Neil Mann, an expert in paleonutrition at RMIT University in Melbourne, Australia, fossil remains from this early period show none of the diet-related chronic diseases that plague us today.[8]

The good times couldn't last. By 11,000 years ago, a warming climate had drawn the big, cold-weather game back northward, away from human settlement. In their place came smaller, faster species, like gazelle, antelope, and deer, which required new hunting skills and weapons. Our Cro-

Magnon forebears adapted: the bow and arrow, for example, was clearly invented to hit a smaller, faster target. (Neanderthal, by contrast, may not have had the ability to update his hunting strategy, and so followed the big animals into extinction.) Yet ultimately, new technologies couldn't save the hunting life. By modern estimates, even well-equipped hunters during this period were probably earning fewer than fifteen hundred calories an hour hunting small game — quite a comedown from the high times of yore. As hunting success faltered, tribes had little choice but to supplement hunting with gathering: nuts and berries, edible roots; peas and other legumes, as well as various seeded grasses such as wild wheat and barley. This hunting and gathering, or broad-spectrum, strategy kept our ancestors alive, but barely. Research suggests that they needed hours and hours of effort, over many miles of territory, to find enough to eat. As the centuries went by, it became clear that the extractive food strategies, built around whatever foods nature made available, were failing. At some point, our ancestors would have to begin *producing* food. They would have to become farmers.

If the shift from hunting to farming was largely involuntary, our ancestors at least had the benefit of excellent timing. The same warming trend that took away the big animals had also expanded the range of certain edible plants and grasses, notably wheat and barley, into human territory. Such expansions had occurred before, but this time was different in several important ways. First, our ancestors now had a deep knowledge of plants, gleaned over several thousand years of foraging; they had certainly learned, for example, that wheat and barley were edible, and that fruit pits left in a garbage pile could sprout. Second, they had the beginnings of the social organization necessary to tackle the complex and large-scale task of farming. Third, and perhaps most important, they were highly motivated: the same weeds that successful big-game hunters had ignored must now have looked very interesting indeed.

This first agricultural revolution began in different places — Central Asia, Central America, and Southeast Asia — and at different times, but the circumstances were probably similar: humans came across patches of some wild food — wheat or barley, or a root crop such as yams — discovered that it could be coaxed into coming back each year, and settled nearby. By between 10,000 B.C. and 6000 B.C., small groups were growing wheat in Asia and the Middle East, corn in Mesoamerica, and rice in Asia. By 6000

B.C., humans had domesticated sheep, goats, pigs, and cattle, although these spindly beasts were kept mainly for their milk and hides, not meat. By 5000 B.C., agriculture had reached every continent except Australia.

Whether our ancestors regarded the agricultural revolution as progress isn't clear. Farming was brutally hard work, less dangerous than hunting, perhaps, but with longer hours and a new set of risks. Crop failures were routine, and even when things went right — when, for example, early farmers learned to use animal labor and were able to slowly increase yields — food challenges only became more complex. The harvested grain needed to be protected from spoilage and pests. It also needed to be transformed into something edible and nutritious, since unprocessed grain is almost indigestible in the meat-adapted human gut.

Eventually, our ancestors learned to grind and cook grain into flat breads and gruels, increasing the grains' nutritional value and, no doubt, its palatability. And yet, paradoxically, this advance probably left our ancestors with less to eat. Gruel could be fed to infants, which hastened weaning and thus halved the average interval between births from four years to two. The result was a phenomenon that would become the hallmark of the food economy: the virtually static population began slowly to grow, from an estimated five million worldwide in 10,000 B.C. to perhaps twenty million by 5000 B.C.

This first population boom was innocuous by modern standards. And yet, as would be the case in later centuries, this demographic surge pushed early food-production capacities to the limit. Our ancestors' meat intake in particular began to fall — with dramatic and quite visible physiological consequences. Whereas skeletal remains from the high hunting period show humans to be tall, robust, and relatively disease-free, the early agricultural period produced a less healthy specimen, one who suffered from various diet-related afflictions and whose height had fallen by a full four inches from that of his more carnivorous predecessors and would continue to fall until modern times.

But our ancestors survived. Their bodies may have craved more calories and certainly more meat, but farming did generate enough food to keep them alive — if only just. And they got better at it, cultivating new fruits and vegetables, domesticating new animals, developing new and often tasty ways to preserve food (by turning grain into beer, for example), and slowly raising output. Further, if farming left individuals smaller, it allowed early societies to get bigger. Whereas hunter-gatherers needed many square miles to collect enough food for survival — and thus were forced to live in small, dispersed, and highly mobile tribes — farmers could produce enough food to feed much larger groups using a fairly small area of land. This more concentrated form of food production meant humans could live in larger, more densely populated communities — villages first, then cities — which became centers for the technical and social innovations that gave rise to civilization.

Farming was also a source of power. By 3500 B.C., Egyptian wheat farmers were routinely producing more grain than they could eat themselves, and these surpluses, the first accumulated wealth, radically transformed society. Surpluses provided food security, something inconceivable to our hunting forebears, and they also gave people something to trade, thus forming the basis of commerce. Surpluses also meant cities could support more people, and some of these surplus people could be excused from the task of *making* food to specialize in other tasks: they could be builders, bakers, beer-makers; they could be potters and smiths, scribes and artisans and entertainers; they could be soldiers and defend the new city-states, and priests and kings, to rule them. Surpluses, note Naomi Miller and Wilma Wetterstrom, somewhat acidly, "provided an impetus toward developments we associate with civilization: urbanization, a high degree of economic specialization, and social inequality."[9]

For all agriculture's civilizing powers, however, it left these new civilizations vulnerable in ways that have never entirely faded. Most obviously, the concentrated populations in these city-states could no longer feed themselves yet were now too large to be fed by local farmers using primitive technologies; for example, because early plows could only cut shallowly, farmers tended to work just a thin layer of topsoil, which was easily depleted of its accumulated nutrients and often had to be abandoned in favor of more remote fields. In a foreshadowing of our own far-flung food economy, cities increasingly depended on production in distant lands, delivered via ever larger supply networks that required more and more resources to maintain and defend. It is no coincidence that the outlines of the Roman Empire followed the borders of the wheat regions in Europe, Central Asia, and North Africa (by 100 B.C., a third of Rome's wheat was coming from Egypt, nearly a thousand miles away)[10] or that once Rome's military power failed, its food system collapsed. In the evolving food economy, food security had become as much a question of military power and political influence as of agricultural capacity; when all three failed simultaneously, as with the fall of the Roman Empire in the fourth century A.D., the Western food economy collapsed so completely that for the next six centuries, global population rose from 300 million to just 310 million.[11]

Further, although the agrarian food system was capable of astonishing adaptation and recovery, it never quite escaped the boom-and-bust dynamic that had tormented the first farmers, despite a series of extraordinary innovations beginning around 1000 A.D. A new plow was invented that cut the soil more deeply and thus allowed farmers to reach soil nutrients buried farther down. Farmers learned to replenish spent soils — by adding animal manure, but also by alternating primary staple crops like wheat, which rapidly depletes soil nutrients, with rotations of cover crops, like beans, that pull nitrogen, a key nutrient, from the atmosphere and biologically fix it back into the soils. They also learned to select seeds from only the hardiest plants for future planting, which gradually improved seed quality. And while these advances were initially discovered through happenstance and trial and error, as the disciplines of biology and chemistry took shape, scientists began to understand and, to a growing degree, control natural systems, especially those systems upon which food production depends, such as soil fertility and plant and animal breeding.

Intertwined with these scientific and technological revolutions was another even more potent booster of output: commerce. As cities recovered, demand surged for the crop surpluses that the new plows and crop rotations were generating, and over time, farmers geared their production to exploit this lucrative new market. Decisions on what and how much to plant were now determined less by individual needs and more by the market: whereas a farmer had once produced the same crops year in and year out, he now raised whatever brought maximum return, and when prices were high enough, he even expanded production to fully exploit the opportunities. Landowners in eastern Europe began growing and selling wheat surpluses to western Europe. Cattle and sheep farmers built enormous herds in Scotland, Denmark, and Poland, which they then drove cross-country to sell in massive slaughtering yards in the large cities, where prices were highest.[12] Enterprising traders set up vast plantations of sugar, coffee, and tea in the tropics, whose output was intended entirely for burgeoning markets in Europe. Food was changing from necessity to commodity, its production driven less and less by sustenance and more and more by a competition for profits. Farming might still be intimately connected to the land, but from that point on, cycles of business would be as important as cycles of nature.

This shift into a more modern mode of food production was hardly easy or automatic. These more commercial food operations were not only larger but more complex, requiring careful husbanding of various inputs — seed, feed, labor, and eventually technology — and thus new organizational structures and a new set of skills: management. Many operations also depended on substantial financial backing (the typical sugar plantation required vast tracts of tropical land, large pieces of industrial equipment, skilled craftsmen, and hundreds of slaves).[13] This growing need for investors, who tended to expect a return on their money, only increased pressure on producers to become more *productive*, that is, to generate more output at a lower cost. Farmers and traders developed elaborate systems to analyze market conditions. They scrambled to boost output or lower costs, using new labor-saving technologies or increasing the scale of their operations, which spread their costs over more bushels or boars. The payoff was impressive. Between 1300 and 1600, grain yields nearly doubled.[14] More to the point, because farmers now had extra grain to feed their livestock, meat production increased, and in a harbinger of our own livestock revolution, meat consumption skyrocketed. By the sixteenth century, the average German was eating half a pound of meat a day (almost as much as a modern American!) and across Europe, even farm laborers ate meat at least once a day.

Yet this prosperity would be only temporary. What few Europeans understood was that their new plenitude did not come primarily from capitalism, technology, or even God's grace, but from a demographic accident: centuries of malnutrition, wars, and disease (not least the Black Death of 1347, which killed one out of three Europeans) had kept population growth from exceeding the production capacities of the food economy.[15] But by 1600, that economy was producing so many extra calories that our numbers again began to grow. Between 1500 and 1750, world population jumped from roughly five hundred million to around eight hundred million,[16] while the number of farmed acres grew hardly at all.

It wasn't simply the greater numbers that overwhelmed farmers' productive capacities; it was also the rediscovery of meat. Raising livestock, it turns out, is a rather inefficient way to make food. A cow, for example, needs to eat seven pounds of feed to gain a pound of weight, and pigs and chickens weren't much better, which meant that as meat consumption increased, the demand for grain, hay, pastureland, and other feed grew at a much faster rate. And while that had been tolerable when Europe had only a few people per acre and lots of unused acres, once population began booming and those unused acres had all been plowed into fields, the food system came under heavy stress. As early as 1600, Italy, France, and the Netherlands all had population densities *greater* than their farmlands could sustain,[17] and China and India were not far behind.

The result of this "demographic tension," as historian Fernand Braudel calls it, was disastrous.* Between 1600 and 1800, France recorded twenty-six major famines, plus uncountable smaller, local shortages. In Florence, harvests were insufficient one out of every four years.[18] In 1696, nearly a third of Finland's population starved to death, and elsewhere in Scandinavia, farm animals became so weak from lack of fodder that farmers had to carry them "like children." Asia fared no better. In China and India, famines in 1555 and 1596 killed millions and emptied entire regions. Armies of hungry people moved across Europe and Asia in search of anything to eat — nuts, roots, grass, leaves, and, by some accounts, one another — prompting towns and villages to enact increasingly stiff poor laws. "In the sixteenth century, the beggar or vagrant would be fed and cared for before he was sent away," notes French historian Gaston Roupnel.[19] "In the early seventeenth century, he had his head shaved. Later on, he was whipped; and [at] the end of century . . . he was turned into a convict."

Even in nonfamine years, much of the population lived in a state of nutritional purgatory. Physical stunting was endemic: it's no accident that doorways, ceilings, and suits of armor from the period all look like they were made for children. But malnutrition's real costs were less visible, at least at first. In unborn and newborn children, when the body's very structure is still forming, malnourishment wreaks a particularly cruel havoc: it retards brain and nerve formation, which leads to learning disabilities and higher rates of schizophrenia; it disrupts hormone production, which interferes with the development of vital organs like the gut, heart, and lungs,

---

* Through the sixteenth century, the European food production system could comfortably handle population densities of around 77 people per square mile. But by 1600, Italy had reached 114 persons per square mile; France, 88; and the Netherlands, 104. Braudel, *Civilization and Capitalism*, 61.

making them weaker and more prone to later failure. Some of the damage is reversible if proper nutrition is restored, but then, as now, malnutrition almost always persisted into adulthood, where it unleashed a cascade of secondary and permanent afflictions: low respiratory capacity; digestive system failure; greater risk of infection; arthritis and other joint diseases; and arrhythmia and heart attack. As Robert Fogel, a University of Chicago economist and expert in ancient nutrition, puts it, malnourished humans "wear out more quickly and are less efficient at every age."[20] In short, for centuries, hunger effectively destroyed the mental, social, and productive capacities of entire populations.[21]

In 1798, in a treatise entitled "An Essay on the Principle of Population," Thomas Malthus, the Anglican parson turned economist, gathered these trends into a single pessimistic forecast: humanity was doomed. Although people were astonishingly clever at finding ways to make more food — and, indeed, the scarcer food became, the cleverer people got — Malthus believed that hunger would never be eradicated because any increase in food served only to make populations even larger. These larger populations then exceeded available food supplies, plunging humanity into famine and strife until scarcity sparked the next round of productivity increases, which then sparked yet another population surge.

In a sense, Malthus was offering a formal explanation for what populations had been discovering and rediscovering ever since the first agriculture revolution, but with a critical difference: where his predecessors had imagined the boom-bust cycle as a permanent condition of human existence, Malthus believed the cycle would soon conclude: because crop yields can increase only linearly (that is, by the same small percentage each year) whereas population grows *geometrically* (doubling every several hundred years), he reasoned that population growth would soon outpace humankind's capacities to feed itself — at which point demographic equilibrium would be fatally out of balance and could be restored only by a cataclysmic famine. "The power of population," Malthus wrote, "is so superior to the power of the earth to produce subsistence for man, that premature death must in some shape or other visit the human race." His own calculations suggested that the "visit" would occur in the middle of the nineteenth century.

Much of the century that followed Malthus's grim forecast was spent desperately trying to prove that forecast wrong. Governments banned merchants from exporting grain. Farmers were urged — and in some cases forced — to plant new, more productive crops, such as the corn and potatoes that explorers had brought back from the New World. More dramatically, new farmlands were carved out of woods, marshes, and other "wastes." Italian nobles spent a fortune filling in their swamps and rivers, and by 1850, half of England's medieval forests had been chopped down and converted into farmlands.[22] As land expansions reached their limits, and food prices rose (between 1750 and 1800, European wheat prices nearly tripled), producers redoubled efforts to increase productivity with whatever new technology or input might help raise yields.

But despite impressive results — between 1600 and 1860, English wheat yields tripled — it still wasn't enough. As Malthus had predicted, whatever extra calories farmers did manage to produce simply allowed more of those who would have starved — young children, the sick, and the aged — to survive, with the result that Europe and Asia had more mouths to feed but not more food for each mouth. By the mid-1800s, German meat consumption was down to two ounces per person per day, and the English working class was living almost entirely on starch — in many households, a fifth of daily calories came from the white sugar added to jam or tea, and many working-class families spent nearly half of their income on bread alone.[23] By the nineteenth century, the average British man stood five foot five inches tall and weighed barely 134 pounds; in France, the average male was five foot three and 110 pounds.[24] Life expectancy remained frightfully low: in 1880, the very peak of the British Empire, the average Englishman could expect to live forty years, while a member of the working poor could expect just twenty years[25] — about the same as our primitive Paleolithic ancestors. Even in the relatively well-fed America of the nineteenth century, men were considerably frailer than today, being nearly three times as likely to suffer heart disease and nearly five times as likely to suffer gastrointestinal problems.[26]

The University of Chicago's Robert Fogel estimates that during this period, a fifth of the population was caught in nutritional purgatory, where they were getting enough calories to escape outright starvation but were so malnourished they could no longer work, and "many of them lacked the energy even for a few hours of strolling."

Despite such widespread evidence of a looming nutritional collapse, many in Europe's political elites remained in a state of denial. Aristocrats had been shielded so long from the nutritional realities of the masses that, in addition to being physically larger and healthier than the average citizen, they continued to blame the epidemic of nutrition-related exhaustion on the moral failings of the poor. But even aristocrats were forced to face facts in 1901; as Britain geared up to fight the second Boer rebellion, it was discovered that one-third of the army recruits were too weak to endure basic training and too short for the army's minimum height requirement of five feet. After twelve thousand years of civilization and progress, this is what humanity had come to: stunted bodies, shorter lives, and, at some point, mass extinction by famine.

What turned our fortunes around? What allowed us to push the Malthusian limits back and move to an era of superabundance? In a word, globalism. What staved off Armageddon was the emergence of an international food system, built on railways, shipping routes, and new preservation technologies, and spurred by a nascent ideology — free trade — that slowly but surely began to connect the starving demand centers in Europe with distant suppliers in Australia, Argentina, and especially the United States — countries that not only possessed surplus land and small populations, but that also were just then undergoing industrial transformations of their food production.[27]

The United States in particular seemed born to superabundance. Land was plentiful and immensely fertile; topsoils in the Midwest were thick and rich with nitrogen and other nutrients from thousands of years of dead and decaying plant matter. The relatively mild climate, and especially the large areas with frequent and reliable rain, provided ideal conditions for grain. And the federal government, eager to encourage rapid economic development and westward expansion (and wipe out the last vestiges of the southern plantation system), gave farmers large tracts of land at little or no cost. Where a European farmer nurtured his small plot carefully with crop rotations and manures, his American counterpart had so much land that he worked a field until the soil was exhausted, then plowed up the next.[28] Moreover, because U.S. farms were larger than European farms, and farm labor was so scarce, Americans were quicker than Europeans to adopt mechanical threshers and other labor-saving technology. As a result, U.S. farmers made rapid gains in their productivity; in 1837, a farmer needed 148 hours of labor to produce a single acre of wheat; by 1890, he was getting it done in just thirty-seven hours.[29] Even low-tech innovations brought revolutionary impacts: barbwire transformed the vast prairie into an equally vast livestock operation, and by 1884, the United States had more than forty million head of cattle — two for every three Americans.[30]

With this rapid mechanization and continued expansion of farmlands, U.S. food production began to outstrip demand, leading to a dramatic shift in American dietary practices and food culture. As food prices fell, the laden table became a national icon. Homemakers and boarding houses competed to offer the most dishes at every meal; a typical breakfast, writes food historian Lowell Dyson,[31] included "steaks, roasts, and chops, along with heaps of oysters, grilled fish, fried potatoes, and probably scrambled eggs, with biscuits and breads, washed down with numerous cups of coffee." Such conspicuous abundance left many Europeans scratching their heads. "Every day at every meal you see people order three or four times as much of this food as they could under any circumstances eat," complained one European visitor after dining at an American boarding house.[32]

In fact, Americans could not even come close to eating all the food that U.S. farms were now producing, and from the 1850s onward, U.S. farmers and ranchers were forced to look to foreign customers to offload their surpluses — much to the relief of starving Europeans. In the late nineteenth century, U.S. exports of food to Europe grew from a trickle to a torrent; grain, of course, but eventually meat. Not only did America have vast cattle and sheep herds, but with the breakthroughs in preservation technologies, such as canning and, especially, refrigeration, that surplus meat could now be shipped around the world cheaply and relatively safely. It was these shipments, joined by similar excess cargoes sent from South America and Australia, that initially slowed Europe's slide toward starvation. And conversely, because Europe's burgeoning industrial centers could now draw meat and grain directly from producers around the world, rising European demand began to drive food production everywhere on the planet. Food was now a truly global commodity. Governments might still be called upon to protect shipping routes and certainly to referee the fairness of trade, but the food trade itself was now almost entirely commercial, driven less by military strength and more by price.

So large had the food trade become (by 1900, Britain was importing almost half its calories),[33] that our political understanding of food changed as well. For centuries, nations had sought to produce as much of their food as possible, importing only reluctantly in times of need. But in a world where population was soaring and centers of demand and centers of supply were often not even on the same continent, the idea of food self-sufficiency seemed obsolete and impractical — and even contrary to the budding notions of internationalism and global fraternity. As one editorialist put it, each new cargo of food unloaded in London or Antwerp furthered "the work of weaving in and blending the interests of remote countries, which is more and more making mankind one great community, with common objects and wonderfully complex interdependency."[34] It was a sentiment that would resurface many times over the next century, albeit in slightly less florid terms.

There was, however, one final step remaining in the conquest of scarcity. For the most part, the great surpluses being shipped to Europe weren't the result of better farming but of *more* farming: in South America, North America, and Australia, farmers could grow steadily more food because they could plant steadily more acres. In terms of actual productivity, per-acre yields had risen little since the Civil War; even in the fecund United States, farmers had hit physical limits as to the total food they could generate from a single acre. If humans were going to truly beat Malthus, we would need to move past such limits with new forms of agriculture whose output was not constrained by acres but could grow as fast as the population. Agriculture would have to become more concentrated or intensified — a leap that would take more than just new plows or commerce; it would take an act of Congress and a breakthrough in chemistry.

By the late 1800s, the leveling-off of yields in the United States caused a panic in Washington. Without the prospect of cheap, abundant food, many policymakers feared America would not be able to keep its new factories and cities booming[35] or allow the nascent middle class to continue on its upward trajectory. Yet policymakers also realized that boosting food supplies sufficiently to meet the nation's economic goals posed a challenge too large and complex for the private food sector alone and would require enormous government intervention in the form of laws, agencies, and a volume of public spending on an unprecedented scale.

To fulfill this vision of a new kind of food economy, during the late nineteenth and early twentieth centuries Congress created a vast system of support for food production: a department of agriculture, whose mission was the provision of affordable food; a system of publicly funded farm programs, meant to maximize output while protecting farmers from harvest failures and market crashes; a construction campaign of dams, irrigation canals, and other reclamation projects to bring agriculture to the desert and semi-desert regions; and a massive railroad network to transport this great bounty from the new areas of production — the midwestern Corn Belt, the California "Salad Bowl," the western-state ranches, and the stockyards in Chicago — to the big urban areas and export facilities. So integral was government in the buildup of the modern food economy that years later, Harvard economist Ray Goldberg described the food system as "the largest quasi-public utility in the world."[36]

But infrastructure was only the start of this agricultural revolution; the improvements in food production that America now sought required fundamental changes in the way food itself was made. At new federal and state research centers and at land-grant universities, researchers began applying rapidly expanding scientific knowledge to the development of new plant varieties and animal breeds that could grow faster and larger. In the 1920s and 1930s, scientists came out with hybrid strains of corn that not only had bigger, more plentiful ears but also grew more closely together in the field — all of which meant more corn per acre; between 1930 and 1940, the number of bushels of corn per acre doubled, and then continued to rise each year.

But there was another parallel development to crop breeding that was in many ways even more important: fertilizers. Traditional methods for replenishing soil fertility — with manure and crop rotations — could no longer replace nutrients as rapidly as the new, fast-growing crops could suck those nutrients out. An acre of modern corn, for example, will pull half a ton of nitrogen and other nutrients from the soil during a five-month season.[37] Without outside replenishment, such soils will become so depleted that the dirt loses not only its productive capacity but its physical integrity; it falls apart and thus becomes highly vulnerable to wind and rain erosion — a phenomenon that contributed to the catastrophe known as the Dust Bowl. The solution came in the form of a new industrial process — known as Haber-Bosch, after its inventors — that could pull nitrogen out of the atmosphere, where it exists in near-endless abundance, and fix it, synthetically, in a convenient and highly concentrated chemical form, ammonia, which farmers could simply mix into their dirt.

Vaclav Smil, a resource economist at the University of Manitoba and an expert in what might be called the global nutrient economy, has described Haber-Bosch as the most important invention of the twentieth century, and its arrival did indeed mark a turning point in food production. Traditionally, nitrogen supplies had been limited to the quantities generated by plants; farmers could grow cover crops, which biologically fixed nitrogen directly into the soils, or farmers could feed the crops to livestock, which then concentrated the nitrogen in their manure. But in both cases, the supply of nitrogen was limited to the amount of land farmers could devote to rotation crops or feed crops. By 1900, for example, farmers were committing as much as half of their total acreage to growing feed or cover crops,[38] which meant that just half their land was available to produce cash crops — a huge constraint in a world where acres are finite. With Haber-Bosch, however, these natural limits no longer seemed relevant. Because atmospheric nitrogen was effectively infinite in supply, there was no limit to how much nitrogen could be produced and used — as long as there was the energy to run ammonia refineries, which, with the rapid growth of the petroleum industry, there was. Farmers could apply as much nitrogen as their high-growth plants would absorb. Thus, for all the importance of the new plant-breeding technologies, it was fertilizer that actually fueled the explosive boom in grain production; by Smil's estimate, nearly half of the extra food produced today (and more than two-thirds of the extra people) is a direct result of the availability of synthetic nitrogen.[39]

Science had revolutionized the raw materials, or inputs, of agriculture, and it soon did the same for the outputs. Breeders began designing plants not just for larger size and faster growth rate but also for uniformity, which would make them easier to mechanically harvest and process. For example, in an old-style cornfield, the corn plants matured at different times during the season and bore their ears at varying points on the stalk — all of which required labor-intensive hand-harvesting. By contrast, the newly developed hybrid corn produced fields of virtually identical corn plants whose ears ripened simultaneously and which were set high enough on the stalk to allow harvest by machine — improvements that allowed farmers to produce more corn faster and at lower costs.

Uniformity quickly became a guiding principle for plant breeding and for all of industrialized agriculture. The tomato was transformed from a sprawling bush with soft, slowly ripening fruit to a compact shrub whose fruits were similar in size and shape and firm enough to endure mechanized picking and long storage.[40] Cucumbers were bred to be straighter, for ease of picking and pickling. As one admiring farm journal noted, because it was cheaper to breed plants for machine harvest than to design a mechanical harvester to cope with nonuniform plants, eventually all commercial produce would be engineered so as to "place the harvested parts in a predictable position in relation to the harvest machine." Or as another journal put it, "Machines are not made to harvest crops; in reality, crops must be designed to be harvested by machine."[41]

But it wasn't enough to improve the efficiency of mere individual crops; the entire operation of farming was ripe for modernization. Just as scientists had revolutionized the study of nature by breaking the natural world into its constituent pieces, such as molecules or cells, agronomists now sought to *rationalize* conventional agricultural systems by breaking them into their constituent pieces, which could then be individually scrutinized and modified for more intensive production and more efficient operations. Farms had once been diverse, one-stop operations, producing a multitude of crops for both food and for livestock feed and using livestock for labor, meat, and fertilizer, but the modern farmer was now encouraged to specialize in a single crop or livestock species and simply *buy* his inputs. By focusing resources and expertise on a single product, a farmer could raise output *and* lower costs; instead of buying different technologies for each of several crops or animals, he now spread the costs of a single set of technologies — for raising corn, for example, or wheat — over more bushels.

To be sure, a more rationalized agriculture was far more reliant on outside players. Synthetic nitrogen, for example, would now be produced from petroleum in huge new petrochemical factories hundreds or even thousands of miles away. Likewise, these new hyperefficient farming operations would no longer process their own outputs; instead, the slaughter of livestock, the milling of grain, or the processing of fruit and vegetables would take place elsewhere. And by the end of World War II, a vast network of commodity buyers and processors had arisen to convert grains, animals, and other farm products into inputs for the food industry.

Yet this outsourcing of traditional farm functions would make the entire food system vastly more efficient. Because these new specialists focused on a single task — milling corn, for example, or synthesizing fertilizers — they could operate far more cost-effectively than farmers could, and they could also focus more intently on improving their technologies through innovation, which in turn spurred even greater yields on the farm. For example, as the petrochemical industry perfected methods to mass-produce nitrogen, the price of nitrogen plummeted, which allowed farmers to use far more of it: between 1950 and 1980, farm use of nitrogen jumped by a factor of seventeen.[42] Importantly, the emergence of defined sectors for inputs and outputs opened the door to new investments in agriculture: farms themselves might be too risky to invest in, but Wall Street was more than ready to gamble on a growing industry that sold those farmers fertilizers and other inputs and on the big grain companies and food processors that bought those farmers' products.

For all intents and purposes, the traditional farm had largely vanished; its operations had been separated and reassembled elsewhere in a new and much larger system — a supply chain that began with financiers and input companies and concluded with grain traders and food processors. The modern farmer resembled more and more a black box that inputs flowed into and outputs emerged from, and he bore about as much resemblance to his old-fashioned predecessors as automobiles did to horse-drawn wagons. So complete was the transformation that in 1957, Ray Goldberg, the Harvard economist, and his colleague, John Davis, proposed that the term "agriculture" be replaced with a new, more fitting one: "agribusiness."

By whatever name, the new uniform, rationalized, and intensified food system that had fully emerged by the late twentieth century was extraordinarily powerful. Whereas farmers in industrializing Europe had been proud to double their yields once a century, farmers now *quadrupled* their output in half that time. By the mid-1980s, the United States was generating 40 percent of the world's corn, and a quarter of that was coming from a single state (Iowa). And all this was happening with fewer and fewer farmers. In 1885, more than half the U.S. population was engaged in farming; in 1985, that share had fallen to less than 3 percent.[43]

But the real significance of hybrid power was what it meant for meat production. When they were fed grain, cattle, pigs, and chickens not only grew larger faster than did their foraging predecessors, they were also produced much more efficiently. Traditional livestock operations had been dispersed to give animals the time and space to forage for sufficient calories, but grain-fed livestock could be confined in pens, barns, sheds, and other concentrated animal feeding operations, or CAFOs, which allowed far tighter control of production. And once the production phase of the meat business had been rationalized, the entire process — from the feed mills to the feed lots to the slaughtering plants — could be integrated into one efficient supply chain under the control of a single company, usually a feed company or a meatpacker.

This factory farming had its complications. Shifting animals to a starchy grain diet caused nutrition problems — for example, protein-starved chickens turned cannibalistic, eating one another's feathers, or worse, to get enough protein — until farmers began supplementing grain with protein-rich soybeans and amino acids, along with bone meal, blood, offal, and other protein-rich waste products from slaughterhouses.

Similarly, the crowded conditions in the new CAFOs created perfect conditions for massive disease outbreaks — until operators began feeding their animals antibiotics (which, as Thomas Jukes, our fish detective, showed, also accelerated growth). As each obstacle fell to some new scientific solution, farmers were able to generate more meat at a lower cost and in a much smaller space; today, the standard index of livestock productivity is "pounds per square foot."[44] And as output rose and prices fell, consumption once again soared. In 1945, the average American ate around 125 pounds of meat a year; by 1980, per capita consumption had reached 195 pounds,[45] an increase of nearly 60 percent.

The impacts of this new high-volume, low-cost production rippled across entire societies. As protein consumption rose, populations accumulated what Robert Fogel calls "physiological capital" — stronger bodies and organs capable of resisting disease; by 1980, life expectancy for the average Western male was around sixty-five years, and he stood five foot ten inches in height — three inches taller than he had as recently as the Civil War — while the average woman went from 63.5 to 65 inches. The average British man was now nearly five foot nine inches. And we got richer. As labor requirements fell, farmers moved to the cities to take better-paying factory jobs, and everyone was paying less and less for food each year. In 1900, the average American family spent half its household income on food; by 1980, that share had dropped to less than 15 percent.[46] Declining food costs meant Americans had more money to spend on other things, like bigger houses, nicer cars, education, and health care — the essentials of the American Dream.[47] In this sense, the modernization of food production marks one of the biggest transfers of wealth in human history. And it wasn't only the industrial West that was benefiting. Between 1950 and the late 1990s, world output of corn, wheat, and the cereal crops more than tripled, which meant that even though world population more than doubled, from 2.5 billion to 6 billion, the volume of food available for each person rose from fewer than 2,400 calories per day to more than 2,700 calories.[48]

In the context of such bounty, one can hardly be surprised that by the end of the twentieth century, official attitudes about agriculture and food production in the United States, Europe, and other industrialized nations began to take on distinctly triumphal tones. In the opinion of most agronomists and agricultural economists, there really wasn't any limit to this industrial plenitude. Under the new agribusiness model, the farm had become an utterly rational enterprise in which ever larger inputs of capital — in the form of seeds, fertilizers, pesticides, machines, fuel, and research — delivered predictable increases in output and profit. In the new food economy, human destinies would no longer be driven by the age-old fear of boom and bust but by, as economist John Perkins put it, "a powerful vision of the nation-state as an industrial economy in which all natural resources, including agriculture, [could be] marshaled by the rational control of modern science." And the undisputed master of this new system was the United States — a nation that, with just 5 percent of the world's population, generated a sixth of the world's meat and nearly half its soybeans and corn[49] and wielded such control over global grain and food markets that by the 1980s, in the words of one foreign diplomat, it had attained "a position similar to that of OPEC in the field of energy."[50]

For all its efficiency and controlled abundance, the new modern food economy was rife with contradictions. Production was booming in much of the world, but it flagged in places like Africa, where antiquated farm methods barely kept pace with rising populations, and even the United States still had large pockets of hunger. Food production might require less labor, but its greater reliance on labor-saving mechanization and chemicals made farming and other food-production jobs among the most dangerous occupations in the world.[51] Global trade was helping to build food security for wealthy importers, but the search for the cheapest production was transforming countries like Brazil, Panama, India, and Malaysia into vast planta-tions of coffee and tea, sugar and bananas — all bound for the tables of middle-class consumers in Europe, Japan, and the United States.

Moreover, there was a growing sense in some quarters, and certainly among agriculture producers, that the machine was out of control. The treadmill effect was becoming more and more pronounced: the more bush-els farmers were able to grow on each acre and the greater the volumes of grain that farmers were able to put onto the market, the lower that prices for that grain fell — a boon for consumers, but a slow-motion disaster for farmers and other producers. In most other businesses, declining prices are a signal to companies to produce *less;* restrict output, and supply will even-tually fall, at which point prices will rise. But agriculture has always been different. Because the biggest and costliest input is land, farmers have far less flexibility in their production. While a factory owner might cope with low prices by, say, laying off half his labor force (labor being his biggest cost), farmers can't lay off their land. In most cases, they must continue to farm those acres in order to make mortgage or rent payments; land is a fixed cost, over which farmers have very little control. Thus, as the price per bushel continues to fall, farmers compensate by spreading their fixed costs over *more* bushels per acre, which they generally accomplish by buying new seeds, better fertilizers, bigger tractors, or some other new technology. Un-fortunately, while growing more bushels maintains their income in the short term, the cost of all that new technology has to be spread over even more bushels, which simply adds to an already oversupplied market and pushes prices down further — a vicious cycle wherein farmers are essen-tially producing more and more bushels each year simply to keep from go-ing under.

Eventually, however, the treadmill became too much; as commodity prices continued to fall, the small and medium-size farms that lacked the scale to keep spreading their costs or the capital to buy new technol-ogy were pushed out, replaced by huge industrial farming operations that could make up in volume and efficiencies what they were losing in price. By the mid-1980s, the U.S. farm system had been so consolidated that more than two-thirds of the nation's entire agriculture output was now coming from less than one-third of the farms.[52]

Social commentators called this the farm crisis, but the same dynamic of price-led consolidation has become a defining trait of the entire food economy. Up and down the supply chain, the smaller, less efficient producers have been pushed out or gobbled up, leaving each sector in the hands of a small number of enormous players with the massive volume and economies of scale to remain profitable. By the 1980s, a handful of petrochemical and pharmaceutical companies controlled most of the seed, fertilizer, and pesticide businesses, while a select group of meat processors handled more than half the meat business.[53] Similarly, the bulk of U.S. grain was being managed by just three grain trading companies, Cargill and Continental (which would later merge) and Archer Daniels Midland. This tiny number of buyers meant farmers had few options as to where to sell their grain, and thus little choice but to accept whatever low prices these buyers offered, all of which simply accelerated the treadmill syndrome.

Farmers had become price takers — that is, forced to accept whatever the market was paying — while the big buyers of grain and other commodities were increasingly the price makers, who could dictate terms. In yet another perverse dimension of the economics of food, farming had become what economists call a "perfect competition," with neighbors fighting to undersell one another by fractions of pennies per bushel — while the market they sold to moved toward becoming the "perfect monopoly."[54] Uniformity and specialization had been the hallmarks of the early modern food economy; consolidation and inequity would be its lasting legacy.

And there were other, less quantifiable impacts of the high-volume, low-cost model. The flood of cheap food that had been so critical in helping societies regain physical and economic health emerged as a liability. Governments struggled to cope with the surpluses generated by their overly successful farm programs: in the United States and Europe, lawmakers had to *pay* farmers billions of dollars a year to grow *less* food. Likewise, as the price of commodities plummeted, food processors had to go to greater and greater lengths to "add value" to those commodities and thus maintain their own profits. As well, by the 1980s, it was clear that many of the populations that had struggled to get enough calories were now struggling to keep from eating too many calories. In the United States, as the physical demands of our work dropped, the amount of fuel available for each of us soared — from 3,100 calories in 1950 to nearly 4,000 calories by 2000.[55] And while much of this excess wound up in landfills, the result of a food chain that had grown incredibly wasteful, our expanding girths suggested that at least *some* of these extra calories were reaching their targets.

And bizarrely, for all the staggering output and perpetual oversupply, there were troubling signs that our capacity for such superabundance was limited. Although demand showed no signs of slowing — in fact, each new breakthrough in yields still spurred a Malthusian surge in population growth — our new massively high output levels were becoming harder to maintain. Our new supercrops turned out to need more protection from weeds and insects, necessitating ever greater applications of pesticides, much of which found its way into the streams, rivers, and water wells of farming regions. (By the late 1970s, federal scientists identified agriculture as the biggest single source of nonpoint water pollution.[56]) Similarly, soils in some regions were being so overfarmed that more fertilizers had to be applied every year simply to keep yields from falling — and in some cases, yields kept falling anyway — a trend that, given the critical link between cheap grain and cheap meat, raised troubling questions about our triumphal return to a more carnivorous diet. In nearly every sector of the food economy, our very success seemed to have caught up with us. After spending centuries to build a food economy that could free us from want, we would spend the next decades coping with the costs of that victory.

2

It's So Easy Now

ACH WEEKDAY MORNING, as most citizens of Lausanne, Switzerland, head off to work, a small contingent of local residents arrives at an elegant, tree-lined corporate campus on the outskirts of town where Nestlé has built its research center. The visitors, most of them Swiss hausfraus with a taste for new foods and a few hours to kill while the kids are in school, sign in at the front gate, pass through security, and are taken directly to the sensory lab, a large, well-lit, spotlessly clean facility with a fully equipped test kitchen and twelve sensory cubicles. The cubicles are small, each furnished with a stool, a computer keyboard, and a stainless steel spitting sink, all carefully designed to minimize distractions. The air is pressurized to keep kitchen odors from seeping in. Overhead colored lights can be switched on to obscure food colors that might affect a taster's food experience.

At preset intervals, a wall hatch opens and a gloved hand passes through one of the day's samples. It may be an existing Nestlé product — an instant sauce, say, or a coffee beverage — that is under "renovation" because sales are lagging. In other cases, the tasters may be sampling some completely new product from one of Nestlé's high-tech and often highly proprietary processes: a frozen pizza crust that stays perfectly crisp; a probiotic yogurt with bacteria designed to improve gut health; a low-fat ice cream engineered to keep its creamy mouthfeel. Between bites, testers key in their responses, and the data are plotted onto a sensory map that details a product's performance on various sensory attributes, such as spiciness or crunch. Once completed, these maps show Nestlé product developers precisely which element of a new food is or isn't working; the map can also show how far a flavor or texture can be pushed before consumers will

reflexively reject it — critical knowledge in a ruthlessly competitive industry where four out of five new products fail. "Human beings were designed
to be hugely conservative when it comes to food," explains Peter Leathwood, head of Nestlé's Food Consumer Interaction Department. "When
you're a hunter-gatherer and something suddenly tastes different, that's a
warning."

And one, apparently, that Nestlé rarely misinterprets. The company is
the world's largest food and beverage maker, with annual sales of $71 billion
and a dominating presence in every geographic market — a position it has
gained by identifying consumer desires and translating them, with near-
pharmacological skill, into tens of thousands of products representing the
full spectrum of the postindustrial cuisine. The company's major brands
are a veritable *Who's Who* of modern convenience: Lean Cuisine and Hot
Pockets, Coffee-Mate and Nestea, PowerBar, Häagen-Dazs, Gerber, and, of
course, Nescafé, the world's third most recognized brand (after only Coke
and Pepsi),[1] which is sold in more than two hundred different demographi-
cally targeted formulations. In the spacious employee canteen at the company's headquarters in nearby Vevey, publicist Hans-Jörg Renk hands me a
cup of Nespresso, a new espresso drink made from prepackaged capsules,
and points to a large, wall-mounted digital counter. "That shows the number of cups of Nescafé that are consumed each year. Assuming ten grams of
coffee per cup, it's about four thousand cups a second." The numbers are
changing so quickly they're a blur, but the year's total is somewhere in the
neighborhood of 44,395,999,000 — enough for seven cups for every man,
woman, and child on the planet. And it's still early May.

If the rise of low-cost, high-volume agriculture marked the birth of the
modern food economy, the food manufacturing industry that Nestlé dominates represents the next step in that economy's evolution. Just as the rise of
agribusiness meant that consumers spent less money on food, now the
emergence of convenience foods meant they would spend less time on it as
well. And in rapidly industrializing societies, time was becoming as valuable a commodity as anything else. When Nestlé was founded in 1867, most
of the world's calories were processed by hand, at home or in local shops,
and the average household spent half its labor hours preparing meals. Since
then, in many parts of the world, consumers have cut that time dramatically by essentially outsourcing the cooking task to companies like Nestlé,
Unilever, Kraft, Tyson, Kellogg, Danone, and tens of thousands of others.

But the success of Nestlé and its rivals represents other, less visible facets of this evolution as well. As much as consumers have come to depend on convenience foods, the industry relies on those products even more. In a food economy geared toward ever lower prices, selling convenience has become the food industry's most important means of making money. Moreover, while convenience has been an economic gold mine — food manufacturers generate nearly $3.1 *trillion* in revenues a year, and the profit margins for leading companies are well above those in other sectors — it has also demonstrated yet again the perverse inadequacy of food as an economic phenomenon.

Before food companies could take on the tasks of preparing and cooking our foods, for example, the ingredients would need to be altered, often significantly, to become more manufacturable — with results that haven't always been in consumers' interests.

Further, just as farmers have been trapped in a cycle of overproduction, companies like Nestlé must sell more and more convenience, by way of a steady stream of new and improved products that is becoming very difficult to sustain. Beyond the enormous material costs of so much processing and packaging, the sale of convenience depends not only on increasingly aggressive promotional strategies (advertising junk food in schools, for example, or selling baby formula to Third World mothers) but on the continued decline in consumers' ability to prepare, or even understand, their own food. If human beings are indeed inherently "conservative when it comes to food," the success of companies like Nestlé marks one of the most radical and potentially troubling developments in the story of food economy.

Although the processing industry that Nestlé dominates is largely a twentieth-century phenomenon, the underlying obsession with convenience would have been quite familiar to our ancestors. It was one thing to acquire large volumes of meat, roots, or grains, but quite another to cut, pound, boil, roast, or otherwise render these raw materials into something edible. And although cooking did provide a context for important social forms and traditions, it remained a job most practitioners would have happily outsourced had there been an easy alternative. There was a tiny "consumer-ready" sector in large cities (for example, one could buy honey cakes and sausages at the Forum in Rome),[2] but most citizens were still rural, without access to such amenities nor the discretionary cash to buy them. Beyond the bread, beer, cheese, and sausage available from local artisans, processed foods remained very much a home-based enterprise, and the food "industry" was really just an extension of the farm system, with small farmers selling their raw surpluses at the town market to local homemakers.*

All this would change in the economic and social tumult of the Industrial Revolution. As labor-saving mechanization remade agriculture in America, Europe, and, to a lesser degree, Japan, it pushed an army of surplused rural workers to the cities, where many — men and women both — worked long hours in shops and factories. Cut off from the means to grow their own food, and lacking time even to cook, this nascent middle class needed (and now had the wages to afford) food that was readily available, easily prepared, and quickly eaten. Humans had always pined for fast food, but this time, there was a real food industry that could respond. The same Industrial Revolution that had emptied the countryside had also transformed food manufacturing. Assembly lines, canning and bottling technologies, mechanical refrigeration, and a growing network of rail, trucking, and shipping lines all meant food could be processed en masse, quickly, cheaply, and with reasonably consistent quality, and then shipped to distant markets. Armed with these new technologies and backed by an emerging capital market eager to invest in a hot new industry, American entrepreneurs like Gail Borden, Henry John Heinz, Joseph Campbell, and William Kellogg introduced a welter of new products, from canned milk and pickles to soups and breakfast cereals. No doubt consumers found these new items less flavorful than homemade, but they were nonetheless grateful for their cheapness, uniformity, relative safeness, and huge convenience.

Prepared food wasn't just an American phenomenon. In Europe, where industrialization was even more disruptive, entrepreneurs offered time-pressed factory laborers an array of fast foods ranging from potato flakes to soup "tablets." In the Swiss town of Vevey, Henri Nestlé, a chemist and fertilizer manufacturer, mixed wheat meal, sugar, and "wholesome cow's milk" to make *Kindermehl,* or "infant cereal," which he marketed to working-class Swiss mothers whose new factory jobs prevented them from nursing

---

* To the extent that anyone controlled the industry, it was the merchant wholesalers who imported bulk quantities of cooking oil, coffee, sugar, or whatever else couldn't be found locally, and packaged it by hand.

their babies. "My discovery has a tremendous future," wrote an elated Nestlé after a positive early response. "All the mothers who have tried it just once, come back for more."[3]

Quite a bit more, as it turned out. As demand for prepared foods grew, companies like Nestlé, Heinz, General Foods, Kellogg, Post, Armour, and Swift introduced thousands of new products, many marketed as time-saving alternatives to home cooking. As early as 1937, Kraft was pitching its Macaroni and Cheese Dinner with the slogan "Make a meal for 4 in 9 minutes."[4] Two years later, Nestlé's Nescafé liberated millions of java-deprived consumers who until then had been forced not only to brew their own coffee but also to roast and grind the beans. During World War II, demand for safe, transportable, easy-to-eat military rations spurred the industry to come up with still more innovations in formulation, preservation techniques, and packaging, and by the 1950s, a busy homemaker could plan entire meals from a growing menu of canned, dehydrated, and frozen dishes.*

As food processing became more competitive and demand for convenience foods more global, ambitious manufacturers realized it was no longer enough to dominate a single product, like chocolate or cheese, or a single country; to protect their share of this growing business and thus maintain their growth in revenues, companies needed a competitive presence across *all* major product categories and *all* geographic markets. American firms such as General Foods and Kraft began exploratory moves into Europe, Asia, and Latin America, as did European firms. Nestlé bought up hundreds of smaller European food companies and expanded its product line from chocolate, dairy, and coffee to include processed foods such as soups, sauces, and frozen fish and vegetables.

These early campaigns to develop a global presence were slow and difficult. Most countries had their own regulations for food product safety, purity, and even package sizes; further, many governments didn't allow foreign companies to invest in domestic markets. To smooth over these obstacles and speed the development of a truly global processed-food business, the big food companies successfully lobbied for universal, or harmonized, food standards, known as Codex Alimentarius, to be administered by the

---

* Some manufacturers, fearing that homemakers might feel guilty about their declining workload, actually re-complicated some of their products — in one case, taking the eggs out of a cake mix so that homemakers could add their own eggs and thus feel necessary to the process.

United Nations. Companies also helped persuade governments to ease restrictions on the movement of capital between countries, making it easier for acquisitive firms like Nestlé to buy foreign companies.[5]

And buy they did. Nestlé was especially keen to break into the United States, the world's biggest market for processed foods. (Even today, a country with less than 5 percent of the world's population accounts for 40 percent of the world's consumption of frozen food, ready meals, and soup.[6]) Beginning in 1970, Nestlé embarked on a string of expensive acquisitions — among them Libby, Stouffer, and Carnation[7] — that gave the Swiss company instant leadership in dozens of product categories and burnished its image as a corporate big spender willing to buy its way to global dominance. "Nestlé's model is to find a way to operate in all markets where they have a competitive advantage and where they are going to win," says John McMillin, a veteran food-industry analyst at Prudential Securities. "They're like the Yankees, buying the best player at every spot." Today, Nestlé has not only its flagship brands, like Nescafé, Nestea, Stouffer's Lean Cuisine, and PowerBar, but also 8,500 national and local brands in more than eighty countries. "A full 95 percent of Japanese are under the impression that Nestlé is an indigenous Japanese company," observes Friedhelm Schwarz in a Nestlé-approved company profile. "All over the world people are buying and consuming Nestlé products without knowing who is behind the brand they have chosen."[8]

By the 1990s, Nestlé was a sprawling if somewhat decentralized global giant. It had tens of thousands of employees and hundreds of factories, generated billions of food items a year, and consumed so much coffee, cocoa, sugar, milk, oil, wheat, corn, salt, and other raw materials that some fifteen thousand square miles of farmland and pastureland — an area almost the size of Switzerland — were needed to produce it all.[9] Yet far from being an encumbrance, its size has been a source of great strength. Nestlé is such a large buyer of inputs that it has considerable power over the chain of suppliers providing those raw materials; not only can Nestlé bargain down prices paid to farmers (who often have no other buyer), but its global status means it can often source its milk, cocoa, coffee, sugar, and other inputs wherever they can be most cheaply produced.

Size also gave Nestlé and its rivals tremendous power at the downstream, or retail, end of the supply chain. As the sole providers of popular brand-name products, Nestlé, Kraft, General Foods, and other companies could dictate how grocery stores shelved, marketed, and even priced their products. By the latter half of the twentieth century, says John Connor, an agricultural economist at Purdue University, "nearly every important decision on product design, advertising, and promotion was being made by the manufacturers."[10] Food companies had in effect become the rulers of a supply chain that stretched from farmers to consumers, or from "dirt to dinner," and through which more than 95 percent of all calories flowed.

Although farmers, grocers, and a few culinary snobs and health nuts found this transformation threatening, its logic dovetailed perfectly with the times. Just as large-scale farmers could grow food more cost-effectively than could the millions of small farms they replaced, there was no reason for tens of millions of separate households to each duplicate the entire process of preparing food and cooking it when much of that task could be done collectively, efficiently, and far more cheaply in centralized factories. And whatever critics might think of the new processed foods, most consumers embraced the change or, at the very least, regarded it as a necessary evil. As the postwar economic boom drew more women out of the kitchen and into the workforce, fewer households could devote so many hours each day to food. Time, not money, was now the scarce resource, and given a woman's growing potential as a wage earner, it made more sense for her to work outside the home and use her new income to pay someone else to prepare the food. (Her new income also made it possible to buy the refrigerators, freezers, ovens, and, later, microwaves that prepared foods required.) It might not be home-cooked. But by making convenience foods available, declared one industry executive, food manufacturers had given the American housewife "the gift of time, which she may reinvest in bridge, canasta, garden club, and other perhaps more soul-satisfying pursuits."[11]

Of course, consumers' growing time shortage wasn't the only factor driving the development of instant coffee or other ready-to-eat products. The reason Nestlé considered instant coffee in the first place wasn't that consumers were desperate for fast coffee, but that coffee beans had become too cheap to sell in their raw form. By the 1930s, Brazil's coffee plantations, like America's grain farms, were so large and efficient that the coffee market had become glutted. Coffee prices fell so low that Brazilians were actually burning the beans as fuel in locomotives. Desperate coffee-industry officials pleaded with Nestlé to develop a more consumer-friendly coffee product in the hopes of spurring demand. Even though Nestlé had no experience processing coffee (it was primarily a milk company at that point), company executives correctly surmised that if the surplus beans were rendered into a more convenient format, consumers would not only drink more coffee but would happily pay a premium well above the price of raw beans.

This lucrative transformation of raw commodities into finished goods is known as adding value, and it's so ubiquitous today in all consumer product industries that it's easy to miss just how central the phenomenon is to the success, and character, of the food-processing industry. Declining grain prices might be killing farmers, but they had the opposite effect on processors like Kellogg and Nestlé, which were turning those cheap commodities into Corn Flakes and *Kindermehl,* and charging handsomely for the service markup. To be sure, craftsmen had been adding value to grain, milk, and meat for millennia; what is wine but grapes with the added value of fermentation? But with the new tools of mass production and marketing, adding value now offered food companies profit potential that was simply unavailable to producers of raw commodities.

The key is what economists call differentiation. A farmer selling wheat, say, or soybeans has few ways to distinguish his product from that of any other farmer and thus can't charge much of a markup. If he raises his per-bushel price by even a fraction of a cent above the market price, buyers will simply go to his competitors — who use the same seeds, fertilizers, pesticides, and machinery, and who, in effect, have an identical product. The only way our farmer can boost profits is by cutting his operating costs (for example, by scaling up his operations) so he can *lower* his selling price and thus increase the number of bushels he sells. This same low-margin, high-volume reality applies to anyone in the commodity business. Companies that trade in grain, or process grain into flour, or mill soybeans into oil and protein have finished products that are virtually indistinguishable from their competitors' and so can demand only a tiny markup — usually just a few percentage points above the costs of labor and raw materials.

But for manufacturers of processed foods — frozen dinners, say, or breakfast cereal — the economics are completely inverted. Where a farmer fixates on cutting costs, the cereal maker does everything possible to *add* to the cost of the raw grain — with more flavor, for example, or more crispness, or more convenient packaging; in short, any new value that might justify an additional markup. Food processors have become so adept at adding value that by the end of the process, the initial cost of the grain or other raw material is only a tiny fraction of the retail price; of the $3.50 you pay for a twelve-ounce box of cereal at the supermarket, less than twenty-five cents represents the cost of the grain itself. Even taking into account the grocery store's cut (20 percent of the retail price) and the costs of cereal's rather expensive production and packaging (another 36 percent), the cereal company is still looking at a markup, or gross margin, of around 44 percent.[12]

The ability to generate such huge margins is why food companies have moved steadily and inexorably toward higher and higher levels of processing: the more a company processes a raw material, and the closer that commodity comes to being a finished consumer product, the more the company can charge. In 1950, about half the retail price of a food product went to pay the farmer or other producer for raw materials, while half went to adding value. By 2000, this farm share had fallen below 20 percent. What this means is that even as farmers and other producers earned steadily less on their products, food processors and manufacturers were able to maintain their own revenues by steadily adding more value. During much of the twentieth century, the prices for manufactured foods held steady or, in some instances, actually rose,[13] and even today, when grain markets periodically glut, causing grain prices to plummet, food processors rarely lower *their* prices. In fact, when wheat prices dropped 40 percent in the late 1990s, bread makers actually *raised* their prices slightly.[14]

Granted, processors' gross margin isn't pure profit. Unlike the farmer, who simply sells his crop to a grain elevator at a set per-bushel price, food manufacturers face numerous distribution and marketing costs, which, due to the nature of consumer products, are quite high. For example, because most food companies use similar manufacturing processes, many ostensibly competing products are essentially identical. Corn flakes, it turns out, are pretty much corn flakes, no matter who makes them; they are, in a sense, little more than the commodities they are made from. Similarly, because breakfast cereal and other food products represent such a small consumer investment (compared with durable goods, such as cars), there's little economic penalty for consumers who switch between competitors' products.

Put another way, food companies must differentiate their products through heavy promotion, bribing consumers with outright financial incentives (like coupons or discounts) or, more often, inducing consumers to associate a product with a cluster of potent and attractive ideas such as "quality," or "health," or "good parenting" — ideas that may or may not have anything to do with the product itself but which, if managed adroitly, can add immeasurably to a product's image, or *brand*.

Successful brands are not easily created. As much as twenty cents of every dollar you spend on food goes to pay for the promotions, clever packaging, celebrity endorsements, event sponsorships, coupons, and relentless advertising in newspapers, on radio and TV, and now online that keep brands strong and consumers loyal. Soft-drink makers alone spend well over $700 million a year in advertising; U.S. breakfast cereal companies, pioneers of the saturation ad campaigns and kid-targeted brand characters, spend nearly $800 million in advertising, and hundreds of millions of dollars more in coupons and promotions. (The advertising budget to launch a new breakfast cereal is typically equal to half of that cereal's yearly sales revenues, and by its third year, advertising is still soaking up a fifth of sales revenues.*[15]) Altogether, the U.S. food industry invests about $33 billion[16] a year in marketing, more than any other sector save the auto industry. Further, where food purchases account for just 10 percent of all consumer expenditures, food ads account for 16 percent of all ad buys — a degree of disproportionate spending, known as advertising intensity, that is behind only tobacco, over-the-counter pharmaceuticals, and cosmetics.

But food companies are *glad* to pay such costs, because the flip side of a high investment in marketing is a huge potential for profits. Historically, the more heavily a food company advertises a product, the more units of that product it sells. And in manufacturing, more is *always* better. The more units a company sells, the further it can spread its fixed costs and thus the faster it can pay off its investment in its factories and other assets. Better still, under the perverse economics of food, the more units a company sells, the more the company can charge per unit. With nearly any processed food (or nearly any consumer product, for that matter), consumers treat the leading brand as if it somehow has more value, and they are willing to pay a premium to possess that value. Specifically, consumers will pay as much as 4 percent *more* for the top-selling brand than for the runner-up, and up

---

* Many food companies engage in what is known as excessive advertising, that is, spending so much on ads that smaller would-be competitors are effectively barred from entering the market.

to 7 percent more than for the number three brand — even if the three products are essentially identical. In the 1970s, recalls Steve Silk, a former vice president of ConAgra Frozen Foods and brand manager at General Foods,[17] "we knew absolutely that a three-ounce box of Jell-O could be two cents more expensive than the number-two brand, Royal Gelatin, and still maintain its 70 percent market share. We knew it, and they did, too."

This so-called pricing power explains food companies' willingness to spend billions of dollars on advertising. The more they spend, the more units they can sell; the more units they sell, the lower their per-unit production costs *and* the higher their per-unit markup, which results in greater revenues, that can then be reinvested in . . . more advertising. "It's the classic 'virtuous cycle,'" says William Leach,[18] a food-industry analyst at Neuberger Berman. Because top brands consistently generate more revenues, Leach says, "companies that own those brands can afford to do even more marketing, which means they can better support their brands, which then leads to further revenue increases."

Pursuing this virtuous cycle has had large, if often bizarre, effects on food-industry practices. Because brand strength is so dependent on heavy marketing, the big food companies are locked in a promotional arms race to defend their brands and market share; even small reductions in ad "spend" can lead to significant sales declines. And because heavily processed products offer the highest potential markups, they tend to be the most heavily advertised: candy, snacks, prepared meals, cereals, and soft drinks make up barely a fifth of all consumer food purchases,[19] yet account for nearly half of all advertising expenditures — which explains much about the state of the commercial landscape today. (By contrast, less than 6 percent of total ad buys goes for fruits, vegetables, meat, poultry, and fish — foods with little added value or branding potential, and thus little potential for markup — even though these account for 41 percent of all consumer food spending.) But again, for food companies that can afford heavy advertising spending, the payoffs have been great. In what became the formula of modern food manufacturing, companies used their massive scale to drive down the costs of raw materials, used their marketing budgets to build brand power, used their brand power to justify higher retail prices, and pocketed the considerable difference.[20] Thus, where the sellers of bulk or low-processed products have profit margins of between 2 and 3 percent, food manufacturers like Nestlé, General Mills, Kellogg, and Unilever boast margins of 8 to 10 percent, and Coke, with the world's strongest brand and an enormous marketing budget, enjoys a margin of 21 percent.

In an important sense, food companies fundamentally redefined prosperity in the food economy. For centuries, producers had competed for *existing* demand in raw commodities, such as wheat flour or corn — demand that could grow only as fast as the population did. With added value, however, producers were selling not just calories but convenience and other qualities as well, which suggested an almost unlimited source of revenues and growth — at least as long as companies could find more value to add.

In a conference room near Nestlé's sensory lab, research director Peter van Bladeren is walking me through one of the greatest technical feats in recent food-processing history: low-fat ice cream that doesn't taste low fat. "If you simply take out the fat, you end up with a sorbet, which has a totally different behavior in the mouth," says van Bladeren, who is blond and slender and seems to relish talking about food science, even with a journalist. "And if you simply replace the fat with some other ingredient, like sugar, the consumer always notices." What Nestlé's researchers realized is that ice cream's mouthfeel comes not from fats alone, but from the way that fats align with the ice crystals, sugars, and proteins that make up the ice cream molecule. By varying the ice cream churning methods, researchers were able to subtly rearrange the molecule's structure so as to halve the fat content while actually boosting mouthfeel beyond that of full-fat ice cream. The breakthrough wasn't cheap; bringing the concept to market took Nestlé's U.S. subsidiary Dreyer's five years and a reported $100 million. But the result, introduced as Dreyer's Slow Churned Light in 2003, just as obesity anxiety was peaking, was a monster hit. Within a year, Nestlé's had taken more than half the $350 million U.S. light ice cream market, mostly at the expense of archrival Unilever — and now seems poised to dominate the global ice cream market.[21]

Such success underscores yet another reality of the modern food business. For all advertising's potency, even the best marketing campaigns will at some point stop working because, sooner or later, even the hottest products lose their power. The pattern is fairly predictable: when a product is introduced, sales initially climb but then flatten out and eventually enter a gradual decline. Thus, in addition to nonstop advertising, food companies must bring out a constant stream of new products to keep sales up and revenues high.

New products come in several categories. The "new and improved" products are intended primarily to revitalize existing lines, with a new flavor, say, or a packaging gimmick; many are brought out mainly to copy a rival's product. But the thoroughbreds of product innovation are the "killers" — items so unprecedented that they create their own categories. Nescafé was a killer — before 1937, there simply was no such thing as instant coffee; so were Kellogg's Pop-Tarts, in 1963, and Oscar Mayer's Lunchables, in 1988. Killer products are important for several reasons. When a killer hits the market, it enjoys a virtual monopoly on consumer attention and dollars until competitors can respond with their own copycats. And even then, the killer tends to retain its first-entrant lead in sales and market share and, thus, its greater pricing power. "First entrants essentially get 'naming rights' for the category," explains Ronald Curhan, professor emeritus of marketing at the Boston University School of Management and a leading chronicler of American food retailing. "Think about it: there may be three brands of toaster pastries, but everyone still calls them 'Pop-Tarts.'" New products are so vital to manufacturers (by the late 1990s, a third of all industry revenues came from recently introduced products) and so unpredictable (just one in three introductions survives for three years) that companies must continuously release new products (currently, around fifteen hundred a month)[22] in the hopes that at least some of them will survive.

To deliver this prodigious output of novelty, food companies have become perpetual product machines, continually translating consumer desires into successive waves of high-margin products. In sophisticated facilities such as Nestlé's research center in Lausanne, scientists pore over sensory maps, sift through sales data, and disassemble competitors' products. Teams of analysts fan out into the field like anthropologists to study how we consumers use our food — in our homes, at work, in our kitchens and dining rooms. After decades of such intensive scrutiny, companies such as Nestlé, and Kraft, and Heinz have not only broken down the mysteries of taste and preference into quantifiable data sets but often know more about what we like to eat and why than we do ourselves. Companies know precisely how the preference for saltiness or crispness varies by gender, age, ethnicity, and nationality: older consumers prefer stronger flavors, in part because their taste buds are worn down; Asians lean toward salty and crisp snacks; Americans are mad for new flavors, yet never stray far from the familiar — macaroni, meat loaf, and "nostalgia driven flavors."[23]

As important, food companies understand how consumers' relationships to food have changed dramatically and how to exploit those changes. In developed countries, where scarcity has been replaced by abundance and endless choices, consumers no longer buy food products solely on the basis of brute calories or even flavor but because the product embodies a particular set of value propositions: it's convenient, it makes some nutritional claim, or it supports a particular lifestyle or identity. For example, one new and lucrative identity is the "foodie," an upscale urban consumer who sees food as a hobby and as a vehicle for socializing and fun.[24]

The food-as-fun concept has also been crucial in developing products for kids, who are drawn to foods that are colorful and provide opportunities for play — a finding that explains much about the neon-colored, toylike qualities of everything from kids' breakfast cereals and energy drinks to candy, lunch-kits, and even condiments. When Heinz wanted to boost ketchup sales in 2000, the company released ketchup dyed bright green and purple — and promptly saw ketchup consumption jump 12 percent.[25] "Today's kids are exposed to lots of vibrant colors and animation, and they expect these same experiences at the dinner table," explained retail expert Gene Grabowski at the Heinz debut.[26] "To kids, food is about more than eating. Color, taste and touch are vital elements for differentiation." Colored ketchup, Grabowski added in a somewhat disconcerting aside, "is a great example of how an item can transcend its food status by first, delighting kids."

And of course, food companies have become especially adept at exploiting our darker food associations. The identification of working-mom guilt led companies to develop the early premade meals for children. "Home-style" products help compensate for the inability to cook at home, while obesity fears have yielded a steady stream of low-fat products. Even the anxieties surrounding the birth of a child can be profitable: new parents are simultaneously time-pressed and obsessed with food safety, and therefore ripe for a variety of new product "relationships." "The period of adjustment and learning of a new [parenting] lifestyle presents opportunities for manufacturers and retailers," market analyst Neil Broome explained in an interview with the trade journal *Just Food*.[27] "A successful strategy targeting new young families will not only reap rewards in terms of direct sales to this group, but will also have positive effects on the ability to target these consumers through their years of parenthood."

In fact, food-safety fears are turning out to be a boon for the big processed-food companies. "As more and more of us move away from producing our own food," says Nestlé's Leathwood, "we've become less and less certain about its safety, its quality or its taste." Such fears, Nestlé believes, are actually making consumers *more* appreciative of today's sophisticated processing technologies, such as tamper-proof and shelf-stable packaging, as well as the familiarity of well-known brands.

For all these sophisticated new categories and tactics, however, many new products target the same consumer desire that the earliest products did: convenience. Food companies know exactly how much time the average household can devote to cooking — around thirty minutes a day, down from an hour in 1970 — and how quickly that number is expected to shrink: by 2030, the ideal cooking time is forecast to be between five and fifteen minutes.[28] Industry analysts have also tracked the decline in cooking frequency (less than half of home meals feature even one freshly made, or from-scratch, item);[29] they have measured the general fading of cooking knowledge (cookbook publishers have begun drastically simplifying their recipes)[30] and documented the demise of the home-cooked sit-down meal. In many households, family members are so overwhelmed by school, work, and conflicting activity schedules that they increasingly eat separate and often different meals — a practice known in industry parlance as flexi-eating. "Today's average home-maker isn't cooking a meal for five people anymore," Steve Silk, the former General Foods executive, told me. "She's cooking five different meals." And "meal" may be an overly generous term; according to one recent U.S. study, sandwiches are now the most commonly served dinner entrée, ahead of beef and chicken dishes.[31]

Social commentators bemoan our shorter mealtimes, infrequent cooking, and fragmented family meals as the cause, or symptom, of family decline. But for food companies, consumers' gradual disengagement from the process of making food has been a windfall. Not only has it meant rising demand for processed foods, but it has put an even greater premium on the added value of speed and thus given companies new opportunities for innovations in food formulations and packaging that let consumers eat in just minutes. One of Nestlé's best-selling recent offerings was Lean Cuisine Frozen Panini, whose specially designed package "grills" the sandwich in the microwave in just three minutes. In the near future, even three minutes may seem overly long. Because millions of office workers now routinely eat lunch at their desks (the lunch hour having been replaced by the "deskfast," or dining "al desko"), food makers are rapidly expanding lines of ready-to-eat vending-machine meals.

Food companies are in fact anticipating the end of meals as we know them, as a growing number of consumers skip more and more sit-down meals altogether. Datamonitor, a British company that analyzes global food sales, estimates that the average American already misses breakfast every third day and is beginning to skip a large number of lunches and dinners. And while such a trend is extraordinarily bad news for consumer health, it is providing yet another opportunity for food makers, because as consumers eat fewer regular meals, they're compensating by eating more of another even more lucrative food category: snacks. In the United States, according to Datamonitor, snacking now accounts for nearly half of all "eating occasions."[32]

Not surprisingly, snacks, which are among the most highly processed of foods and thus have among the highest profit margins, are increasingly the focus of product innovations. Companies are betting heavily not simply on candy, chips, cookies, and other traditional snacks but on a new generation of portable, or "on-the-go," products, designed to be consumed anywhere, anytime, with zero preparation. For example, when researchers at Skippy realized that the traditional mode for peanut butter consumption — the sandwich — had become too complex for time-pressed families and kids, the company introduced single-serving tubes of peanut butter, called Squeeze Stix, that kids empty directly into their mouths. Kellogg, meanwhile, has successfully recast its breakfast cereals as portable snacks by repackaging them in a mobile "bagged snack format."[33] And according to Datamonitor, as companies continue to tailor the snack concept to specific demographics — working parents, for example, or teenagers whose hands are full of iPods or cell phones — the critical thresholds for product development will be whether it "can be consumed one-handed, and whether packaging causes a mess."[34] The future of food is as an accessory.

In Singen, Germany, a small industrial city on the Rhine, motorists on Lange Strasse can glimpse the state of the art in industrial food: the Nestlé Product Technology Centre. Built in 2003 at a cost of $27 million, the sprawling complex develops everything from frozen meals and pasta to chilled and "wet" culinary foods (sauces, mayonnaises, and baby food) for the rapidly growing markets in Europe. Like the research center in Lausanne, the Singen facility boasts extensive laboratories, test kitchens, and tasting rooms, as well as something else: a pilot factory, a miniature production plant where technicians pretest each new product's manufacturability — ensuring that it can be made in mass quantities, cost-effectively, from readily available ingredients, and by using industrial machinery, without suffering too much in quality. "Anyone can have a great idea in the kitchen," Nestlé's Leathwood tells me. "But in scaling it up to full production, things that worked on a small scale just don't work, or don't work in the same way."

In fact, in the modern food business, *most* traditional food practices, and most traditional foods, just don't work. As production has become almost entirely automated, with vegetables diced, meats ground, batters mixed, doughs extruded, and ready-to-serve dinners assembled all by computer-controlled robots at rates of thousands of units per minute,[35] the food itself has had to be amended, often significantly, in order to tolerate the process.

In some cases, the amendments are relatively mild: perishable ingredients like milk must be dehydrated or homogenized; meats are frozen; vegetables canned. (In fact, cans and other packaging, a critical if rarely acknowledged piece of the modern food system, make up the industry's second biggest cost, after labor.) In other cases, however, the rigors of the manufacturing process have required the development of an entirely new job description — the food engineer — to actually change the molecular structure of the food. Companies found they could thicken and preserve vegetable oils by injecting them with hydrogen atoms — thus, hydrogenation. They learned to stop flours and powders from clumping with anti-caking agents; they found that humectants like glycerin and sorbitol kept moist foods from drying out, and emulsifiers prevented fats from separating.

Companies also began using additives to repair the damage done to food during manufacturing. In flours, the vitamins and minerals that were destroyed by bleaching were replaced, a process companies called fortification. Colors that had faded from vegetables and meats during cooking or pulverizing needed to be restored, and so the foods needed brightening; chlorophyll put the green back into canned peas, while caramel lent a more wholesome, home-cooked brown to pressure-cooked meats.

Flavor also needed resuscitating. Natural flavor is easily damaged by heat and other processing rigors; as little as 3 percent of a cookie's flavor remains after baking, and much of that gets absorbed into the packaging. Some flavor losses can be minimized by changes to the manufacturing process; for example, many commercial bakeries now spray flavorings on cookies and other confections *after* baking, to avoid heat damage. But the easiest method is simply to boost the flavor synthetically. Although natural flavors are made up of hundreds of molecules, the dominant note for any flavor usually comes from just one or two molecules, or character-impact compounds, which are often quite easy to synthesize: vanilla's loudest note, for example, derives from a single compound — 4-hydroxy-3-methoxybenzaldehyde, better known as vanillin; the essence of chicken broth can be re-created by the amino acid L-cysteine. And these are merely the primary colors of flavor; with the right chemistry, companies like Givaudan, IFF, Firmenich, and Symrise, leaders in a global flavor industry worth $18 billion a year,[36] can duplicate nearly any flavor experience imaginable. Breads can be made to taste more breadlike; orange juice orangier; canned meats meatier; pork more porklike.

Beyond their ability to soften the impacts of processing, flavorings and other additives offer food manufacturers other valuable services. Because additives are generally made from industrially available materials (vanillin, for example, can be synthesized from paper-manufacturing residues), they're usually much cheaper than their natural counterparts. Beverage makers spend only a fifth as much money flavoring their strawberry soda artificially as they would using natural strawberry.[37] Bulking agents such as pectin and xanthan gum let companies add heft and mouthfeel to fillings cheaply.[38] In fact, one has to be impressed by the way food manufacturers now routinely re-create flavor from the cheapest of commodities: corn, for example, provides meal for breads and crackers, starch for adding bulk to processed meats and burgers, hydrogenated oils to replace butter in fillings and baked goods (and even the cocoa butter in chocolates) and, of course, high-fructose corn syrup, or HFCS, a cheap substitute for sugar in many processed foods.*

Additives also let manufacturers skirt not only the costs of natural ingredients but their finite supply. Demand for grape flavor by makers of so-

---

* U.S. sugar prices are about twice as high as world prices due to restrictions on cheaper imports.

das, gum, candy, and other foods now exceeds the quantity of grape flavor produced naturally — that is, in actual grapes — by a factor of ten to one, according to Gary Reineccius, a flavor engineer and professor of food science and nutrition at the University of Minnesota. Similar disparities exist for many flavorings, like vanilla and coconut, which are now used in both food and cosmetics. Even certain textures are hard to come by: because demand for creaminess and crunch now exceeds global supplies of dairy fats and lard, these textures are increasingly duplicated with hydrogenated vegetable oils and tropical oils, like palm.[39] Even the demand for more complex flavor experiences exceeds the industry's capacity to supply them naturally. Beef gravy, for example, is traditionally made from flour, butter, milk, and the pan drippings of a beef roast. "But if you're a company making ten thousand gallons of gravy in one batch, how do you flavor it?" asks Reineccius. "You can mix massive quantities of flour, butter, and milk, but you don't have pan-drippings, not for ten thousand gallons, and you're not going to grind up that much roast beef, so you need to add a flavoring" — usually an additive like monosodium glutamate, or MSG, which the mouth reads as meat flavor.

Food additives, and food engineering generally, have allowed companies to dramatically simplify what was once a very complex process — cooking — and thus gain a considerable measure of control over costs. In homemade foods (or classic food, in industry parlance), attributes like flavor and texture all depend on traditional and very specific ingredients and cooking procedures: a traditional apple pie, for example, can be made *only* from apples, sugar, butter, flour, shortening, salt, and spices, and baked *only* in an oven — requirements that are very costly to replicate en masse in a factory setting and from industrially available materials. By contrast, in the reengineered version of an apple pie, or any food product, a company is free to create flavors and textures by whatever ingredients and processes give consumers an acceptable food experience while also meeting the company's cost and operational imperatives. And although some consumers object strenuously to artificial flavors, many of us have become so accustomed to the synthetic version that we actually prefer it to the original.[40] Benzaldehyde, which creates cherry flavor, is now more familiar than natural cherry, while the compound diacetyl became "butter" for many consumers of microwave popcorn — that is, before it was taken off the market in mid-2007 as a possible cause of lung disease.[41]

In fact, because food engineering and ingredient substitution are such routine practices and because consumers seem increasingly willing to accept synthetic and processed products, companies have removed a large measure of the risk from food production. Not only is preparation itself much simpler and more rational, but companies can more easily switch among ingredients and suppliers and thus protect themselves from price increases, crop failures, or many of the other perils that have traditionally beset food production. Using the now standard disclaimer "May include one or more of the following . . . ," a food company can sweeten its products with corn syrup from Iowa or sugar from Brazil, depending on which is cheapest.[42] A company can switch from cottonseed oil to soy oil, from cornstarch to potato starch, from soy protein to wheat protein or to whey, or make any number of other substitutions depending on availability and price or some new consumer concern, and yet mask these scattered origins, ingredients, and methods under the aegis of a single unifying experience and a single unifying brand. A Nestlé's Crunch is still a Nestlé's Crunch, no matter where the sugar or cocoa comes from. Indeed, as long as the brand remains strong and the products themselves continue to deliver their explicit and implicit promise of taste, texture, convenience, status, health, purity, cost, or any of a myriad of added values, companies are free to produce those products in the most economic and profitable way possible.

The relentless need to expand markets and issue a steady stream of new products and additional value has frequently led manufacturers into questionable territory. Early food companies routinely enhanced their products with an assortment of inedible and occasionally toxic materials (extending flour with chalk, for example, or using lead to brighten candy colors) and promoted their products with fraudulent claims. And while the more egregious of these practices were curbed by reform legislation in the early twentieth century (at least in the United States and Europe — China's food industry, it seems, is just now entering its own food-safety jungle), food companies continue to stretch the boundaries of what is permissible to add to, or say about, a product.

In the 1980s, for example, Nestlé and many of its rivals were excoriated for aggressively marketing infant formula in developing countries. The companies worked assiduously to persuade mothers to switch from breastfeeding to formula — handing out free samples and paying local doctors to recommend the formula and denigrate breastfeeding — even as evidence mounted that the practice could be deadly. In Africa in particular, many mothers prepared the formula using contaminated local water; others, hoping to economize, so diluted the formula that their babies essentially starved to death. Nestlé and other formula makers reined in their aggressive promotional practices in Africa and have since become loud promoters of breast milk. But the controversy remains one of Nestlé's most enduring public images — not least because Nestlé and its rivals, despite their professed support for breastfeeding, continue to earn billions of dollars a year selling infant formula in the emerging markets of Asia.

Moreover, food companies continue to stir controversy with their heavy reliance on entirely legal ingredients — chemical additives, but also fats, salt, and sweeteners, which manufacturers use, often copiously, as a means to boost flagging flavor. (As one industry executive told *Business-Week* in 2005, "[A]ll those food technologists know that the more fat and the more sodium you add — and it's cheap to do that — the more you help the flavor. It's much cheaper to put fat and salt into food than real expertise."[43]) Sweeteners are used so routinely to enhance flavor and texture in cookies, crackers, candies, breakfast cereals (which now contain almost as much sweeteners as candy does), and even breads, sauces, and canned vegetables that many consumers now regard unsweetened foods as unpalatably bland. And yet it was precisely such practices that by the 1980s were being scrutinized as possible contributors to a variety of health problems, including an increase in obesity rates. Critics were also raising concerns about the industry's impacts elsewhere, such as its excessive and energy-intensive packaging.

Food companies, of course, have tried to change their packaging and promotion strategies, and the industry now speaks constantly about its ongoing efforts to develop healthier fare and become more sustainable. But even as the industry reassessed its practices, it came under pressure of a different sort: after decades of near-monopoly control over the food business, the big food companies found themselves in an economic battle that required them to shift many of those questionable practices into even higher gear.

Competition from restaurants, especially fast-food chains, has been steadily taking more of the consumer's food dollars,[44] often by appropriating manufacturers' cherished concept of convenience and advancing it to a level the traditional food companies haven't been able to copy. Companies such as McDonald's have embedded the idea of time savings and ease of use into every particle of their products and processes. Menu items are engineered to be not only quickly prepared but also easily consumed; many are meant to be eaten one-handed while driving. Store locations are carefully chosen for ease of access: suburban outlets allow consumers to drive to the restaurant, eat, and get home in less time than it takes to cook a meal. Similarly, with fast-food outlets in gas stations, feeding the family is now no more difficult than fueling the car. The goal of this convenience strategy, boasted McDonald's in its 1994 annual report, "is to monitor the changing lifestyles of consumers and intercept them at every turn. As we expand customer convenience, we gain market share."[45] Indeed. In 1962, U.S. consumers were spending just twenty-eight cents of every food dollar in restaurants;[46] today, nearly half of the $840 billion Americans spend on food annually is spent "away from home," and of that, more than half is for fast food. Harry Balzer, a market analyst with the NPD Group, is only half joking when he calls the power window "the fastest-growing food appliance in America today."[47]

Food manufacturers haven't been idle, as a walk down the aisles of any grocery store makes clear. The frozen pizzas, single-serving yogurts, snack bars, energy bars, breakfast bars, and pastries; the instant oatmeal, instant soups, and instant noodles; the string cheese, microwave popcorn; lunchkits, and canned lattes — all represent a counterattack by food manufacturers to take back convenience. And yet walk a few more yards and you'll come to an in-store deli, with a long line of customers and a long display case full of ready-to-eat meals — salads, noodle dishes, and rotisserie chickens. Grocery stores too have been cashing in on dining out.[48]

In fact, everyone now wants a piece of the convenience market. Gas stations and convenience stores sell jo-jos and burritos. Starbucks sells breakfast. Even the big commodity companies, weary of the low-margin commodity business, are inching down the supply chain into consumer space and processing their raw materials directly into their own value-added foods. Archer Daniels Midland, more famous as a dealer in raw soybeans and a maker of high-fructose corn syrup, now blends those ingredients into finished foods for restaurants, while another commodities giant, ConAgra, turns its raw materials into consumer-ready products under such familiar brand names as Butterball and Chef Boyardee.

More crucially, grocery stores have diluted and even usurped manufacturers' lucrative brand concept. Working with private-label food manufacturers such as Ralcorp, most grocery chains now offer generic versions of everything from instant coffee and breakfast cereals to fully prepared frozen meals that not only rival national brands in quality but are also considerably cheaper, dealing a serious blow to the price power leading brands once commanded. "Manufacturers lost control of their brand," says Boston University's Ronald Curhan. "Here they'd spent the last hundred years building up the value of their brand, and now it becomes another homogenized commodity being sold indiscriminately against other commodities."[49]

These are not inconsequential developments. Because food manufacturers are such an integral part of the entire food economy, the setbacks they experience have broad ramifications for the food system as a whole. Once manufacturers lost the ability to differentiate their products by brand alone — and as their products began behaving less like premium items and more like commodities — manufacturers themselves began behaving more and more like commodities producers. Just as farmers responded to declining prices by getting bigger, many food companies embarked on desperate acquisition sprees to gain market share and greater economies of scale. The 1989 mega-merger of tobacco company Philip Morris with Kraft, General Foods, and Post — which brought names like Oscar Mayer, Philadelphia, Nabisco, and Maxwell House under one enormous roof — was only the start of a consolidation wave that left the food business under the control of a much smaller number of much larger businesses. Whole sectors have been all but monopolized by single companies. Nestlé's 2002 purchase of U.S.-based Dreyer's Ice Cream, for example, not only made Nestlé the largest ice cream maker in the world (with 17 percent of the global market and nearly a quarter of all U.S. sales)[50] but also gave it more than half of the premium ice cream market in the United States.

More important, in the same way that farmers responded to declining margins by growing more bushels, food makers moved toward a strategy of ever more massive output — more ads,[51] more marketing campaigns, and, of course, more new products, which are even more critical today. Food companies are now so big and operate across so many product lines that they can no longer survive on just one or two killers a year; they need dozens to, as one analyst put it, "drive the needle" on continued growth.[52]

This is one reason that the number of product introductions has soared — from about fourteen thousand a year in 1995 to nearly nineteen thousand in 2005.[53]

And yet despite the increased numbers, the prospects of a product becoming successful are slimmer than ever. Most traditional food categories are already saturated with products, and many new food trends — the fresh-foods craze, for example — are difficult for food manufacturers to exploit. Companies keep trying, of course. The industry's recent discovery of the Hispanic market has food companies frantically bringing out somewhat less processed items to entice a population that still does a lot of home cooking. Most food companies have also made peace with the enemy and now supply restaurants with a wide variety of premade foods (which are then presented to consumers as freshly made). Food manufacturers are also leveraging growing fears of obesity, hypertension, and other nutritional ills by offering lines of healthier, lighter products: Nestlé, for example, is recasting itself as a "health and wellness" company and offering some products with less salt and fat and more so-called nutraceutical ingredients, engineered to deliver a quasi-medical value.

But for all the talk of a new healthier industry, companies have been just as quick to exploit demand for the not-so-light fare. Most have brought out lines of comfort-food products, including meat loaf, lasagna, macaroni and cheese, and, of course, pizza, one of the most popular prepared foods ever invented. (Americans, according to one industry researcher, "love and are comfortable with cheese."[54]) Many companies are trading in their home-style products for products done up in something called restaurant style, which seems to mean larger servings and a richer ingredient list. And premium or upscale desserts remain a key market: the same day that Unilever bought Weight Watchers in 2000, it also bought Ben & Jerry's.[55]

Food makers have also made increasing forays into snacks, one of few remaining growth opportunities left in mature markets such as the United States, western Europe, and Japan. Whereas sales of nonsnack foods are growing at barely 1 percent a year or less, sales of candy, chips, cookies, crackers, various mini-meals, and other snack foods are climbing by around 5 percent a year and generating more than $65 billion in annual sales in the United States alone.[56] Further, because snacks are among the most heavily processed of all processed foods, their large measure of added value and associated large margins help companies compensate for poor sales in other categories.[57] Indeed, in mature markets, snacks may offer the most profitable avenue still open to food makers — which means that for companies like Nestlé, future success now depends even more on the continued decline of the sit-down meal as a significant part of Western food culture.

Although the United States and other mature consumer markets are hard for the big food companies to profit in, the story is quite different in the emerging markets of eastern Europe, South America, and Asia, where the same economic and social trends that launched convenience in the West a century ago are taking root — and where the big food companies are now moving aggressively. Industry excitement is especially keen over China. Its five hundred million urban consumers are growing wealthier and are also working longer and longer hours and showing less inclination to cook. Sales for premade dinners in China, though currently just a fifth of that in the United States, are expanding almost ten times as quickly, and the Chinese ready-meals market is expected to top $6 billion by 2009.[58] In thriving urban centers such as Shanghai and Beijing, consumers have embraced everything from breakfast cereals (Raisin Bran, Fruity Pebbles, Oreo Extreme, and Special K are especially hot) to energy drinks, foods with herbal additives, and, surprisingly, coffee products: Nescafé is doing exceedingly well here, and when research indicated that consumers would also buy an instant milk-tea mix, Nestlé had one designed, tested, and in grocery stores in just seven weeks. "It's a consumer renaissance," says Josef Mueller, who runs Nestlé's China operations. China is evolving "from the 'iron rice bowl' to a fledgling consumer culture, even more competitive than the U.S."

Nestlé is betting heavily on that renaissance. Near the old airport in Shanghai, the company has built a research center, with tasting booths, a pilot plant to test new recipes, and a staff of food experts who carefully scrutinize local food patterns for product opportunities. When I visited the center, its director, Chris Brimlow, showed me a map of China in which the nation had been divided not by political affiliations or topography but by flavor preferences. Consumers in the predominantly Muslim western provinces, for example, tend toward heavily spiced meat dishes; in Beijing, by contrast, consumers prefer strong flavors, wheat-based foods, and plenty of salt. Another diagram described the Chinese system of herbal medicine, showing how consumers here have always regarded food not simply as calories but as a means of treating illnesses and restoring spiritual balance.

Studies like these have helped Nestlé bring out products that mix Western attitudes with Chinese flavors — for example, a black pepper–flavored
snack bar called *Yo* that Nestlé hopes will appeal to China's emerging on-
the-go consumer who still wants traditional flavors. "We're deconstructing
a cuisine," Brimlow tells me. "We're describing the fundamentals, seeing
where we can add value." Indeed, in return for its initial $1 billion invest-
ment, Nestlé generated sales in 2004 of $1.3 billion and has seen annual
growth of around 20 percent.

Critics of this Westernization of traditional food cultures may take
some solace in the fact that the sheer inertia of these older cultures will
slow the conquest. In China, few homes have conventional ovens (baking
was never part of traditional cuisine), and microwaves are not yet widespread. More fundamentally, Chinese food preferences and habits are in
some ways completely antithetical to the model of prepackaged convenience foods. Although younger consumers embrace high-tech foods, many
middle-aged and older consumers still put a high premium on fresh food
and local producers; wet markets, where grain, produce, and meat are sold
fresh from the farm and fish swim live in tanks, remain popular, and many
housewives will travel to the market several times in a single day to get food
for a particular meal. Further, many of these potential emerging markets
are in fact a collection of markets — a patchwork of local cultures, languages, customs, and cuisines, which are harder to reach with a single
product or a single national marketing campaign. Perhaps most problematic, in many emerging economies, cooking is still widely practiced. In India, despite rising incomes and a middle class larger than America's, the
penetration of prepackaged food is tiny: more than 90 percent of Indian
food is processed in the home or at local shops.

And yet companies like Nestlé are finding opportunities to break
down even these barriers. In Asia, ice cream is proving surprisingly popular
among a people that aren't supposed to tolerate dairy products; in fact,
Nestlé's researchers now contend that Asians aren't any more lactose intolerant than any other ethnic group. The problem, Brimlow told me, is that
cow's milk has historically been so scarce and expensive in China that most
Chinese never developed the enzyme needed to digest dairy foods. If Chinese children are introduced to milk products early on, says Brimlow, they
have no trouble tolerating lactose — a finding that has spurred Nestlé's
China operation to launch a wide range of dairy products aimed at the

youth market.[59] "Even as adults, it takes only three months to develop the enzyme," Brimlow says. "They may feel a little sick for a while, but they get used to it. Yogurt is a great way to reintroduce dairy." And dairy is a great way to generate high margins. Nestlé is especially gratified to see demand grow for its high-end, premium products, whose fat margins cover rising raw materials costs and which accounted for more than half of Nestlé's product launches in 2006. "If we can convince the consumer to pay more for the higher margin, impulse ice cream products," said Frank Li, marketing manager of Nestlé's ice cream business in China, "we can deal with higher costs."[60]

And in China, as elsewhere, Western food companies have shrewdly exploited consumers' fears of unsafe food — a fear that has become something of a national passion in China. Long before the melamine scandal alerted Western consumers to China's problems, consumers there witnessed a series of food-safety scandals involving tainted food, including black-market infant formula that left dozens of babies with permanent brain damage, and they are now far more likely to seek out prepackaged foods with familiar brand names. "The Chinese perceive us as offering quality and security," says Nestlé's Mueller. "We have a very powerful image here."

For many locals, the image is too powerful. Although Asian governments have officially been quite welcoming toward Western food companies and Western methods, seeing them as a source of much needed investment and expertise to upgrade their own overwhelmed food sectors, many in government now privately say the same things their Western counterparts have been saying about the Western food industry. Health officials in China and elsewhere in Asia and the developing world see the arrival of premade convenience foods, snacks, and sweetened beverages (China now accounts for one in every ten carbonated beverages consumed worldwide)[61] as contributing to a larger trend toward obesity and poor nutrition. Health officials and private watchdog groups also say that the Western food industry hasn't quite abandoned its old, unsavory marketing tactics. According to the United Nations, the big makers of infant formula have gradually intensified their marketing efforts in Asian countries and, by some accounts, are using a few of the same deceptive methods as in the past, paying doctors to tell new mothers that formula is better than breast milk, or that breast milk alone cannot meet an infant's nutritional needs. In one case, according to a former Nestlé employee interviewed by the *Christian Science Monitor*, Nestlé itself advised Chinese employees attending a company-sponsored postbirth class that its Good Start baby formula was superior to breast milk. "The company told the staff that infant formula is better," the woman, Ding Bing, told the *Monitor*. "It didn't tell us that mother's milk is enough."[62]

But ultimately, neither these complaints nor any of the other challenges inherent in entering emerging markets will deter the industry's expansion. With declining sales in their home markets, big companies like Nestlé and Pepsi and Kraft cannot afford to miss opportunities in the developing world, whatever the costs. And slowly, inexorably, the barriers are falling. During the 1990s, international food companies began the process of unifying these chaotic emerging markets in Asia, eastern Europe, and Latin America by buying up hundreds of local food and beverage brands. Consumers too seem to be playing their part. As one ebullient food-industry consultant noted recently about Latin America, "maturing demographics, rising urbanization, dual income households help explain why Latin Americans continue to abandon the tradition of home cooked meals in exchange for fast food and convenient prepared foods. The trend is expected to accelerate over the next decade."[63]

3

Buy One, Get One Free

O N   T H E   G R A Y   R U B B E R   F L O O R of a meatpacking plant in northwestern France, surrounded by pork in various states of disassembly, a certain Monsieur M. is giving me an impromptu lecture on the anatomy of the postindustrial pig. M., who is president of a company that makes premium "chilled" meats for upscale consumers and is somewhat shy of publicity, has invited me to tour his plant and just now is standing near a production line, watching two employees reconstruct a pig "leg" from pieces of pork. The pieces, deboned, trimmed of all fat, and brined for twenty-four hours, lie in small metal bins around the workers, like components for some factory assembly line. Every few seconds, one of the men selects a piece, checks its appearance, and then lays it carefully alongside other pieces inside a meter-long, loaf-shaped stainless steel mold — with the result being a stack of meat that mimics the muscular strata of a pig leg. "Of course, in a real leg, you would have six muscles," admits M., a fortyish man whose athletic build and stylish suit are hidden beneath his white factory coveralls. "Here we use five, to keep the shape as natural as possible." He shrugs. "Or even better than nature."

M. is only partly joking. When the mold is pressure-cooked, the separate pieces of meat will bind into a single chunk of pork that, when sliced into thin cross-sections, looks just like real ham — albeit without a bone and, more important, without the normal fat, connective tissue, or color variations that now are increasingly unsettling to the French consumers who frequent the big grocery chains. "Customers want ham to be pale pink and homogenous," M. tells me almost apologetically. "And this is a problem, because the muscles you find in a real pig leg are sometimes dark, sometimes light, and sometimes next to fat, which is not welcomed by the customers. Or retailers. So now we must make our hams homogenous."

That consumers in France, bastion of traditional *porc*, are now fright-ened by the look of the real McCoy is an irony not lost on M., who despite years in the processed-meat business is something of a gourmet. Yet given all the other changes in the meat trade, the decline of the French palette is the least of M.'s worries. When he launched his company, a joint venture with U.S.-based Smithfield Foods, in the early 1990s, the French meat-packing business was brisk and margins fat. Since then, the food economy has been turned upside down. The big French grocery stores and food-service chains, like their counterparts in England, Germany, and especially the United States, are so powerful that they now dictate everything about the products they buy from M., from how the meat looks to how much it costs. Beyond their demands for "homogenous" hams, French retailers routinely insist on discounts of 15 percent or more, squeezing M.'s margins to the bone. In such a market, M. says ruefully, "you just watch your profit melting in the sun."

M.'s complaint is hardly unique. After a century in which manufacturers like Nestlé and Kraft essentially pushed calories through the supply chain, with products and prices that satisfied their own strategic requirements, today's food economy is driven far more by the pull of demand. Demand from consumers, who expect the food they buy to be better and cheaper every year, but, even more important, demand from retailers — mega-grocers such as Wal-Mart, French-based Carrefour, and Britain's Tesco, as well as food-service giants such as McDonald's, Burger King, and Wendy's, huge companies that have channeled consumer expectations in ways that have put the *sellers* of food, not the producers, firmly in charge of the food chain.

For most consumers, this retail revolution has been a largely beneficial acceleration of the historic trend toward food that is more convenient, more varied, and certainly less expensive. To stroll through the grocery store and see heaps of fresh produce in January, fresh fish from Chile, or economy-sized packages of giant-size boneless chicken breasts — all at ab-surdly low prices — is to realize that the food economy of today is light-years from where it was even a decade ago.

But as M.'s melting profits aptly illustrate, such improvements at the retail level come at a huge cost to those farther up the supply chain and are changing the rest of the food economy in ways that make even the "chemogastric" revolution of the last century seem mild by comparison. Whereas in the past, manufacturers and processors profited by charging a premium for convenience, pleasure, or other added values, now retailers make their profits by offering *more* value — more freshness, for example, or more year-round availability, or simply more food — while charging consumers *less,* a contradictory proposition that retailers have achieved only by squeezing margins in the global supply chain to the point of non-existence.

To survive this relentless downward price pressure, food manufacturers have embarked on desperate cost-cutting strategies, building even larger production facilities, finding even cheaper materials, but mainly by obsessively seeking out more and more efficiency — more efficient equipment, more efficient workers, and even more efficient food, especially fresh produce and meat. With rapid advances in breeding, production, and shipping, a ham or a chicken can now be manufactured as uniformly and precisely as the frozen pizzas or fast-food burritos they wind up in.

In an important sense, food, even high-end food, is being transformed back into a commodity, a reversal that has accelerated the trend of declining prices but is generating other, less tangible costs. Our hyperefficient, growth-dependent system now overproduces more readily than ever and is flooding the food economy with even more unnecessary calories. In addition, the relentless focus on price has resulted in food that is of lesser quality and nutritional value; a food culture that is increasingly defined by value pricing and portion size; and a global production system so lean and just-in-time that it is simultaneously more likely to be disrupted (by foodborne illness, for example, or a spike in energy prices) and less able to absorb the impact of a disruption. And as with the earlier transformations of farming and processing, even as our confidence in this new retail-driven system falls, its momentum increases.

Ten thousand miles to the west of M.'s plant in France, at the refurbished Albertsons grocery store in Wenatchee, Washington, the retail revolution that has discomfited manufacturers takes a few moments to come into focus. Albertsons is the fourth largest grocery retailer in the United States, after Wal-Mart, Kroger, and Safeway, and its stores, with their massive floor space, enormous product selection, and famously gracious staff, create the impression not of revolution but of supplication and ease: everything the hungry shopper could possibly want or need is within easy reach. Here is what must be an acre of produce; hundreds of products, from baby watermelons to big bags of salad greens, all fresh, all gorgeous, and all surprisingly cheap, despite having traveled from farms hundreds and even thousands of miles away. Here is the enormous meat department, with an equally large selection of products — immaculate hams, of course, but also piles of steaks and roasts, huge Styrofoam trays of boneless chicken breasts, shrimp from Thailand, fillets of tilapia from China. Here too is the delicatessen with dozens of salads, side dishes, and hot entrées, and in the center of the store, aisle after aisle of processed and packaged goods, most of which seem to be on sale.

For a few moments, the sheer magnitude of the bounty almost persuades me that I'm not here for anything so base as buying food but rather to partake in a casual and festive celebration. And yet, there is little about this or any other modern grocery store that is truly casual. Every element of the experience — from the cheeriness of the employees to the neat piles of produce, from the color of the walls to the size of the shopping carts and layout of the products on the shelves — has been carefully chosen to maximize sales, minimize costs, and defend Albertsons' share of a retail-food market that over the last decade has become one of the most brutally contested economic battlegrounds in global history.

The opening salvos in that battle were heard two decades ago when grocery retailers recognized that the food system was ripe for a putsch. It wasn't just that food manufacturers kept raising prices (despite the fact that consumers' wages were stagnating);[1] it was that manufacturers arrogantly saw themselves as immune to competition because they alone could deliver the food values consumers desired. Such misplaced confidence created a huge opening for retailers, who, being much closer to consumers, were actually much better positioned to deliver food value. In Europe, upstart grocers like Carrefour and Aldi introduced the hard-discount concept: ultralow prices and private-label products, which French business leaders scorned as anticapitalist but which French consumers loved. In the United States, warehouse stores like Sam's Club, Costco, and Price Club, which were acutely aware of consumers' inflation anxieties, offered similarly deep discounts. And in 1989, the retail rebellion hit critical mass when Wal-Mart, the king of the low-cost model, began selling groceries in its first supercenter, in Washington, Missouri.

As it had with its nonfood business, Wal-Mart turned the conventional grocery pricing model on its head. Whereas traditional grocery stores had profited by charging high markups (on top of manufacturers' high markups), with an occasional sale to bring consumers into the stores, Wal-Mart offered the sale price *every* day, which in turn generated enough volume to compensate for its substantially lower markups. Second, to keep its everyday prices low, Wal-Mart systematically, and often ruthlessly, cut costs and inefficiencies at every point in its supply chain. Traditional grocers relied heavily on expensive on-site storage; Wal-Mart squeezed its inventory to the bare minimum by taking product deliveries only as products were needed — the so-called just-in-time model already pervasive in other industries. In place of the union labor that traditional grocers used, Wal-Mart, which now accounts for 1 percent of the U.S. labor force,[2] has largely crushed unionization and lowered its labor costs to about a third of that of its competitors' — although, admittedly, not without a lot of negative press and other costs: many of Wal-Mart's twelve thousand new hires each week are replacements for employees who have quit.[3]

Less controversially, the company is a master at exploiting economies of scale: its grocery sections are almost twice the size of traditional grocery stores' — 61,000 feet versus 35,000 feet[4] — which means Wal-Mart can more than double consumer traffic and sales without doubling costs due to its more efficient operations. Moreover, where traditional grocers rely primarily on food products for their revenues, Wal-Mart enjoys a much broader base: roughly two-thirds of a supercenter's floor space is devoted to nonfood items such as clothes, housewares, and cosmetics. Because these items have higher profit margins than food items do, they effectively subsidize Wal-Mart's food sales, allowing the store to sell its food for even *less*.[5] Wal-Mart also pioneered the use of scanning data and other information technology, which allow store managers to measure precisely how fast each product "turns," or sells, and thus to ensure that the store carries only those products that generate the greatest revenue for each inch of shelf space.

More famously, because Wal-Mart enjoys such a large share of the grocery market — today, twenty-one cents of every food dollar spent in the United States is spent at Wal-Mart (and some experts say it may be fifty cents by 2010)[6] — the retail giant has unrivaled bargaining power over its suppliers, who must keep prices flat or even lower them to retain Wal-Mart's business. "Wal-Mart never talked about taking a chunk from Kraft or Heinz or General Mills" when it began selling food, says Prudential's John McMillin. "But the reality is that Wal-Mart is now 22 percent of General Mills' business and once a customer gets that big, you're going to have to justify any price increase. Fifteen years ago, General Mills sure didn't have to justify any price increases."

All told, Wal-Mart can generate a dollar of sales far less expensively than competitors can. And because Wal-Mart passes most of those cost savings on to its customers in the form of lower prices, the company essentially uses its cost savings to generate even more volume, and thus even more cost savings, in what Wal-Mart calls a "never-ending virtuous circle." So self-feeding is Wal-Mart's low-price, high-volume formula that the company, already the largest commercial entity in the world and generating nearly 2½ percent of the U.S. gross domestic product,[7] can open new stores in the United States at a rate of two hundred a year, giving it an even larger market share and thus even more bargaining power with its suppliers. Wal-Mart has recently taken its model abroad, building or buying stores in Mexico, the United Kingdom, China, and India, and exploring opportunities in eastern Europe and Russia — moves that are being shadowed by its two main rivals, Carrefour and UK-based Tesco, in a high-stakes race to control the emerging global retail-food market.

As dramatic as Wal-Mart's power over its suppliers seems, Wal-Mart's more significant impact on the food system has come indirectly, in the way that its success has transformed rival grocery retailers. Unable to match Wal-Mart's[8] radically lower cost structure, many traditional grocers either went out of business or merged with other chains — with the result that more than half the entire U.S. retail grocery market is controlled by six U.S. retailers — Wal-Mart, naturally, followed by Kroger, Albertsons, Safeway, Costco, and Dutch-owned Ahold. (Compare this to twenty years ago, when the top six retailers controlled just a fifth of the market.) In urban areas, just four chains control three-quarters of all grocery sales.[9] Even the smallest of these mega-chains now has enough market share to enjoy a Wal-Mart-like degree of leverage over manufacturers' prices. Simply to get their products into the stores, manufacturers must actually *pay* the big retail chains anywhere from $75 to $300 per item[10] per store for shelf space, and even more to have their products positioned in premium spots, such as on eye-level shelves, aisle end caps, and at checkout stands. By some estimates, these slotting fees and other kickbacks add $2 million[11] to the manufacturer's cost of marketing a new product and collectively cost the food industry about $16 billion a year.[12] And sometimes, even fees aren't sufficient. Because retailers allot shelf space to strong sellers and will drop, or delist, poorly performing products, manufacturers are under constant pressure to promote their products or develop new killers. As Nestlé's Renk puts it, "[I]f you are third or fourth in a category, consumers won't want you, and retailers will kick you off the shelf."[13]

As retailers have demanded steadily more price cuts, the big food companies have been forced to reengineer their operations for better efficiency and cost-effectiveness by installing labor-saving automation, shutting down older plants, seeking cheaper sources for raw material (China, for example), and boosting production scales through acquisitions. In fact, much of the last decade's consolidation in the food industry, when large companies spent billions of dollars buying smaller ones, was driven by the need to cut costs in order to survive retailers' incessant price pressure. (Today, in the categories of breakfast cereals, snacks, and beer, three-quarters or more of all products are generated by the top four companies.[14])

All told, Wal-Mart has so successfully squeezed the supply chain — either directly, with its own buying power, or indirectly, by forcing rival retailers to become similarly aggressive — that since 1985 it has driven down U.S. grocery prices by a stunning 9.1 percent.[15] In other words, Wal-Mart has provided massive and tangible benefits to consumers — an assertion often made by the company or its boosters whenever Wal-Mart is accused of laying waste to some small town, hapless supplier, or Western culture generally. Less often does one hear the other half of that marvelous statistic: much of that price drop has come from Wal-Mart's success at cutting labor costs, which, according to the same study, has driven down average U.S. wages by 2.2 percent[16] during the same period.

But retailers are changing the supply chain in other, even more significant ways. Unable to beat Wal-Mart on prices, many new mega-retailers are instead trying to differentiate themselves from the low-cost behemoth. Where Wal-Mart stocks a narrow number of products so as to keep inventory maximally lean, other retailers have dramatically expanded their offerings — in part because studies show that consumers presented with a greater number of choices tend to buy more items on each shopping trip. "The bigger the store, the larger the average market basket — by as much as around 20 percent," says Roger Betancourt,[17] a retail expert at the University of Maryland. This is one reason that the number of individual items, or stock-keeping units (SKUs), in the average grocery store has risen sharply, from around fourteen thousand in 1980 to around forty-five thousand today[18] (which, of course, has only added to the pressure on food manufacturers to develop new products).

Similarly, while Wal-Mart targets a lower-middle-class shopper with a household income of less than $30,000 and dresses its stores in a blue-collar décor, many traditional grocery chains are aiming for a more affluent demographic, with softer lighting, simulated wood floors, and smiling greeters. Retailers are also offering lots of upscale value-added products and services, like bakeries, espresso bars, and delis, usually found on the store's high-traffic perimeter. Such amenities account for an increasing share of retailers' sales and profits (the margin on hot takeout food is around 41 percent, almost twice that of cosmetics)[19] and are among the biggest drivers transforming the retail supply chain.

Consider the recent evolution of the produce section. To capitalize on consumers' rising demand for fresh fruit and vegetables (and their willingness to pay substantial markups), retailers now offer vastly larger selections: the average produce section features 350 SKUs, almost double that of 1987 — again, because greater selection induces higher sales.[20] Shelf stock is turned constantly to keep displays looking perfect, which means stores now substantially overbuy and then toss large amounts into the dumpsters. Similarly, because consumers get locked into buying patterns and will often buy a produce item as long as it is available, retailers have steadily eliminated seasonal variation in availability: today's fresh produce supply chains source from farms in multiple climate zones to ensure that new crops are being harvested more or less continuously. So carefully orchestrated are freight logistics that Chilean raspberries can be picked, packed, shipped, and in the grocery display case in four days.[21]

The benefits of such advances have been substantial. Produce sections are now so lucrative, they generate about a sixth of the average store's total profits.* Consumers too have a vastly greater selection and are eating about

---

* Retailers have extended this fresh premium to other foods, including store-baked breads and cookies, deli foods (even though most are manufactured by food-service companies), and many refrigerated items. A study by the market analyst Hartman Group found that "consumers think shelf-stable soymilks displayed in refrigerated cases to be 'significantly fresher' and 'higher in quality' than their identical counterparts in non-refrigerated center [part of the] store." http://www.hartman-group.com/products/HB/2005_02_10.html.

30 percent more produce than they were in 1980.[22] The produce revolution has also been a boon for countries such as Chile, which, being blessed with a geography and topography that spans hundreds of microclimates, is emerging as a produce powerhouse, capable of delivering nearly any product to nearly any market — North American, European, or Asian — in nearly any season. Even smaller countries are exploiting their tiny growing zones to target gaps in the global produce flow. Berry farmers in Guatemala, for example, have found they can profitably grow and export fruit to the United States during a narrow window in the spring when output from Chile has subsided but California growers aren't yet on-line, and then again in the fall, after California peaks.[23]

But there have also been considerable costs. Growers have had to develop hardier varieties of fruits, vegetables, and berries; these can tolerate the longer shipping times and the new shipping systems, but they are often far less flavorful and, according to some studies, can be less nutritious as well. To ensure freshness, suppliers must now ship more and more of their produce by air, in costly climate- and atmosphere-controlled containers. And although all this extra travel and technology adds substantially to delivery expense, in the new retail-dominated food business, such costs cannot be passed along to retailers. Instead, suppliers must find ways to shave costs at their own end — typically by moving their growing operations to regions that are blessed not only with favorable climates but also with low land and labor costs. Chile's emergence as a great produce powerhouse derives as much from its large pool of cheap labor as from its versatile climes.

There are other, more subtle downsides to this new model of global produce. Because consumers have come to expect their produce to be as uniform and blemish free as packaged foods, retailers insist that fruits and vegetables meet exacting criteria for quality, visual attractiveness, size, and weight. Avocados headed for the United Kingdom, for example, must come within a half ounce of a target weight. Green beans bound for France must be straight and precisely 100 millimeters long. Further, because retailers refuse to keep expensive inventories of produce on hand and want their produce just in time, suppliers must replenish their customers' stocks with constant deliveries, usually overnight, despite the fact that these cargoes may be traveling ten thousand miles or more. And woe to a supplier that fails on promised volume or quality. "If you come up short for a delivery," one African exporter told me, "the buyer will let your competitor fill in and take your orders, and you won't get them back" — a financial disaster, given how much growers must invest in modern produce operations. Indeed, so high are the costs of failure that the big export-oriented commercial farms now routinely overplant simply to ensure they'll have enough acceptable product to make their promised deliveries — a practice that results in adequate supplies but is incredibly wasteful. "Some of the kit we throw away is obscene," says another tropical exporter who deals with European retailers. "We get fifteen tons of running beans, and only eight tons can be exported, because the others aren't straight. Maybe 30 percent of that can be sliced and processed, but the rest — two or three tons — is gone."

And even these precautions are sometimes insufficient. As retailers continue to demand lower and lower prices, only the largest suppliers, with the greatest economies of scale, can survive. For example, more than half of the global banana trade is now handled by two U.S. companies — Chiquita and Dole[24] — in part because ongoing price wars among European retailers drove wholesale banana prices down by more than 30 percent[25] and pushed smaller suppliers out of business. Even among the survivors, there is the grim recognition that they can look forward to a future of ever smaller margins and a relentless search to cut costs. "It's depressing," one farm manager told me with some rancor. Farming, he said, now has almost nothing to do "with production, with the quality of the product. It's all about the price coming down, down, and down, and anything else you focus on is irrelevant."

A few minutes' walk from M.'s pig assembly line, in a part of the plant devoted to cooked meats, I see another byproduct of the retail revolution: rack after rack of tubes made of processed pork. The tubes are about three feet long, about three inches wide, and, when sliced, will yield a pink circle of meat that fits perfectly between two slices of bread in a sandwich on the menu of an international fast-food chain. The chain's French outlets began selling the ham product a few years ago after an outbreak of mad cow disease made beef unpopular and forced the company to develop a cheap alternative protein. M. jumped at the chance to work with such a huge customer and was soon churning out tube after tube of pork. But as the fast-food chain, too, has begun demanding steep price cuts, M. now questions the wisdom of the venture. An attempt to cut costs by increasing his output, or "line speed," brought quality complaints from the chain, underlining the Catch-22 dynamic that makes it harder and harder to turn a profit in the meat business. "When they ask you for a 15 percent discount, there is no way you can get it back," M. says.

In some respects, meat processors were the earliest victims of the Great Retail Squeeze and the other half of the story of the retail revolution. In the mid-1970s, well before Wal-Mart took on the big food manufacturers, the other big retail food sector — food service — went on a cost-cutting crusade of its own. The protagonist was McDonald's, which even then was desperately searching for a new meat for its U.S. menu. Americans still loved beef — in 1976, per capita consumption was a phenomenal ninety-two pounds a year, more than twice that of chicken — but the gusto was fading. Medical authorities had criticized red meat as cardiovascular poison, and beef prices had been driven up sharply by a global shortage of grain that made it expensive to fatten up corn-fed cattle. Between the higher prices and the health fears, beef consumption was falling — bad news for a fast-food empire built on cheap burgers. McDonald's needed a new and inexpensive meat in a hurry, and the animal it turned to — the chicken — would change protein history.

Chicken fit McDonald's new program in several ways. The meat was leaner than beef, and thus healthier; it was more broadly acceptable among global ethnic groups; and it was much cheaper to produce, in part because chickens convert grain into meat three times as efficiently as cows do. Plus, with the advent of then new meat-processing technologies, chicken could be transformed into a huge number of consumer end products. Excitement was especially high over a process known as mechanical separation, wherein chicken meat is forced through a sievelike screen and made into a slurry, which is then filtered and reformed, with the help of chemical binders, into virtually any shape, from hot dogs and patties to, eventually, finger-size nuggets. In 1980, McDonald's tested a product it called Chicken McNuggets — basically, mechanically separated chicken that had been pressed into bite-size bits, breaded, deep-fried, frozen, and then reheated at the franchise and served with dipping sauces. Launched in 1983, the heavily marketed McNuggets were a monster hit. Although the product offered no true health advantage over beef (ounce for ounce, McNuggets have more calories, fat, salt, and cholesterol than a Big Mac),[26] Americans lined up outside franchises to get them. Within a year, McDonald's, the quintessential burger joint, was the world's second-largest seller of chicken.[27]

The poultry craze had officially begun. Rival fast-food chains scrambled to launch their own cheap chicken entrées, while traditional chicken outlets like KFC and Bojangles' went into expansion frenzies. Food manufacturers such as Swanson and Campbell's brought out processed chicken products for the freezer case, and supermarkets stocked more fresh and frozen chicken. By 1993, per capita chicken consumption in the United States hit sixty-eight pounds a year, catching and passing declining beef consumption[28] and sparking a gold-rush mentality among chicken farmers, who built massive new broiler barns all the across the South. By the early 1990s, the U.S. chicken flock was growing by nearly a third of a billion birds a year. "This thing is going to get even bigger," Bill Haffert, a poultry-industry analyst, excitedly told the *New York Times*.[29] "The broiler industry will never be the same."

Haffert had no idea. To satisfy America's new appetite for chicken, poultry producers had to reinvent every step of the production process. For example, because most fast-food chicken products were made from *boneless* meat (not the whole or cut-up fryers that poultry processors had been raising), poultry breeders needed to develop a new, more manufactureable chicken. The bird had to be larger, of course, both to give more meat and because big birds are easier to debone. It had to be of a more consistent size, to better fit the new automated processing equipment that companies like Tyson, Foster Farms, and Perdue had begun installing. It also needed to have substantially larger breast portions; American consumers prefer white meat to dark, and white meat forms more readily into patties and nuggets, which meant the demand for white meat was now growing about twice as fast as that for dark meat — and far faster than a traditionally proportioned chicken could supply.

Above all, this new industrial-strength bird needed to be exceedingly cheap. From their earliest days, fast-food chains regarded low prices as a strategic necessity — first, to draw consumers out of their own kitchens, and then, in the 1980s, as a weapon in the brutal price wars that broke out between McDonald's and its fast-food rivals. Thus, even before Wal-Mart began beating up food manufacturers, fast-food chains were demanding massive price cuts from Tyson, Pilgrim's Pride, Perdue, and other poultry suppliers. And where these big-name processors might once have balked, times had changed: by the 1980s, the fast-food chains had become such important customers (selling a third of all chicken consumed in the United States) that processors had little choice but to slash their own production costs.

Some of these cost reductions came from the newly designed birds. Breakthroughs in genetics let commercial breeders like Aviagen and Cobb-Vantress manipulate most of the factors that govern a bird's growth, from the tendency to distribute, or partition, muscle mass into the breast region (critical for optimizing white-meat yields) to the efficiency of the digestive tract (which lets the bird convert grain into muscle faster). The resulting broiler was a walking meat machine: twice as big as its 1975 predecessor, with breast portions weighing more than a half pound each, and the ability to reach this sumolike stature with freakish speed. Whereas a 1970s-era broiler needed ten weeks to reach slaughter weight, today's model does it in forty days, which means an enterprising chicken farmer can raise two more crops each year and thus increase his annual output by 40 percent.

Faster growth also translates into lower feed costs: because the birds reach slaughter weight in fewer days, they consume fewer meals. In the 1960s, a broiler needed two and a half pounds of feed to put on a pound of weight; today's designer bird needs just 1.9 pounds of grain to gain a pound of body mass — no small improvement, given that feed is 70 percent of broiler production costs. The modern chicken, says Paul Aho, a longtime poultry-industry analyst, has metamorphosed "from a lean barnyard racer that was all skin and bones to a slower moving meaty animal that fully utilizes its internal organs."

But more efficient birds alone couldn't satisfy this new demand for low-cost chicken. Like the big grocery retailers, fast-food restaurants were so price obsessed (McDonald's, Wendy's, and KFC all offer a chicken meal for a dollar or less) that they turned on their suppliers with a Wal-Mart-like ferocity.

Buyers not only demanded lower prices but, under cost-plus contracts, insisted on examining poultry companies' operations and accounts to be sure that the chicken was being produced as cheaply as possible. The fast-food chains "basically forced us to open our books and show what it *really* cost to produce a chicken," Blake Lovette,[30] the former CEO of ConAgra poultry, told me. "They wanted to be involved in all our major decisions, like when we purchased our feed grain, and how much, and at what price." So great is the price pressure now that poultry processors average just two and a half cents profit on each pound of chicken they sell.[31]

To survive on such thin profit margins, the poultry industry has had to reinvent itself in ways that go beyond a bigger bird.* Feed is now mixed by computers to obtain the ratio of starches, amino acids, antibiotics, and protein (much of it in the form of slaughterhouse offal) that maximizes growth. Processing has been increasingly automated: at most big slaughtering plants, chickens are killed, plucked, eviscerated, and split by machines, lowering labor costs and boosting line speeds from around fifty birds per minute to nearly one hundred.[32] Scale of operation is vastly larger. In 1980, a poultry processor was considered competitive with two slaughtering plants and a combined output of 32 million broilers a year; at that volume, says Aho, the processor's costs per bird were low enough to be profitable. Today, after nearly three decades of dwindling margins, a processor needs four much larger plants and a combined output of 260 million birds to stay in business. "In 1980, the biggest poultry processing plants in the United States handled 16 million birds a year," Aho says. "Today, you need 1.25 million birds a *week* just to break even." And 1.25 million is the minimum; the biggest plant in the United States — the Choctaw Maid plant in Carthage, Mississippi — handles more than two million chickens per week.

And, of course, poultry processors have turned on their own suppliers. Because the poultry industry is such a large user of grain — consuming a seventh of the U.S. corn crop and almost a fifth of all the soybeans[33] — processors like Tyson have bargained down feed prices substantially. They've also put enormous pressure on the chicken farmers who raise the birds, to the point that, according to a Purdue University study, half of all chicken farmers now have farm debt of more than $100,000.[34] Processors have slashed costs for another key input: labor. Although slaughtering is increasingly automated, chores like deboning are still done largely by hand, and as high chicken demand has forced processors to add thousands of new line

---

* Poultry wasn't the only industry to feel the pressure from the fast-food market. The cheese for Pizza Hut's pizzas consumes nearly 3 percent of the total U.S. milk production, giving the chain and its rivals immense bargaining power with dairy manufacturers; when Pizza Hut introduced its cheese-stuffed-crust pizza in 1995, demand for mozzarella jumped by 17.5 million pounds — or more than half the nation's supply at the time.

workers, the companies have worked assiduously to keep labor costs from bruising their already thin margins. Most poultry processors not only aggressively oppose unionization[35] but have steadily moved operations into southern states, where unions are simply less popular and where economically depressed towns are willing to overlook the social costs of massive processing plants. And, as in other food sectors,* poultry companies frequently rely on immigrant workers, many of whom are here illegally and who will accept poor working conditions and wages that average eight dollars an hour. Adjusted for inflation, poultry-industry wages are 24 percent lower than they were in 1977,[36] which is surely one reason poultry workers are five times as likely as other manufacturing workers to quit their jobs.[37]

In return for such relentless attention to cost and efficiency, the industry has managed to keep its customers supplied with an enormous volume of meat. Since 1980, U.S. chicken output has more than tripled, from 11.3 billion pounds to 37 billion pounds, while prices have plummeted. Boneless, skinless breast meat sells now for about $1.40 a pound wholesale, or less than a quarter of the inflation-adjusted 1980 price. There is so much chicken on the market that the new challenge is finding places to sell it all. Processors rely heavily on the export market: today, a seventh of all U.S. chicken production, some 2.5 million tons a year,[38] much of it the less desired drumsticks, are sold abroad, especially to Asia and Russia. At home, retailers and food processors are continually looking for new poultry vehicles — everything from pizza topping and microwavable nuggets to buffalo wings, which Americans consume at the rate of eleven billion a year. All told, the average American consumer eats eighty-seven pounds of chicken a year — twice the amount eaten in 1980, and nearly twice that of beef — and the trend shows no signs of slowing, despite mounting concerns we're eating far too much. "Clearly, there's a limit to what the U.S. stomach can hold," Aho told me. "But I've been predicting we'd reach that limit for the last ten years, and it never arrives."

Because the food economy is such a tightly interconnected system, chicken's meteoric rise has had a domino effect in other meat sectors. Pork produc-

---

* The practice isn't limited to chicken. In December 2006, federal agents apprehended 1,282 illegal alien workers at six facilities owned by Swift and Company, one of the nation's largest processors of fresh pork and beef. See http://www.ice.gov/pi/news/newsreleases/articles/061213dc.htm.

ers, seeing their market share eroded by chicken, quickly adopted the same high-output, low-cost tactics. Smaller, less efficient pork slaughterhouses were swept away by massive new facilities: the largest in the world, built by Smithfield Foods, in Tar Heel, North Carolina, processes two thousand pigs an hour. Such massive "throughput" has in turn made necessary an uninterrupted flow of hogs from farmers, who, to meet the demand for high volumes and low prices, are also consolidating into large-scale operations specializing in a single stage of a pig's life: piglets are raised at "nurseries"; weaned pigs go to "feeder operations" and then on to finishing facilities to be fattened to slaughter weight. The pace of the consolidation is breathtaking: in 1980, the United States had 667,000 hog farms, with an average of 101 hogs each. Today, there are fewer than 50,000 hog farms, with an average size of 1,173 head.[39]

And as with the chicken, the pig itself has been made far more productive. With better breeding, new feed formulas, and new additives, the animals mature faster, carry 20 percent more meat, are more uniform in size and weight, and are far more prolific. A quarter century ago, a breeding sow averaged fourteen piglets a year. Today, says industry analyst John Nalivka, the average litter is twenty piglets, "and at some of the bigger facilities, like Smithfield's Murphy Farms, in North Carolina, they're getting twenty-three pigs per sow per year." Between the larger pigs and bigger litters, a single sow will generate nearly two tons of meat a year — more than double the 1980 output.[40]

With so much pork rolling out of these new processing facilities, pork companies, like their poultry rivals, have been aggressive in finding new marketing vehicles. Bacon, for example, has been heavily and successfully promoted as an über-topping, for burgers, salads, and pizzas. Companies have also worked to improve pork's flavor and convenience. Because the new, leaner pork dries out easier when cooked (meat's juiciness derives largely from the fat marbling), nearly half[41] of all pork products are now "pumped" with a proprietary brine containing flavorings, salt, and other chemicals that help retain moisture and enhance taste.[42] And of course, processors have worked to make their product homogenous: after decades of fast food, frozen dinners, and boxed food, consumers are now entirely accustomed to, and have come to expect, meat products that don't vary from one purchase to the next. As former Smithfield chairman Joe Luter once put it,[43] "McDonald's may or may not be the best hamburger in the world. But it is consistent, consistent, consistent."

Even the beef industry, long the contrarian of the food economy, has been brought to heel under the low-price, high-volume retail regime. Feed-lots and slaughterhouses are larger. The average steer weighs 1,350 pounds, up from 1,000 pounds in 1980, and the meat is more tender, more marbled with fat, and more partitioned into the high-profit regions: the ribs, the loins, and the rounds.[44] The supply chain has been leaned out for maximum cost savings, and production has been fully rationalized: hamburger is now ground in multiton batches and then injected into tubes for bulk buyers or machined into patties that are shaped to appear handmade. Nothing is wasted. With so-called advanced meat recovery, a series of rollers and screens clean the bones of even the tiniest scraps, which are then added to everything from hot dogs and sausages to taco filling and pizza topping.[45]

Even steaks, chops, and other "intact" beef products are now produced in ways that meet retailers' need for cheap, efficient, and uniform products. Where grocers traditionally took delivery of large pieces of cow, known as primals, and hired butchers to break them into individual cuts (the process is, somewhat bizarrely, called fabrication), Wal-Mart is changing this practice, insisting that its suppliers deliver beef in a new format known as "case-ready." At a central packing plant, the beef is cut, weighed, sealed in a special modified-atmosphere package, and labeled, then trucked to Wal-Mart and placed directly in the meat case.

Wal-Mart's primary motive was, as usual, cost savings; because case-ready beef requires no special skills to handle, Wal-Mart was able to get rid of its eighteen-dollar-an-hour union butchers. But case-ready beef also provided Wal-Mart with new ways to add value. The special, low-oxygen atmosphere inside the case-ready package not only retards spoilage for days but also encourages the beef to "bloom" with the ruddy coloring of a freshly butchered steak — a key feature, given that color and appearance[46] are the most important factors in consumers' meat-buying decisions. And because Wal-Mart knows its shoppers lack the time and, often, the skill for careful cooking, case-ready beef, like pork, is now pumped with a proprietary solution to enhance flavor and ensure that the cut, even if it's overcooked, will be as juicy as the more expensive meat grades sold in higher-end grocery stores. "Cooking at home is becoming a lost art," Bruce Peterson, Wal-Mart's head of perishables, told *Beef* magazine in 2003.[47] Wal-Mart's customers are "expecting to buy a piece of meat that has a particular flavor profile, stick it in the oven for 15–20 minutes,

and yet have something they remember eating when they were growing up."

To be sure, shifting to case-ready beef was enormously costly for beef
packers, who were required to invest hundreds of millions of dollars building or renovating packing plants. And because cattle are nowhere near as
uniform as pork or chicken, workers must often laboriously hand-trim the
steaks, roasts, and other cuts so they fit neatly in Wal-Mart's containers. But
as elsewhere in this retail-dominated food economy, such additional costs
had to be absorbed by suppliers. Because Wal-Mart is the nation's biggest
buyer of beef (a million tons a year, almost twice as much as McDonald's),
"you just can't say no," says Nalivka, the beef-industry analyst.[48] "You may
not like what you have to do to keep doing business with Wal-Mart, but if
you step out of line, there are a whole lot of other suppliers ready to take
your place."

In reality, it's not clear how long the industry will have a "whole lot of
other suppliers." As in farming and produce, retail price pressure has cut
out the smaller meat producers, to the point where just four companies —
Tyson, Cargill, Swift, and National Beef Packing Company — control 80
percent of the beef market, up from 40 percent in 1980,[49] a degree of concentration that gives packers enormous pricing power over the ranchers
and feedlots they buy from. Half of all chicken and 60 percent of all pork
are also under the control of just four companies.

Industry critics complain that such concentration is undermining the
food system: as these big meat processors gain market power, their suppliers have been forced to cut their own costs so dramatically and squeeze so
much slack from their own operations that they are slowly losing any margin for error — that is, the ability to deal with disruptions such as a labor
strike, or a crack in their supply chain, or, worse, an outbreak of disease.
But few in the meat industry see an alternative. When asked by a reporter
for *Meat & Poultry* magazine if industry consolidation posed a problem,
Robert Peterson, CEO of the IBP meat company, then one of the largest in
the country, snapped: "You want to go back to the nineteenth century? You
want to have a packing house in every little town and deal with twenty-
first-century marketing? There's no way! . . . There's no stopping it. This is
an evolution that's going to take place in spite of whoever is in the way."[50]
Peterson's comments were prescient: in 2002, IBP was bought by Tyson, a
move that made Tyson, which began as a small poultry processor, not only

the world's largest protein supplier but also the largest maker of food of any type in the United States.[51]

On the fourth floor of the Meat Sciences Building at Iowa State University, researcher Dong Ahn is showing me the delicate metal probe he uses to test the elasticity of chicken muscles. Elasticity is important because it determines the texture of the meat when cooked and, lately, is a source of concern. The problem, says Ahn, a soft-spoken Korean native, is that chicken breeders have been too successful: commercial broilers now put on muscle so quickly that the rest of their anatomy can't keep up. Breast muscles grow so swiftly that muscle cells don't fully form and cannot completely relax, with the result that breast muscle often remains in a state of semi-contraction, which affects meat quality. Likewise, capillaries in the breast muscle fail to develop extensively enough to nourish the entire muscle. "Blood supply gets too low," Ahn says. "And sometimes, an area of muscle tissue actually dies."

But the industry's biggest concern is something called PSE, or "pale, soft, exudative" meat. Breast muscle is made up of fast-twitch fibers, which produce the rapid contractions needed to flap the wings. When a chicken is killed, these fast-twitch fibers contract rapidly (this is why chickens convulse so famously when killed), which squeezes a cellular waste product, lactic acid, into the muscle tissue, where it causes damage. All chickens have this postdeath reaction, but because modern birds come with such big breasts, the volume of lactic acid released is huge, and the impacts on meat quality are massive. The acids denature the proteins in the meat, which causes the meat to turn pale, lose its ability to retain moisture (hence the bloody residue at the bottom of the grocery package), and become so soft that it crumbles when cooked.*

PSE is only one of a multitude of tradeoffs the industry makes in order to deliver the low-cost birds that consumers now take for granted. Because chickens are slaughtered as juveniles, while their growth curves are still profitable but before their bodies fully develop, their bones are still

---

* Interestingly, PSE also occurs among the new breeds of fast-growing hogs, whose meat has poor texture, turns an alarming pinkish gray, and exudes so much water (processors call it "purge") that it results in a lower selling weight and lower processor profits.

soft, so when they're cooked, the bones often leak a blood-red and quite unattractive fluid into the surrounding muscle.[52]

Food companies are acutely aware of such deficits. Livestock breeders are trying to control genetically for problems like PSE. But as with food manufacturers and additives, the cheapest solutions are often merely remedial. Meat companies treat PSE after the fact, pumping the meat with the salts and phosphates to make the meat retain water. Such water enhancement is not only easier and cheaper than switching to a slower-growing breed of chicken but, as a bonus, increases the meat's selling weight by anywhere from 10 to 30 percent, thus allowing the processor to charge a higher price for only a small additional production cost.[53]

Sadly, even though the effects of overly rapid growth can be hidden from consumers, this is hardly the case for the birds themselves, for whom this new Goliathlike stature is no small burden. Although breeders have labored to give the new meat-type birds stout skeletons, hearts, lungs, and other organs, many birds are so meaty they can't walk or even stand after they're about five weeks old. One study by the University of Bristol found that one in four commercial broilers suffers crippling leg problems, and many die prematurely from cardiac arrest and congestive heart failure as their hearts struggle to feed the massive breast muscles. Even healthy specimens cannot survive long enough to reach sexual maturity. Some of these shortcomings are academic, since few birds ever live to see adulthood: a chicken converts feed most efficiently during its first few weeks of life, when its growth is fastest. Once a broiler's rapid growth tapers off and its per-pound costs begin rising, the bird is said to be at its economic endpoint and should be slaughtered. "In meat birds, you're not interested in what they look like as adults, but as juveniles," Susan Lamont, an expert in the molecular genetics of poultry at Iowa State University, told me. "In fact, most of the commercial customers who buy these birds never see them fully grown."

But, of course, commercial growers and the birds themselves must still deal with consequences of unbalanced growth. Because so much of the chicken's intake of energy and protein is shunted into muscle growth, the bird has less energy for other bodily functions, such as immune response: well-muscled birds produce fewer antibodies and are thus far more prone to infection by the diseases that are endemic to close-packed commercial flocks. Breeders are working to develop disease-resistant birds, but in the meantime, the steady increase in bird size has been accompanied by a steady rise in the use of subtherapeutic antibiotics and a subsequent rise in bacteria that are resistant to those antibiotics — an effect that is already showing up in the form of human diseases that are harder and harder to treat.

There are an almost endless number of examples of the way the retail-led drive toward maximum productivity, efficiency, and everyday low prices has had unintended side effects throughout the food economy. The huge streams of food that issue from these giant, hyperefficient facilities have added to a global calorie glut. The massive supply chains that link these producers to their retail buyers are more attenuated, and more vulnerable, than ever. External costs — that is, impacts not directly paid for by the producers — have skyrocketed. In California, the state's huge dairy herd produces twenty-seven million *tons* of manure a year, the particulates and vapors from which have helped to make air quality in the agriculturally intensive San Joaquin Valley worse than it is Los Angeles.[54] And cows are relatively benign crappers: the typical hog produces three gallons of feces and urine every twenty-four hours, and the typical hog CAFO, or concentrated animal feeding operation, generates as much sewage as a midsize city; this outflow is stored in enormous lagoons that not only taint the local breezes with unwholesome miasmas but also can pose serious threats to surrounding people and property. On June 21, 1995, an eight-acre hog lagoon in North Carolina gave way, unleashing twenty-five million gallons of excrement in what one account described as a "a two-hour, knee-deep stream that destroyed the cotton and tobacco crops of a neighbor's fields, crossed the highway, and drained into the New River, where it killed all aquatic life for 17 miles."[55] Even absent such spectacular failures, livestock CAFOs represent such a concentrated source of nitrogen and other nutrients that, paradoxically, they actually cause harm: if nitrogen in particular escapes into surrounding water systems, its presence in drinking water can contribute to human cancers, while its potent fertilizing powers so disrupt ecological systems that most fish and other animals die off.

Legislators in North Carolina and other states have passed laws restricting poop lagoons, but as lawmakers everywhere have discovered, problems like these aren't so easy to fix. In response to stiffer regulations, many meat producers simply move their operations to places where laws are less

onerous. This is one reason pork processing has migrated from its traditional midwestern locations to more welcoming spots in Missouri, North
Carolina, Oklahoma, Texas, Utah, and even outside the country entirely.
"We try to go to places where we can succeed," Smithfield's Luter told an interviewer. "Some countries offer more opportunity than we see in the U.S.
at this time. If regulations become too tight in this country, we'll invest in
Canada and Mexico."[56]

More fundamentally, poop lagoons, bad air, and attenuated supply
chains are simply the more visible manifestations of a business model that
is effectively locked into a race for lower costs and larger scales. To meet retailers' price requirements, producers have scaled up their operations in order to spread their costs over the largest possible number of units. Yet because these new farms and production facilities are so large and so costly to
build (the typical high-speed hog plant goes for $100 million),[57] and be-
cause the profit margin per animal is so narrow, these operations must be
run at full capacity continuously to provide a sufficient return on the enormous investment. Overproduction is, in effect, financially embedded in the
system. "If your lines are built to run 8,000 birds an hour, then you want
8,000 birds running through that line," says Aho. "You've got to keep your
shackles full."

And yet, as with farming, the benefits of maximum output are short-
lived. As prices continue to drop, processors have little choice but to go for
another round of efficiency — with new equipment, or by pushing down
labor costs even further, or, most often, by finding some way to pump more
volume through the same plants — thus trapping themselves in a cycle that
not only narrows profit margins to the point of nonviability but renders
the entire meat sector so lean and just-in-time that it is losing its ability to
rebound from any kind of shock. As Charles Olentine, a poultry-industry
veteran, told the trade publication *Watt Poultry USA* in 2003, no matter
how fast poultry companies improved their productivity, retailer price
pressure has increased even faster, so that even at greater volumes, companies' margins haven't increased at all. "If this is success," Olentine quipped,
"I would hate to see failure."

Perversely, despite the possibility that the retail-driven food model is accruing so many unpaid costs that it will eventually collapse, that same model is
rapidly becoming the global standard. Just as food manufacturers were

forced to migrate from mature Western markets, so the big grocery retailers, finding their home markets saturated, have had to aggressively go after opportunities in developing countries[58] whose middle classes are becoming large enough, wealthy enough, and busy enough to support a supermarket culture. Studies show that such a culture can thrive when annual per capita incomes reach $6,000, which suggests that there are forty-five million potential retail grocery shoppers in Mexico, a hundred million in India, and a whopping three hundred million in China.[59] In fact, expansions by mega-retailers are already following such fertile demographics as closely as the Roman Empire once chased after wheat production. Wal-Mart, for example, is blasting into Mexico (where it opened 120 new stores in 2006 alone)[60] and China (where its acquisition of domestic Trust-Mart chain for $1 billion made it the top Chinese food retailer)[61] and is actively scrutinizing eastern Europe, Russia, and India, as are Carrefour and Tesco.

To be sure, the big retail chains are not solely responsible for the retail revolution. Typically, the arrival of a big multinational like Carrefour or Wal-Mart in a developing country marks the final stage in a retail evolution that began with smaller domestic grocery players,[62] who are eventually bought out by some multinational retailer. But once the big boys arrive, the retail evolution shifts into high gear. In Mexico, Argentina, South Africa, Chile, and the Philippines, countries where supermarkets were almost nonexistent a decade ago, the supermarket format now accounts for more than half of all food purchases.[63] As significant, these big players tend to shake up local supply chains and suppliers, whose operations are generally far below Western standards and practices. Local producers of vegetables, fruit, meat, and dairy products who wish to become preferred suppliers for the new high-end retailers[64] must often abandon traditional practices and schedules for a more consumer-oriented Western model — promising, for example, to harvest and deliver a product seven days a week, twelve months a year, even if it means skipping religious holidays and other traditional activities. And whereas traditional grocers in developing countries once tolerated inconsistent quality or late deliveries, the new retailers simply reject late or low-quality deliveries and eventually delist nonperforming suppliers. In response, farmers have scrambled to upgrade their operations and equipment and to focus on one or two crops so as to raise product quality and consistency.[65]

As in developed countries, such changes have often meant cheaper and better food for consumers who previously were forced to tolerate spoiled food and the massive markups of innumerable middlemen.[66] But the flip side of this new efficiency has been serious dislocation. In many evolving retail markets, retailers' strict new discipline (coupled with the standard Western practice of delaying payment to suppliers by as much as ninety days) have pushed tens of thousands of smaller, less efficient producers out of the supply chain — a serious change, given that farming in these countries is still a major source of jobs and wages. In Malaysia, according to a report by the United Nations' Food and Agriculture Organization, a new retail chain that was buying from two hundred vegetable suppliers in 2001 winnowed that number down to just thirty two years later.[67] Such a trend may indeed result in a food industry that is more efficient and lower cost. But given that much of the developing world is still confronting major challenges in food security — and in many cases lacks the roads, rail lines, warehouses, and other infrastructure to distribute food to all consumers who need it — the steady loss of small, geographically dispersed farmers seems a step that may be ominously premature.

This winnowing impact is not limited to the developing world or to emerging markets. As retail price pressure continues to build, even producers in mature markets are feeling the heat. For example, to supply their retail customers in Europe, the big meat companies are anxiously searching for ways to cut their costs. Many are building processing plants in eastern European countries, such as Poland, where cheap grain, minimal environmental regulation, and low wages (due largely to massive unemployment) make the perfect base from which to ship cheap pork into wealthy western Europe. "We've got people in Western Europe who make 20 Euro an hour [and] people in Eastern Europe who make one and two Euro an hour," explained an excited Larry Pope, Luter's successor, at a Smithfield shareholders meeting in 2006. Whereas land costs in western Europe were high, Pope continued, "land in Eastern Europe they will virtually give you. [Processing] plants in Western Europe are very expensive. Plants in Eastern Europe, they will virtually give to you for small dollars."[68]

And, of course, American meat companies are bringing the American model of high-volume, low-cost meat production to western Europe as well. When M. partnered with Smithfield in the late 1990s, the hope was that his local experience and least-cost expertise, backed by Smithfield's financial muscle, would allow the new venture to pursue the classic American strategy: by buying up his competitors and consolidating operations, M. was supposed to gain enough scale, cost-cutting efficiencies, and market power to deal forcefully and profitably with the big French retailers. Instead, those big retailers have kept the upper hand, shrewdly throwing enough business to M.'s rivals to keep them financially strong enough to avoid merging with him. Unable to expand his market share or get the dramatic cost reductions that might have come with consolidation, M. has seen his profits slowly squeezed by retail buyers. A few months after I visited him, M. explained that he'd been "replaced" by Smithfield. "They wanted to find someone with more marketing skill," M. told me in dejected tones. "Someone who didn't offer all his profits to retailers."

M. acknowledges that he didn't meet Smithfield's financial targets. But he's not sure how his successor will do any better against the increasingly powerful retailers, who continue to wage price wars against one another and insist on lower and lower prices from their suppliers. "It is a model that backs you into a corner," he says. A big grocery retailer "wants a huge order, four thousand tons of ham, and they want it as natural as possible and as cheap. And if prices they can pay collapse by 20 percent due to the price war they're having with their competitors, they're passive about it, because in their minds, they were paying 20 percent too much to begin with." The situation, he told me, is unsustainable. "Competition is getting tougher, margins are getting smaller. It has been like this for the last three years and it will end when the retail industry is even more concentrated and when French suppliers are either more concentrated or dead."

4

Tipping the Scales

O N THE MORNING OF March 22, 2006, at the Florida state capitol in Tallahassee, Bob Barrios, head staffer for the powerful health-care committee in the House of Representatives, began getting calls from anxious executives in the food industry. The day before, a House subcommittee had been scheduled to vote on a bill banning the sale of foods containing high-fructose corn syrup, or HFCS, in schools. Few observers gave the bill much of a chance: its sponsor, thirty-nine-year-old Representative Juan Zapata, was making the usual claims — that the sweetener makes children obese — but in a state whose economy had benefited so handily from sweeteners generally, that argument rarely gained much political traction. In the subcommittee hearing, however, Zapata, a jocular, youthful-looking banker who favored dark suits and an Elvis pompadour, was quite persuasive. He explained how food companies began adding HFCS to processed foods in the late 1970s — just as U.S. obesity rates exploded. He showed how HFCS was used in everything from soda and ketchup to candy and bread. He offered research suggesting that HFCS inhibits the body's ability to know when it is full. High-fructose corn syrup "is the crack of sweeteners," declared Zapata. "You get addicted, and you want more." The measure passed unanimously, and by morning the offices of the House heath-care committee, which now had to deal with Zapata's bill, were in an uproar. "Barrios told me his phone just lit up," Zapata recalled with some humor. "They were worried."

To say the least. Within days, corn refiners, for whom HFCS is a two-billion-dollar market,[1] launched a lobbying campaign attacking the science of the bill. Zapata was "trying to link HFCS as a unique contributor to obesity, and research has found the opposite to be true," insisted Audrea

Erickson of the Corn Refiners Association. Companies whose products contained HFCS joined the fray. Coca-Cola flew a company nutritionist down to meet with Zapata. Conservative pundits, meanwhile, assailed the bill as another example of government intruding in an area best left to parents. Zapata, a loyal Republican, was unrepentant. "In a home, the parents can make choices for the child," Zapata told reporters after defiantly sending his bill on to the health-care committee. "But in the school, we are the caretakers for these kids and we're providing them with something that can be potentially harmful."

Juan Zapata is hardly the first person to suggest that consumers need protection from the modern food system. If Americans and other rich-world consumers are no longer haunted by scarcity or blandness, the cost of such varied bounty has been a food supply that is in many respects ill suited to our physiology. Our scientifically bred produce grows so quickly that it contains measurably fewer micronutrients. Our processed foods are often packed with large quantities of salt, fat, and sweeteners, not to mention hundreds of chemical additives, some of which, such as the preservative sodium benzoate and yellow food coloring, are definitively linked to medical problems, such as hyperactivity.[2] And where the wild animals our ancestors gnawed on were naturally lean, our grain-fed livestock is specially bred not only to put on lots of fat, but to partition that fat *inside* the muscle: indeed, today's premium cuts are those with more marbling.*

More fundamentally, plenitude itself is rapidly shifting from a sign of progress to one of our top health risks. Each year, according to the U.S. Centers for Disease Control, complications from obesity and related problems, such as diabetes and heart disease, cause 112,000 premature deaths and account for $75 billion in extra medical costs in the United States, and there is every reason to believe that this is just the opening round: because children are growing obese in greater numbers and at earlier ages than before, we can expect a much larger wave of obesity-related medical problems to strike the country just as our medical system is buckling under the needs of the present generation. And by then, what was once a distinctly Ameri-

---

* Also, because wild animals feed largely on grass, which is one of the few natural sources of omega-3 fatty acids, their body fat has a higher proportion of these "good" monosaturated and polyunsaturated fats, and less of the "bad" saturated fats; corn and soybeans, by contrast, do not contain beneficial fatty acids.

can problem will be global; already, obesity afflicts a billion people world-wide, or roughly the same number as those who are underfed — a gro-tesque symmetry that before the 1980s would have been as inconceivable as Zapata's legislative proposal.

Not all of this can be blamed on our food or our food industry. Obe-sity, as food and beverage companies never tire of pointing out, is the result of numerous factors, not least genetics, a general decline in physical activ-ity, and a food culture that oscillates between Bacchanalian abandon and Calvinistic self-reproach. But even industry defenders acknowledge that in making food cheaper and easier to use, they've removed two of the biggest natural restraints on overconsumption. And although consumers are ulti-mately responsible for what they eat, they're getting plenty of help in how and what they choose. It's probably not pure coincidence that obesity rates turned up sharply in the 1980s, just as plunging food prices and heavy re-tail pressures left food companies needing to move more and more value each year with promotions and advertising that grew more creative and aggressive.

The last decade has seen numerous well-publicized initiatives meant to curb the obesity problem: laws like Zapata's, or proposals to tax junk food, or bans on soft drinks in schools; but also a flood of healthier, low-fat food products from an industry desperate to avoid the costly government sanctions suffered by the tobacco industry. Yet such initiatives seem des-tined to fail because they miss the underlying problem — namely, the sys-temic more-is-better economic model that not only defines the modern food system but also exceeds anything that system's primary customers — us — were built to withstand.

Humans were designed for scarcity — this much seems clear. In the prehis-toric calorie economy, when feast was far less likely than famine, those who survived did so because they had adaptations that let them make the best use of whatever foods were available. Some of these adaptations ensured that we took in the right inputs; because we use animal foods to build our bodies, we were programmed to crave the flavor of fat and protein. Like-wise, because we burn carbohydrates for fuel, we're wired for the sweet taste of starches and sugars. But it wasn't enough to simply *want* fat, pro-tein, or sugars; we needed a mechanism to make sure we consumed these foods in the right measure. Too little and we'd starve, but too much could be just as bad, since we'd waste time and effort getting calories we didn't absolutely need — an expense that in an environment of intense competition could have lethal consequences. Thus our bodies adapted to the external calorie economy by developing an internal version of that economy, a complex accounting system that uses hormones and neurotransmitters to ensure that the calories we consume (our inputs) match the calories we burn (our outputs).

Consider, for example, the biochemistry of hunger. When your stomach has been empty several hours, a chemical, ghrelin, is released into the bloodstream. As levels of ghrelin rise, the chemical reaches a part of the brain called the hypothalamus, where ghrelin initiates a physiological chain reaction you experience as hunger; you look for food and start eating. But hunger is only the first step: how does the body know when to *stop* eating? As it turns out, the body regulates intake in part by controlling the length of the meal. When you eat fat, for example, your gut stimulates the release of another chemical, cholecystokinin, or CCK. CCK acts on the hypothalamus as well, but instead of promoting hunger, CCK fosters a sensation of fullness, or satiety. Protein and carbohydrate in the stomach cause the release of their own respective satiety messengers. The process is linear: as food accumulates in your stomach, more chemical messengers are released into your blood, and more messages reach the hypothalamus. When this combined chemical signal becomes loud enough, the hypothalamus initiates a satiety reflex throughout the brain, and you consciously feel full enough to stop eating — at least, until your stomach empties and another round of ghrelin is released, touching off yet another cycle of hunger and satiety.

Of course, the job of maintaining energy balance goes beyond a single meal. Along with the short-term cycle of hunger and satiety, our ancestors' bodies were also striving to maintain a long-term reserve of energy, mainly in the form of fat, to draw on between meals or when food was scarce. When a person misses a meal, the body begins metabolizing, or breaking down, stored fat into components called fatty acids, which can be burned as fuel. But once fat stores are depleted — for example, after a few days or weeks of poor hunting — the body goes into crisis mode. Energy is withdrawn from nonessential bodily functions, notably reproduction, and conserved for the brain and vital organs. If fat depletion continues, the body turns cannibalistic, breaking down muscle tissues into metabolic fuel. This last-ditch process isn't sustainable, which is why the body does everything it can to keep enough fat in reserve.

Each of us has an optimum level of fat storage, genetically tailored to body size and metabolism and probably influenced by environment. The body also has ways of detecting when these fat stores are at, above, or below optimal levels and it adjusts hunger or satiety accordingly. One mechanism uses leptin, a hormone that is secreted by fat cells and makes the hypothalamus more sensitive to satiety signals from the gut. When fat stores are high, more leptin is secreted and finds its way to the hypothalamus, which then needs only a small satiety signal (and, by extension, only a little food in the stomach) to initiate fullness and end the meal. In other words, when fat stores are high, we're inclined to eat less.

The leptin mechanism is self-correcting; when fat levels begin to fall — after a person has missed several meals — blood levels of leptin naturally fall off, which makes the hypothalamus increasingly deaf to satiety signals from the stomach. As a result, one has to eat more food (and release more CCK) before the hypothalamus finally hears that the stomach is full. In effect, a person will stay hungry and keep eating larger meals until the fat stores recover, at which point leptin levels rise, the hypothalamus regains its hearing, and appetite is shut down.

The leptin-satiety reflex is fairly brilliant in and of itself, but like most biochemical systems, it's also connected to a host of other related functions designed to help humans survive in an environment marked by scarcity. For example, leptin affects not only appetite but also metabolism — that is, how fast calories are converted into energy for muscles and other bodily functions. As leptin levels fall, so does the metabolic rate; this means that when there's not enough food and fat stores fall, the body automatically tamps down the muscle engine so it burns fewer calories, thus conserving energy for vital brain and organ functions. This so-called starvation response, which can reduce energy usage by 20 percent,[3] will also shut down other nonessential functions. Bone growth stops, which helps explain stunting in malnourished children. In women, reproduction function ceases and will not return until fat levels are restored to a minimum level and the body is again strong enough to carry a fetus.[4]

Leptin is one of many hormones in a system of metabolic checks and balances that for most of human history allowed the body to match energy inputs to energy outputs with extraordinary accuracy. Over the course of a decade, the average human will eat about ten million calories; until recently, that human could reasonably expect to gain or lose less than a pound over that entire ten-year period. What that implies, says Jeffrey Friedman, a molecular biologist who studies weight regulation at the Rockefeller University in New York, is that energy intake and energy expenditure are normally within 0.17 percent of being perfectly balanced: that is, of the ten million calories our average human consumed in a decade, only 1,700 of them might wind up as unnecessary padding. "This extraordinary level of precision," writes Friedman, "exceeds by several orders of magnitude the ability of nutritionists to count calories."[5] In other words, the body is programmed to do automatically what none of us could do consciously. "You couldn't regulate your energy balance that precisely even if you weighed everything you ate," Randy Seeley, a neuroscientist at the University of Cincinnati, told me. "You couldn't find a scale that sensitive, and if you did, the crumbs you accidentally dropped on the floor would completely throw off your calculations."

Unfortunately, for all that precision, this remarkably fine-tuned mechanism is marred by several major flaws. First, because our ancestors lived with scarcity and were far more likely to encounter too few calories than too many, the human systems for maintaining energy equilibrium tend to err on the side of encouraging overconsumption. Consider leptin's effect on appetite. When fat levels fall and leptin levels drop, appetite increases proportionately: less leptin equals more hunger. But the reverse isn't quite so automatic or proportionate. As fat levels return to normal and leptin levels rise again, appetite tapers off slightly but doesn't disappear completely. Why not? To initiate satiety, leptin molecules must physically reach the hypothalamus, which means crossing a membrane separating the blood from the brain. This blood-brain barrier is a dumb filter; its tiny passageways can pass through only so many leptin molecules, or any other substance, before a molecular traffic jam ensues. Thus, even if lots of leptin is in the blood, it may be some time before the hypothalamus is sufficiently aroused to hear satiety signals, which means that even after fat levels are restored, a person will probably continue eating more heavily.

In other words, appetite lacks an effective upper limit — which for most of human history wasn't a problem; beyond a tiny group of elites who commanded surplus calories, our ancestors' risk of overeating was zero. In fact, any kind of upper limit on appetite would have been deadly, since it would have prevented people from rapidly regaining weight after lean times and from rapidly storing fat in times of plenty. In this sense, the internal energy economy operated as a complement to the external economy. The body provided a lower limit for intake, while nature provided an upper one.

Not only does the human body lack any effective internal upper limit on food intake, but also, once weight is gained, the body will do everything possible to keep that weight on. Again, the culprit appears to be leptin. If fat storage falls even slightly, leptin levels actually plunge, which sends the appetite soaring. A person will eat more, even though his or her fat stores aren't even close to being fully depleted — a disproportionate response that, again, makes perfect sense from a survival standpoint: eating larger meals *before* fat stores are completely exhausted is the biological equivalent of stopping at a gas station *before* our tank is empty. But in a world of abundant calories, such a mechanism makes losing weight very hard. "Humans have evolved excellent physiological mechanisms to defend against body weight loss," writes James Hill, director of the Colorado Clinical Nutrition Research Unit at the University of Colorado, but "they have only weak physiological mechanisms to defend against body weight gain when food is abundant."[6]

And in fact, nearly every system involved in energy regulation has a similar bias — either toward weight gain or against weight loss, and often overlaps with other systems to create a kind of fail-safe redundancy. If one system fails to defend the fat stores, another takes its place. The human body "is designed to defend itself against starving to death," says Rudolph Leibel, a geneticist at Columbia University College of Physicians and Surgeons.[7] "You can bemoan the fact that we're set up this way, but it's what has gotten us here."

We all know what happens next: having "gotten us here," having kept us alive long enough for us to build a modern, high-yield food economy, a metabolic system designed to defend against starvation was not only obsolete but overwhelmed. In the United States, as the farming boom of the early nineteenth century produced food surpluses and lower food prices, the human energy equilibrium began to falter: we got larger. At first, our increasing size was more of a restoration: we grew not just heavier but taller, regaining our prehistoric stature. But eventually, as gains in weight overtook gains in height, we grew fat. By the 1890s, the rotund phyique had become commonplace, even among public figures. As historian Lowell Dyson notes, the femme fatale of the 1890s was the two-hundred-pound actress Lillian Russell, while leaders like J. P. Morgan and Grover Cleveland "set the standard for both the upper and middle classes, with their huge bellies accentuated by fashionable vests and heavy gold watch chains."[8]

Public health officials, worried by the changes, launched vigorous campaigns to dissuade consumers from rich diets, and these early efforts, abetted by the economic hard times of the Depression, managed to reverse Americans' gastronomic excesses — but only temporarily. By 1942, the Metropolitan Life Insurance Company[9] warned that what it called the body mass index, which measured the relationship between weight and height and which seemed to be directly related to a person's longevity, was creeping upward. By 1960, 13 percent of American adults — roughly three in twenty — were classified as "overweight"[10] (the term "obese" was not yet in use).

During the 1960s and early 1970s, the trend leveled off. But starting in the 1980s, for reasons that are still being debated, the obesity rate not only resumed its climb but did so with unprecedented sharpness. By 1990, 23 percent of all adults met the criteria for "obese" (the term that had replaced "overweight"), while 11 percent were classified "overweight," which now defined a person who, though not yet officially obese, had a body mass index that was still too high for good health. By 2000, the percentage of obese adults had jumped to 31 percent — nearly one in three — and the percentage of overweight was at 16 percent, which meant that altogether, 47 percent, or nearly half, of all Americans now weighed too much.[11] Children were gaining too: in 1960, almost no children had been classified as obese; in 2000, the number was one in seven.[12] Nor did the increases show any signs of abating; by 2000, the American male was twenty pounds[13] heavier than he'd been in 1980 and was adding nearly two pounds a year.

The consequences of a heavier, larger population are apparent at every level. Clothing sizes have been adjusted upward; what passes for a woman's size 10 today would have been a size 14 in 1940. Our infrastructure is being redesigned for larger, heavier users: office chairs must be built more sturdily, mattresses must be more load-bearing. Airlines have had to raise their per-passenger weight allowances[14] and spend $275 million a year more on fuel now than they did in 1990 simply to lift heavier passengers.[15] Even the funeral industry has had to adjust and now offers larger coffins and wider cremators to fit the larger coffins.[16]

Of course, if larger coffins and bigger jet-fuel bills were the worst of obesity's costs, society could fairly easily adapt to a heavier population. But this is not the case. Obese people are more likely to suffer a wide variety of ailments: sleep disorders, blood clots, leg ulcers, pancreatic inflammation, and hernias. Heavier people put more stress on bones and joints, especially the knees, and the extra padding in the chest cavity prevents the lungs from fully expanding, leading to lower blood levels of oxygen and shortness of breath. Obesity also makes medical treatment harder, since body fat hides lumps and other symptoms, dilutes medications, and extends hospital stays by 50 to 130 percent.[17] And these are relatively minor complaints. Obesity is clearly linked to higher rates of heart disease,[18] both because the heart must work harder and because people with more fat tend to have higher blood levels of triglycerides and LDL (bad) cholesterol and lower levels of HDL (good) cholesterol, all of which, as the American Heart Association website points out, are contributors to heart attack and stroke.

Obesity is also strongly implicated as a cause of adult-onset diabetes, possibly owing to the link between fatty acids and insulin. Insulin is a hormone that helps the body regulate the amount of glucose in the blood. Glucose is produced when starchy foods are digested in the gut. It is one of the body's primary fuels — and the only nourishment that can be absorbed by the brain (muscles and organs, by contrast, can also burn fatty acids). Because the brain functions best when blood glucose levels are constant, there is a system to keep those levels steady. Whenever glucose enters the blood, the pancreas produces insulin, which signals the muscles, liver, and fat cells to absorb some of that glucose from the blood and either burn it as fuel or store it for later use, thus keeping the blood glucose level from rising too high. If levels fall too low, other hormones pull glucose out of storage (from the liver, for example) and into the blood.

Unfortunately, insulin's vital regulatory role can be diminished by high blood levels of fatty acids,[19] which make liver and muscle tissue less sensitive to insulin's signal, and thus less able to pull excess glucose from the blood. The more fat a person carries, the more fatty acids circulate in the blood, and thus the less sensitive to insulin the liver and muscle tissues become, in a progressive condition known as insulin resistance. At first, the body responds to insulin resistance by secreting more insulin, which overcomes the resistance and clears the blood of excess glucose. But if fatty acids levels remain high — which is the case with overweight and obese people — insulin resistance also increases, forcing the pancreas to secrete even more insulin. An individual with a robust pancreas can tolerate the situation indefinitely, but in others, the pancreas gradually loses its ability to produce insulin, leading to type 2 diabetes, a disease than can lead to blindness, loss of feeling and circulation in the extremities, amputations, and death.

Not all obese people develop type 2 diabetes; the predisposition appears to be genetic. But among those so predisposed, obesity substantially raises the risk. As more children become obese, type 2 diabetes is found in younger and younger age groups, and this is why researchers like David Ludwig, with Boston Children's Hospital, regard childhood obesity as a demographic time bomb. Because obese children have a high risk of growing into obese adults, the increase in childhood obesity has essentially locked the population into a future with high rates of adult obesity and its related health problems, such as diabetes. "When this generation of obese children carries the increased health risks into adulthood, it will mean the end of the trend toward increased lifespan that we have seen in this country for the last century," Ludwig told me. "And it may in fact actually shorten lifespan by two to three years, which is more than the effect of all cancers combined."[20]

As these and other gloomy findings accumulated, government health officials began to speak in earnest about an obesity epidemic. In 2001, the U.S. surgeon general reported that obesity was causing 300,000 premature deaths a year. Three years later, the federal Centers for Disease Control upgraded the number to 400,000, while other CDC studies found that obesity was responsible for more than $61 billion a year in direct medical costs, or nearly 5 percent of the nation's total health-care expenses,[21] and another $56 billion in lost wages and other indirect costs, making obesity more expensive than smoking.[22] Such research suggested that obesity was on track to become the largest health issue in the United States, raising the ironic prospect that the most serious threat to health would very soon be . . . food.

It wasn't just government agencies that were up in arms. In the minds of many health advocates, obesity hadn't simply *happened* to humanity; it was being encouraged by a food industry whose bottom line depended more and more on processed foods and snacks. These foods contain lots of sweeteners and fat, ingredients that are not only calorie dense in themselves but that also, by some accounts, actually stimulate us to eat *more*. And while that claim was, and is, hotly debated, even the industry had to acknowledge that the national diet was indeed changing. Per capita consumption of sugar, HFCS, and other sweeteners had climbed by more than a third between 1970 and 2000. Cheese consumption was up by 50 percent, in part because of the rising popularity of pizza.[23] All told, the per capita intake of calories, after falling steadily between 1965 and 1987, had rebounded, rising 17 percent by 1995.[24]

In the context of these trends, advocacy groups like Center for Science in the Public Interest (CSPI) and industry critics Marion Nestle and Kelly Brownell began to argue that traditional approaches to preventing obesity, which focused on changing individual behavior, were no match for the rising tide of calories. What was needed were interventionist policies that confronted the environmental, cultural, and economic factors driving the crisis. Lawmakers began offering a raft of proposals ranging from bans on junk food to taxes on fat. Attorneys and advocacy groups, unwilling to wait for legislation, filed lawsuits based on the highly successful antismoking litigation: McDonald's and other fast-food chains were sued for serving supersize portions; school districts were threatened with legal action for selling sugary soda pop and other high-calorie junk. In early 2006, Center for Science in the Public Interest threatened to sue Viacom because SpongeBob SquarePants, the title character of a very popular kids' cartoon show, was pitching Kellogg's Pop-Tarts and other less-than-wholesome fare.[25]

As litigation and legislation loomed — and as public ire was piqued by a gaggle of investigative reports about the food industry and by movies like *Super Size Me* — food companies came to realize they were heading down the same expensive road the tobacco industry had marched a decade before. "Despite the fact that [food] companies are saying this is all bogus and personal responsibility is what counts, I think the lesson we've learned, and that I would preach from my experience, is that you have to take the plaintiffs' bar at their word," Joseph Price, a veteran product-liability attorney, told the *New York Times* in 2004. "They've said they're going to make this the next tobacco."[26]

Predictions like that are what gets Rick Berman out of bed every day. Six foot three and broad-shouldered with a smoothly shaved head, piercing blue eyes, and a gravelly voice, Berman runs the Washington, D.C.–based Center for Consumer Freedom, an industry-backed think tank that seeks to counter the influence of groups like CSPI by attacking the "steady diet of obesity myths" being foisted upon consumers — starting with the myth of an obesity epidemic itself. In Berman's view, the only significant thing in the food economy that changed in the last two decades was that advocacy groups, scientists, attorneys, journalists, and the weight-loss industry discovered obesity as the hot new vehicle for political and economic gain. Obesity research, Berman told me as we sat in his office on K Street, is often funded by entities that would financially benefit from obesity, most notably pharmaceutical companies hoping to capitalize on antiobesity drugs and who "see a marketing opportunity in an obesity epidemic."[27] Berman doesn't deny that obesity is a problem. But he argues that other factors, notably genetics and lack of exercise, are the real culprits, not food or food companies.

Berman's Center for Consumer Freedom is itself hardly pure of motive: most of its funding comes from food companies who would very much like obesity *not* to be seen as an epidemic — among them Coca-Cola, Wendy's, Tyson, and Outback Steakhouse. In fact, Berman honed his skills defending the restaurant industry against antitobacco legislation, and his many critics say his "center" is merely the current incarnation of a PR operation that specializes in defending the consumer products industry. (The guard in the lobby downstairs had never heard of Center for Consumer Freedom but was able to direct me to the offices of Berman & Company.)

Berman does raise some important issues. Obesity's health risks have been overstated: the U.S. Centers for Disease Control has lowered its original estimate of obesity-related deaths from 400,000 down to 112,000. And obesity has indeed become a lucrative business for a great many companies and organizations, not least the five-billion-dollar-a-year diet industry. Perhaps most usefully, Berman and his fellow skeptics have brought attention to the underplayed but surprisingly important role that physical activity plays in our expanding girth.

Americans are not physically fit. Not only do we exercise less — not even half of U.S. adults get the minimum recommended level of exercise, which is around thirty minutes of moderate activity five times a week — but our entire lifestyle is now significantly less energy intensive than it was even thirty years ago. The layout of the modern suburban community — huge streets, no sidewalks, vast distances between homes, shops, and schools — discourages walking and all but requires a car. Recreational activities are increasingly immobile, centered more and more on TV, gaming, and the Internet, and also increasingly indoor, since rising fear of crime means we're less inclined to walk or jog or allow our kids to propel themselves to friends' houses, playgrounds, or school. From 1980 to 1990, the share of children walking to school dropped from four-fifths to less than one-third — a decline a Centers for Disease Control report attributes to the missing-children scare of the 1980s and 1990s,[28] when a generation of parents simply stopped letting their children out of their sight.

Similarly, our jobs, once dependable opportunities to burn calories, are less and less physically demanding: in the early part of the twentieth century, the hard work of the industrial era began to give way to lighter clerical and service jobs, which in turn gave way during the 1970s and 1980s to information jobs, which, with their reliance on computers, phone, and e-mail, meant people could labor for hours without leaving their desks. According to a Mayo Clinic study, the apparently small difference between a sedentary office job, such as data processing, versus one that requires periodic movement, like that of a shipping clerk, can be as much as 350 calories a day.[29] And again, because we're talking about energy balance, even a small reduction in exercise can yield a significant weight gain. According to James Hill, with the Colorado Clinical Nutrition Research Unit, America's weight gain over the past several decades can be explained by a caloric imbalance so small — around 100 calories a day — that we could burn it off with a twenty-minute daily walk.[30]

Conclusions like Hill's have led obesity skeptics (and a great many food companies and lobbyists) to argue that our rising weight has far less to do with food and far more to do with lifestyle changes, and especially changes that became more pronounced during the 1980s, just as obesity rates began to rise. Yet such factors never attracted much attention, because instead of implicating a single highly visible (and quite wealthy) culprit — the food industry — they pointed to consumers themselves. Obesity, in this view, is really just the unintended but unavoidable consequence of economic progress — the natural evolutionary endpoint for a species programmed to find the most calories for the least physical exertion. As Tomas Philipson, a University of Chicago economist specializing in obesity, writes, "the obesity problem is really a side effect of things that are good for the economy." We may be fatter, Philipson concedes, "but we would rather take improvements in technology and agriculture than go back to the way we lived in the 1950s when everyone was thin. Nobody wants to sweat at work for 10 hours a day and be poor."[31]

Thus, obesity is less a manifestation of corporate greed than an expression of rational decision making: we no longer have to work as hard for our calories, so . . . we don't. And given that obesity isn't really any more complicated, or nefarious, than rational consumers *choosing* to work less (and perhaps eat more), the solution will come not from costly pharmaceuticals or (especially) intrusive government regulation but from consumers *choosing* to change how much they eat or exercise. "When you're ready to start dealing rationally with your weight problem," writes Todd Seavey, with the American Council on Science and Health, another food-industry-backed group, the solution is simple: "Eat less and/or exercise more."[32]

The problem with blaming obesity entirely or largely on our burgeoning sloth is that sloth wasn't the only calorie-related factor that changed in the 1980s. What professional skeptics like Berman rarely acknowledge is that many of the same technological and economic changes that contributed to the ease of modern life, and thereby allowed us to burn fewer calories, also made it much, *much* easier to get those calories in the first place.

For example, while innovations in food production and retail strategies have cut overall food costs, the greatest price impact has been on foods with the greatest caloric impact — starch and fat. Andrew Drewnowski, a researcher at the University of Washington, has found that the per-calorie costs of the two most common processed-food ingredients — starch and fat — have fallen faster than the per-calorie costs of less caloric ingredients, such as fresh fruits and vegetables. Potato chips, for example, now cost consumers about a tenth of a penny per calorie, whereas carrots cost about four times that much, mainly because fresh produce is still more labor intensive to grow and more expensive to handle and store. Thus, says Drewnowski, not only can we afford to buy more potato chips than carrots, but because chips are more calorie dense (about 150 calories per ounce of chips versus about 13 calories per ounce of carrots), when we eat more chips, we're consuming even more calories.[33]

Such a disparity may help explain why poorer consumers tend to be more obese than consumers at higher income levels are: cheaper foods tend to be more caloric. It may also shed some light on the so-called Zip Code effect, wherein people in poorer neighborhoods suffer higher rates of obesity. Poor neighborhoods tend to have more fast-food restaurants and more convenience stores (which stock more heavily processed and higher-calorie food) and fewer grocery stores, (which offer far more fresh produce and healthier options) than wealthier, whiter neighborhoods have. One study of all food stores in three low-income Zip Codes in Detroit found that fewer than one in five carried a minimal healthy-food basket — that is, food products representing all strata of the food pyramid. The study also found that perishable items weren't as fresh as they were in richer neighborhoods, and that, in the cruelest twist, basic staples like bread and milk were actually more expensive in poor Zip Codes than in wealthy ones.[34]

As important, food has not only become dramatically cheaper in the last thirty years but vastly more accessible as well. Before the 1970s, most meals were prepared by a small minority of experts — mainly the housewives, cooks, and others who possessed the skills, tools, and time to prepare food* — a culinary tyranny that played a huge role in how much and how often we ate (and probably in our attitudes about eating, hunger, and satisfaction). Today, that tyranny has been vanquished by ubiquitous restaurants, snack machines, and delis, as well as convenience foods so cleverly packaged they can be made in minutes, by children, in a microwave.[35] Granted, convenience foods bestow considerable benefits on a time-crunched culture, but because they tend to rely on Drewnowski's low-cost but calorie-dense sweeteners and fats, they also tend to bestow extra calories: Biing Hwan-Lin, an investigator at the federal Economic Research Service, has calculated that if restaurant and convenience-store food had the same average caloric densities as food typically cooked at home, Americans would eat 197 fewer calories per day — more than twice Hill's caloric

---

* So many of the traditional limits on consumption — eating three meals a day, for example, or the requirement of eating together as a family or group at set times — were dictated in large part by the cooks, whose main interest was in minimizing their workload and whose monopoly over cooking tools and cooking knowledge gave them much authority over what and how often people ate. In such a world, people didn't necessarily eat when they were hungry; they ate when food was prepared — a reality that must have influenced attitudes about eating and appetite. Earlier generations almost certainly felt hunger between meals, but chances are they'd been brought up to regard such feelings as routine, normal, expected — not a reason for panic or anxiety, nor a signal to head to the candy machine.

gap — while substantially reducing their intake of fat and saturated fat.[36] Our increasing reliance on away-from-home food is probably a big reason that Americans' total per capita intake of fat, which fell from the mid-1960s to the late 1980s, has since climbed by nearly 10 percent.[37]

This disproportionate reliance on high-calorie ingredients is affecting consumption and weight gain in other ways. Fat, for example, is not only more than twice as calorie dense as protein or starch, but it also appears to have some capacity to stimulate appetite. Research by Columbia University's Rudolph Leibel shows a striking connection between our rising consumption of dietary fat and changes in the body's capacity to monitor its own fat stores. In newborns, a high-fat diet can impair the sensitivity of the hypothalamus to satiety signals. If this loss of sensitivity leads to weight gain — a plausible outcome — the larger fat stores that result may initiate a vicious cycle of weight gain: as a person's fat levels rise, so do their blood levels of leptin, and if leptin levels remain high for too long, the brain eventually gets desensitized, or loses its ability to respond, to leptin's effects. And since leptin's job is to help the brain hear satiety signals from the stomach, Leibel says, a diet high in fat — which, increasingly, describes what Americans eat — may gradually be reducing our ability to monitor our caloric intake. "The brain becomes desensitized to its own signal as to how much body fat it has," he says. "It essentially loses its ability to 'read' that signal."

Less is known about the effects that sweeteners have on consumption. For decades, some researchers have argued that because humans evolved in an environment almost totally lacking in sugar and other simple sweeteners, these foods overwhelm our natural abilities to monitor intake. Scrutiny has been especially high for high-fructose corn syrup, or HFCS, which entered the food system in a big way in the late 1970s just as obesity was picking up steam, and which possesses several unusual nutritional features that have been linked to obesity. First, where glucose moves easily from the blood into the brain, fructose lacks the biochemical password to get through the blood-brain barrier. As a result, the brain, which senses blood levels of glucose (and alters appetite accordingly), is essentially unaware of the amount of fructose circulating in the blood, and thus doesn't trigger a satiety reflex, no matter how much fructose is consumed. Second, unlike glucose, fructose doesn't stimulate the release of insulin, without which there is no leptin, which the brain needs to hear the I'm-full satiety signals

from the stomach. In short, a person can consume fructose without trip-
ping many of the main mechanisms for curbing appetite. Third, according
to work by Sharon Elliott, while most sugars are digested in the gut (and
converted into glucose), fructose isn't fully digested until it reaches the
liver. Here, fructose's unique molecular structure — particularly, the way
its carbon atoms are arranged — acts as a kind of backbone for the construction of long-chain fatty acids; fructose, in other words, converts to fat
more easily than other sugars do.[38]

In recent years, other researchers have raised doubts about HFCS's
unique obesity-causing qualities. The main objection is that high-fructose
corn syrup actually has only slightly more fructose by weight than table
sugar; sucrose is made up of one molecule of glucose and one molecule of
fructose. This suggests that the switch from sugar to HFCS wouldn't have
had that large an effect on human physiology. Yet even those researchers
skeptical of fructose's unique effects worry that the debate over HFCS has
served mainly to distract us from a more fundamental, if less sensational,
reality: whether fructose alters appetites or not, its growing presence in the
food chain represents a significant volume of calories that were largely absent from the food supply forty years ago. "The problem is that the consumption of all sweeteners — sucrose plus HFCS — has greatly increased
in the last 15 years, much of it in the form of soda," Walter Willett, head of
the nutrition department at the Harvard School of Public Health, told me.
"And almost certainly, this has contributed to obesity. We need to reduce
both of them."[39]

Willett's admonishment goes straight to the economic paradox at the heart
of the obesity crisis. In the last half decade, the food industry has been on
the offense, rolling out numerous reduced-fat/low-calorie products (four
thousand since 2002) and launching campaigns to improve consumer nutrition awareness. But such efforts are hard to take seriously. To begin with,
the four thousand healthier products the industry touts represent just 7
percent of the roughly 56,000 food products introduced between 2002 and
2006.[40] More to the point, for all that the industry claims to support better
health, sensible eating, and a vision of America as lean and mean, the
reality is that food companies cannot afford to have us eat any less than
we already do. On a percentage basis, if Americans cut their daily intake
by the 100 calories that Hill says would suffice to rebalance our internal

economies, it would cost the industry between $31 billion and $36 billion in U.S. sales alone.[41] Such losses would be painful at any time, but they'd be especially unwelcome today, when food companies face growing competition, dwindling margins, and a saturated U.S. market. Put another way, food companies may dispute the claims about HFCS, the glycemic index, or anything else about the *quality* of the food they sell, but they can't deny that they're doing nearly everything possible to sell us greater *quantities* of food.

Consider how diligently the food industry has worked to increase the number of opportunities for consumption. Grocery stores are designed to maximize our exposure to potential impulse buys. Vending machines are ubiquitous, as are convenience stores and especially restaurants. Between 1972 and 1995, as the U.S. population increased by barely a third, the number of restaurants nearly doubled, and the number of fast-food outlets nearly tripled — in part because food companies know more opportunities lead to greater sales. McDonald's, for example, strives explicitly "to have a site wherever people live, work, shop, play, or gather," according to company literature.[42] "To build pervasiveness of our products," declared Coca-Cola in its 1997 annual report, "we're putting ice-cold Coca-Cola classic and our other brands within reach, wherever you look: at the supermarket, the video store, the soccer field, the gas station — everywhere."

And although we can't fault food companies for wanting greater sales — that is, after all, how businesses make money — the fact that the number of consumption opportunities has grown significantly faster than has the population of consumers suggests rather pointedly that the industry is counting on each consumer to buy *more* food. Nor can it be coincidence that food companies create those opportunities in environments specifically chosen for their proconsumption demographics — in malls, in airports, and in poor neighborhoods. Krispy Kreme is certainly not the only purveyor of high-calorie goodies to open new branches in neighborhoods dominated by Hispanics, whose cultural inclination toward sweets (and relatively low emphasis on nutritional education) boosts sales. As the manager of a newly opened store in a heavily working-class Latino community in California told a *Harper's* reporter in 1999, "We're looking for bigger families . . . Yeah, bigger in size."[43]

Food companies counter that they're no longer in the calorie business; in the modern era, food profits come not from pushing raw calories but by adding value and charging a huge markup. Yet the distinction between quantity and quality was never really that clear cut; even in the industry's early days, one of the easiest and cheapest ways of adding value was to add sugar and fat. More recently, the distinction between caloric quantity and quality has become even fainter. With the rise of the Wal-Mart more-for-less model, manufacturers and restaurateurs have moved from a value-added model to a low-cost, high-volume commodity model, in which the consumer must be enticed not simply with added value but with more and more added value, for lower and lower prices. And while this extra added value can be offered in the form of something noncaloric — a new flavor, for example, or upscale ingredients, or more convenient packaging — once the product is established in the marketplace, the easiest and cheapest way to add more value to that product is simply to offer *more* of it — selling two for the price of one, say, or offering larger serving sizes.

Put another way, given the declining costs of food and given the American obsession with "getting better and better deals, that is, getting more for less," as Colorado's James Hill puts it, food companies and restaurants have found that the most economical way to add value and boost sales is simply to supersize. Rather than spend money developing some entirely new product or package based on some truly innovative (and expensive) idea, you just need to figure out how to boost package or portion sizes. Thus, although supersizing has received loads of bad press lately, it remains a standard business practice, and not just with fast food. Cookies, muffins, pizza, candy bars, bagels, meals at traditional restaurants — all are between two to seven times larger[44] than they were during the 1980s, according to research by Lisa Young and Marion Nestle, and in most cases are far larger than federal recommendations for the portion size.

In fact, most of American food culture now reflects the supersize model. Restaurant dinner plates are larger; muffin tins and pizza pans come in larger sizes; automakers are installing larger cup holders; even cookbooks are written with larger servings in mind. "Identical recipes for cookies and desserts in old and new editions of classic cookbooks such as *Joy of Cooking* specify fewer servings, meaning that portions are expected to be larger," write Young and Nestle.[45] And in case you missed it, this trend's timing is quite familiar: after rising slowly but steadily during the 1960s and early 1970s, supersizing emerged as a full-blown marketing tool in the 1980s and 1990s. McDonald's introduced its Extra Value Meals in 1991;

7-Eleven first offered its forty-four-ounce Big Gulp in 1983 and a sixty-four-ounce version in 1988.[46] Since then, the number of new food products that are brought out in larger sizes or portions has doubled every decade. The net result, some nutrition experts suggest, has largely erased many of the gains made in healthier eating. "We've made tremendous success in getting people to eat less saturated fats," says Mark Pereira, who studies the links between nutrition and physical activity and obesity, diabetes, and heart disease at the University of Minnesota. "The problem is, they've replaced it with carbohydrates."[47]

Food companies, although they no longer deny that larger portions are a key marketing strategy, vigorously resist any suggestion that these larger portion sizes actually encourage consumers to eat or drink more — a denial that has to qualify as one of the most laughable claims in the entire obesity debate. Not only have numerous studies* shown that large portions *always* induce greater consumption, but it would be hard to understand why else the food industry would offer them. Given that consumers register the value of food primarily by *eating* it, if bigger portions *didn't* increase consumption and thus cause consumers to feel they were getting greater value for their dollars, no food company would bother offering larger portions in the first place.

On one level, the supersize trend is hardly different from the trends throughout the modern retail business: as competition drives down margins, all manufacturers and all retailers must sell more and more product to maintain their profits. Indeed, in the nonfood sector, consumers can now afford so much stuff that the average home, though 45 percent larger today than it was three decades ago, can't hold it all.[48] The crucial difference is that where consumers who buy too many nonfood items can effectively hide that problem by putting the surplus in the garbage or in off-site stor-

---

* In one study conducted by Nicole Dilberti and colleagues in a cafeteria-style restaurant, customers who were covertly served thirteen-ounce pasta entrées ate 43 percent more calories than did customers who were served nine-ounce versions of the same pasta. In another study, Barbara Rolls of Penn State University fed subjects large portions of food three times a day for eleven days. Over the course of the study, these subjects ate 16 percent more calories — or about four hundred more calories a day — than did subjects served smaller portions of the identical food. Similar results have been reported with potato chips, sandwiches, and even soup; in a Cornell University study, subjects who ate soup from bowls cleverly engineered to surreptitiously refill automatically consumed 73 percent more soup than did subjects eating from normal soup bowls.

age, there are no off-site options for surplus food; all excess calories are kept decidedly on-site.

Beyond what it says about industry marketing tactics, portion size points to another way in which food companies are working very hard to make us better, or at least bigger, consumers. In studies of children's eating behavior, Barbara Rolls of Penn State University has found that very young children are far less affected by portion size than older children. When three-year-old preschoolers were given lunches with a small, medium, or large portion of macaroni and cheese, portion size did not influence consumption: the kids ate a certain quantity and then stopped, regardless of how much food was on the plate. By contrast, Rolls found that with five-year-olds, greater portion size led to greater intake.[49] Such findings, argues Cara Ebbeling and her colleagues at Boston Children's Hospital, suggest that "as children grow older, they become less responsive to internal hunger and satiety cues and more reactive to environmental stimuli." Or, more plainly, as children grow, they stop listening to their own systems and begin to take their cues from their surroundings. Historically, these external cues about food came mostly from family and friends at the dinner table or during meals in communal settings such as schools or churches. In modern times, these traditional signals have been supplemented, and perhaps supplanted, by another set of signals — those of food makers, who spend tens of billions of dollars to fill the cultural landscape with messages for and about food.

Much of this saturation takes place on TV: more than three-quarters of the total advertising budget of the U.S. food industry is spent on television ads, and within certain categories, such as fast food, the share is closer to 95 percent. And although TV is losing some of its luster for advertisers as audiences are fragmented by new media offerings, food makers have been quick to adapt, branching out to nearly every possible venue, from video games (which McDonald's uses to reach the eighteen- to thirty-four-year-old males who are watching less TV)[50] to sports sponsorships (as when Yum Brands sponsored the Kentucky Derby) to food-industry-backed websites for kids.

Food companies deny that such saturation marketing contributes to obesity, arguing that food ads don't make consumers do anything but instead provide information so consumers can make informed choices. This, too, is a bit rich, given the amount of money food companies spend on advertising, given the well-known fact that sales go up as advertising rises, and, more to the point, given that many of these campaigns are predicated on the hope that consumers will *not* make informed choices, since consumers who were truly informed about food would almost certainly buy fewer of heavily processed but quite lucrative products.

Consider the way companies market their increasingly strategic snack products. Companies are not only spending more money advertising snacks but are using those ads to quell consumer concerns that snacking is an inappropriate way to eat. Some manufacturers are doing this by making snacks heartier and more filling, in the hopes of persuading consumers to see snacks as "a meal occasion."[51] Others are going to the opposite extreme, marketing snacks as so-called me-time products — full-blown indulgences for consumers who need a break after a long day of work, parenting, or, apparently, breathing.[52] According to *Food Navigator* magazine,[53] me-time marketing reflects a new industry strategy aimed at the "emotional fulfillment of busy, stressed out consumers," whose anxiety, disappointments, and sheer boredom can be used to induce purchases of foods that might otherwise be shunned as inappropriate. Beyond the cynicism underlying this trend (according to Datamonitor, two top target markets are office employees, who snack out of "psychological need for stimulation and reward,"[54] and women, whose need for emotional fulfillment "should not be underestimated"),[55] it makes hash of industry's stated interest in helping consumers eat healthier. One of the main reasons the industry is so keen on me-time foods is that category's apparent imperviousness to health concerns. As a report by Datamonitor notes, me-time products won't be hurt by the new surge in health consciousness because consumers aren't "generally prepared to abandon the pleasure that they derive from treating themselves" — particularly, Datamonitor notes, if companies are successful at "blurring consumers' perception of what constitutes a need and what is merely a desire."[56]

Again, one could argue that adults aren't complete idiots when it comes to marketing and are thus responsible for their own food choices. Even food libertarians quail, however, when such marketing power is directed at young consumers, who cannot be expected to make informed decisions about food. And food companies are undeniably interested in reaching young consumers. Kids and adolescents are not only the consumers of tomorrow but, by conservative estimates, are already responsible directly or indirectly for some $500 billion in food purchases every year, up from $295 billion in 1993[57] — in large part because children heavily influence parents' buying decision through the so-called nag factor or pester power. Nor is there any doubt that food companies try to reach children as early as possible, when eating habits and preferences are developing. Food companies not only advertise heavily during kids' shows (on average, an hour of children's programming features ten food commercials)[58] but also link the shows themselves to products, as when SpongeBob SquarePants shills for Pop-Tarts, Oscar Mayer's Lunchables, Kraft Macaroni and Cheese, and other high-margin, high-calorie fare.

Granted, adults also see a lot of food ads. Yet adults presumably have the cognitive capacity to judge the accuracy and intent of ads; young children do not. Development experts say that before age eight, children lack the ability to understand the persuasive intent of ads and instead take their claims as truthful: children are, in other words, highly vulnerable to marketers' messages. And when these messages are about food, they are most often about products and eating practices associated with obesity — again, because the most profitable products, and thus the products that companies market most heavily, also tend to be the most processed and the most caloric. According to research by Kristen Harrison at the University of Illinois, more than 80 percent of the foods advertised during kids' shows are convenience foods, fast food, and sweets. Harrison also found that "snacktime eating was depicted more often than breakfast, lunch, and dinner combined." When Harrison analyzed the nutritional content of the products advertised, most exceeded the recommended daily values for fat, saturated fat, and sodium and represented about a cup of added sugar.[59]

Industry executives reply that it is parents' responsibility to monitor and limit their children's TV viewing — another fairly hollow argument, since, given the amount of money the industry spends for ads on kids' TV shows, it is clearly hoping parents do no such thing. But even if parents can be reasonably expected to filter their children's TV viewing, the food industry has worked hard to reach children when parents aren't around, most notably by building a substantial presence in schools. Thousands of cash-strapped school districts have turned over their lunch programs to fast-food vendors (one study found that more than half of public high schools in California offered items by Taco Bell, Subway, Domino's, Pizza Hut, and other brand-name vendors),[60] which provides these companies with a large, captive, and impressionable audience of future consumers.[61] Beverage companies also exploit sagging school budgets by offering lucrative "pouring rights" to schools that allow the placement of vending machines. Even individual teachers are recruited for the cause: in 2001, General Mills paid ten elementary school teachers in Minneapolis $250 a month to serve as "freelance brand managers," a task that involved commuting to school in cars plastered with huge ads for Reese's Puffs sweetened cereal.*[62]

And apparently, kids are paying attention: consumption is rising for many of the most aggressively marketed foods.[63] In 1965, U.S. boys between the ages of eleven and eighteen drank an average of six and a half ounces of soft drinks a day; by 1996, that had risen to eighteen ounces.[64] That's an increase of 143 calories — which could more than explain overall national weight gain during that period. Nor is it simply soda pop consumption that has increased. By age two, the average American child has begun demanding specific products (most often breakfast cereals). Between the ages of three and eleven, the most commonly requested items are snacks and desserts (accounting for 24 percent of all requests) and candy (accounting for 17 percent). By comparison, fruits and vegetables are asked for 3 percent of the time. All of which may help explain the Americanization effect on diet that is observed among many immigrant groups. For example, first-generation Latino immigrants, though poorer than whites, tend to be healthier and slimmer than whites, but their children, who grow up in the United States and thus spend their formative years immersed in U.S. commercial food culture, tend to be much heavier.[65] Give us your poor and your underweight, and we will fix them right up.

For all the debate over the cause of obesity, there is broad agreement that this trend will be hard to reverse. The body's natural defense systems effectively fight any conscious effort to lose weight, which helps explain why people who do manage to lose weight through dieting typically put it back on within one to two years. Exercise, while it seems effective at slowing or even halting weight gain, does surprisingly little on its own to help people

---

* The degree of food-company penetration into children's lives seems boundless. According to the *New York Times,* McDonald's pays thirty-one thousand schools to feature Passport to Play, a physical education program designed by McDonald's in which every piece of literature carries the Golden Arches logo.

lose weight and must be paired with serious calorie reduction. Some pharmaceutical treatments seem promising: many researchers believe we are close to treating, at the molecular level, the body's tendency to defend its weight. But others argue that the very nature of the human energy-regulating system — a complex and redundant series of checks and balances — makes it unlikely that pharmaceutical treatments will be effective for any but a small number of patients. Trying to cure obesity, says Randy Seeley, the University of Cincinnati neuroscientist, "is fundamentally different from trying to treat something like cancer. The tumor may fight you, but the rest of your body doesn't want the cancer, and doesn't keep trying to stop you from attacking the tumor. But your body *wants* to be fat. Getting fat is what it is supposed to do, and trying to persuade your body not to be fat is going against everything your body is designed to do."

Nor have any of the likely external factors changed in ways that will make obesity less likely. Lifestyles are becoming even more indolent, work even less physical. Food companies continue to depend on selling more calorie-dense foods, and consumers continue to eat them. Although fast-food companies have brought out numerous lower-fat items, the industry is nowhere near abandoning the marketing magic of large portions: in 2004, Hardee's proudly launched its 1,420-calorie Monster Thickburger.[66]

Moreover, outside of the insurance industry, which must bear the brunt of obesity-related illnesses, most of the business world has few problems with fatter consumers. William Weis, a management professor at Seattle University, calculates that revenues for what he calls "obesity industries" — everyone from the fast-food restaurants who feed us to the medical and nutrition experts who treat us — will exceed $315 billion a year, or almost 3 percent of the nation's gross domestic product, which makes them unlikely to act in any way that might reduce the problem. "Put simply, there is a lot of money being made, and to be made, in feeding both oversized stomachs and feeding those enterprises selling fixes for oversized stomachs," writes Weis in the *Academy of Health Care Management Journal*. "And both industries — those selling junk food and those selling fat cures — depend for their future on a prevalence of obesity."[67]

Indeed, American culture is busily remaking itself not simply to accommodate a more obese population, but to normalize it. Beyond the bigger clothes and wider seats, we have movies, TV shows, and commercials that are reconfigured for a larger demographic, with big characters played by big actors to connect with this increasingly significant demographic. And the food industry is closing ranks. Although food and beverage companies recently volunteered to remove some vending machines from schools, the industry is pushing ahead on other fronts. Industry lobbyists have successfully pushed through laws in several states that bar lawsuits over obesity or any nutritionally related claims. And in Florida, despite Representative Juan Zapata's command performance, his bill outlawing HFCS in schools was killed in committee.

Even as Americans resign themselves to living large, the rest of the world is waking up to the realities of caloric surplus. Obesity rates in Europe, which once trailed those in the United States, are finally beginning to catch up and have tripled since the 1980s.[68] And if Europe seemed ripe for an outbreak — with consumers who are affluent, busy, and surrounded increasingly by fast food, snacks, and other American-style eating opportunities — developing countries are also showing signs of obesity problems, as falling food prices put a wealth of rich new foods before even poor consumers. A quarter of all Middle Easterners, for example, are overweight or obese, as are 40 percent of all Moroccans, and a third of all South Africans. In Kenya, a country where one in seven people is malnourished, one in eight is now overweight.[69]

There is, in fact, no shortage of irony in a global obesity epidemic. In cultures around the world, economic success now goes hand in hand with physiological failure: to be rich is to be fat. No country illustrates this paradox more dramatically than India, where obesity is now growing faster than either the government or traditional culture can respond. Rates of obesity and related ailments such as diabetes are soaring, as is demand for gastric bypasses, gastric banding, and other obesity surgeries — a perverse reality in a country where nearly half of children under the age of five are malnourished. The reasons are all too familiar. With industrialization and rising incomes, Indians are exercising less and eating more, especially value-added, processed foods; McDonald's, which has localized its menu with such offerings as the Chicken Maharajah Mac, has stores in Delhi and is expanding to other cities. The result, says Anoop Misra, a researcher with the All-India Institute of Medical Sciences, is a dramatic change in eating habits. "People are snacking in a new way," Misra told the *Observer* of London.[70] "Many children no longer take lunch-boxes to school. They drink cola and eat burgers. There is no awareness among parents that this is a problem."

In fact, say Misra and other medical experts, obesity is becoming a massive problem. The combination of dietary changes and nutritional illiteracy means obesity will be much harder to control, as will its side effects, such as heart disease and especially diabetes, which already afflicts twenty-five million Indians and is expected to grow to fifty-seven million by 2025. Despite such bleak forecasts, Misra says, India's government remains in a state of denial. "Politicians still ask, 'How can people here have obesity when they are dying of malnutrition?' They think malaria and TB are much more serious. You can treat TB with six months of treatment. But diabetes needs to be treated until the patient dies. This is going to be disastrous."

In a sense, obesity has emerged as an early warning about rising problems with the modern food system, and the most visible sign that with a product as complex as food, the traditional measures of economic success may in fact be indicators of looming failure. This catastrophe is hardly new; for a century now, farmers and food executives have struggled to transform food into a commodity like any other, and many have fallen by the wayside as their margins have become too thin, their risks too high, and the treadmill of this new economic model too exhausting. But in the rising obesity crisis, we see that the risks of a high-volume, low-cost system aren't limited to declining farms, failing companies, or the consolidation of entire sectors under the control of a handful of transnational giants. The risks have spread beyond the conventional economic arena and indeed are now wreaking actual physical havoc on the very people that food industry was ostensibly designed to help, and in whose name it is continually defended.

To be sure, cheaper, more efficiently produced food is clearly still needed in hungry nations — and, as we'll see in the next section, its absence in the developing world poses a substantial challenge. But in the United States, Europe, and other developed regions — which is to say, advanced food economies presumed to be operating at or near their optimum levels — that very proficiency has started to kill us.

Thus far, the most obvious damage has been to our individual health. But as we'll see in coming chapters, the disconnect between the economic drivers of the food system and the biological limits of our bodies points to a more egregious gap between the food economy and larger world around it. As we explore the global food trade, the persistence of world hunger, the rising threat of food-borne disease, and the precipitous decline in irreplaceable natural systems, we will see that the high-volume model is now disrupting the equilibrium not only of our own internal systems but of larger global systems, with consequences that will be profound and long lasting. What will be unmistakably apparent is that the very logic of the food economy is increasingly at odds with the biological systems, both human and natural, upon which that same food economy ultimately depends. In that sense, obesity may be the perfect metaphor for the modern food crisis: having escaped one set of limits, we now seem destined to grow until we hit the next.

PART II

5

Eating for Strength

A T TEN O'CLOCK on a hazy spring morning, the vast central plaza of the Shouguang exposition center in the Chinese coastal province of Shandong has the look and feel of a county fairground just before opening day. Shandong is China's biggest grower of farm produce, and in a few days it will play host to the International Vegetable Science and Technology Fair, a prestigious event that draws thousands of buyers, generates billions of yuan in sales, and is promoted each year with near patriotic fervor. Dignitaries of every stripe are invited. Journalists are courted with farm tours and sumptuous banquets showcasing local delicacies, both solid and liquid. ("The people of Shandong Province drink liquor," my government-appointed translator, Lin, had warned as we'd left Beijing. "So you better be prepared for that.") The pièce de résistance, however, is the exposition center itself, which is laid out along the lines of a People's Vegetable Theme Park. Around the central plaza, where workers are setting up thousands of folding chairs, billboard murals depict China's sundry agricultural miracles and the products that yield them. Closer in, massive, frighteningly realistic statues of fruits and vegetables — among them a cabbage, a bunch of grapes, a butternut squash, and a silo-size bok choy — loom above us like pieces from some gargantuan Warhol exhibit. Near the entrance, a red archway blazes with the motto "March Toward Science and Technology Frontier Developing High & New Agriculture."

The reality of Shandong's "high & new agriculture" is much more subtle and complex. At the rear of the plaza, Lin and I are herded by a small entourage of officials onto a long boulevard lined with willows, flowering shrubs, and enormous greenhouses. Inside one, standing waist high among green pepper plants, a handsome middle-aged man named Ren Quing

Hun, who is introduced as "the farmer," but who, with his stylish leather jacket, slacks, and cell phone, looks more like a venture capitalist, which in a sense he is. Like many local farmers, Ren grew traditional grain crops before switching to peppers, cucumbers, and other hothouse produce. The move wasn't cheap: a greenhouse costs 6,000 yuan, or about $750, in a country where rural annual incomes are below $1,000. But the returns can be substantial. With urbanites in Beijing, Shanghai, and even Tokyo and Seoul demanding fresh fruit and vegetables — a trend accelerated by the arrival of supermarket chains, including Carrefour and Wal-Mart — Ren makes around $420 from each of his greenhouses. This is several times the amount he could expect planting grain on the same acreage, which explains why Shouguang city alone has more than 400,000 greenhouses and why Shandong officials are so bullish on the province's future in the global food economy. As we leave, our hosts rattle off the names of countries whose trade delegations have come to see Shandong's "high & new agriculture" — among them, Holland, Mexico, the Philippines, and the United States, whose emissaries, I am assured, were "deeply impressed."

No doubt the American delegation was impressed by Shandong — although not, perhaps, for the same reasons my hosts are. More greenhouses here have meant fewer acres for wheat and corn. Since 1995, provincial grain output has fallen 20 percent, even as grain demand has soared, especially for corn, to feed the burgeoning pork and poultry industries. So quickly are supply and demand diverging that Shandong, China's second-biggest corn producer, must now *buy* corn from neighboring provinces. With similar horticultural shifts under way in other provinces, China, once a major corn exporter, will soon be forced to import — a forecast that delights American farmers and trade officials, who see China and its 1.3 billion people as the ideal market for American surpluses. To be sure, the United States isn't the only exporter eyeing the China market — Argentina, Russia, the Philippines, and Indonesia have been wooing Beijing for years. But American experts such as Mike Callahan, president of the very export-focused U.S. Grains Council, contend that once China begins to rely on imported corn, Beijing will quickly realize that American farmers, with their dependable output and low costs, are China's only realistic option. "Argentina can sell them corn only eight to ten months a year," Callahan told me. "China could get some corn from Thailand, and some small quantities from the Philippines and Indonesia. But if China needs large volumes, it's going to be the U.S."

But in a global food economy, one nation's opportunity is another's anxiety. If the prospect of corn imports delights the American Midwest, it terrifies the central government in Beijing, which has never trusted the international market, with its volatility and its susceptibility to political manipulation. In fact, for decades, even as China endured massive famine and tens of millions of deaths, its leaders still clung to a policy of food self-sufficiency and shunned imports — and many resist the idea still. As Zhang Hua Jian, a top agricultural official in neighboring Anhui Province, told me, "When 1.3 billion people must rely on imported grain, that is *not* food security."

Within China's internal argument over the merits of imported food, we can see the elements of a much larger debate about the future of the food economy. For decades, proponents of free trade argued that the key to continued progress in the food economy, as in the larger economy, was to take the low-cost, high-volume model that was so successful on a local and national level and allow it to operate on an international scale. By importing your corn, or chicken, or cherries from producers who could grow them more cheaply, you not only lowered food prices for your consumers, you also freed up your own farmers to use their land for whatever *they* could produce most efficiently, whether that be corn or condos.

This is the basis of the theory of comparative advantage, whose author, a nineteenth-century economist named David Ricardo, was the first to show that nations are better off economically when they specialize in the few products they grow best and trade freely with other countries for everything else. Just as farmers would later be urged to focus on one or two crops, this global rationalization of food production, in which crops were to be grown wherever they could be produced most efficiently, was seen as the best way to liberate the world's scarce economic resources for other critical endeavors, such as building factories, schools, roads, and other bits of civilization.

For all the very real power of comparative advantage, however, Beijing is not alone in questioning how the theory should be applied in a modern, rapidly changing world. As numerous experts have pointed out, by enlarging the low-cost, high-volume model to a global scale, we expand not only the benefits of that model but its considerable costs as well — among them the rapid consolidation of farms and food companies; damage from "externalities," like livestock sewage and farm chemical runoff; the collapse of tens of thousands of local food traditions into a handful of meta-cuisines; and a flood of excess calories. In China, a country where anyone over fifty can recall the famine that left thirty million dead, more than a hundred million people are now overweight or obese.

Free-trade proponents counter that such costs may be unavoidable: only by expanding the Western model of rational, fully industrialized food production to a global scale have we any hope of feeding the extra three to four billion people the planet can expect by the middle of the century — a not altogether dismissible argument, as we'll see in coming chapters. Yet as we'll also see, this eventual need for massive new food supplies highlights the enormous problems inherent in trying to internationalize a food system that is already fast approaching its economic and natural limits. Even in its evolving form, the global food system often ignores the poorest nations that would benefit most from its alleged wealth-creating dynamism. And in regions where the Western model has taken hold, the results are almost as problematic: the successful "export" of this hyperefficient, low-cost, year-round format has left us with a global food system so interlocking and thinly stretched that the risk of outbreaks and other disruptions rises even as our capacity to respond to such disruptions (to say nothing of climate change or declining energy supplies) is falling.

Compounding matters, even as the swift expansion of the global food trade is highlighting concerns about how long we can sustain our massive output, these concerns are already feeding back into, and sharpening, the politics that govern that very system of trade. For most of the last half a century, the trade disputes between nations occurred in a context of overabundance: under the auspices of supranational institutions such as the World Trade Organization, or WTO, agricultural overproducers like the United States and the European Union, along with the international food companies that actually make and ship those surpluses, maneuvered and negotiated for the right to sell to the developing world. But that context is changing. Not only is the United States, the former OPEC of the food market, losing global market share to emerging export powerhouses such as Brazil and Argentina, but the era in which global food markets could be imperiously managed by the big food exporters is fading. Increasingly, the global food economy will be driven by an emerging class of mega-importers, among them India and China, as they compete for access to the acres, water, and good soils they themselves no longer have in sufficient quantities.

In an important sense, how we manage the globalization of the food system over the next decade will determine the complexity and scale of the food challenges we will confront over the next half a century, as well as our prospects for success or failure.

Although food has been traded for thousands of years, a truly global system didn't emerge until the early twentieth century, and even then, only with the grave misgivings of many of the participants. Whereas countries such as England, which depended hugely on imported food, became ardent champions of free trade with minimal government interference, the United States and other big exporters were skeptical. Dependent though they were on the world market to absorb their surpluses, these countries also understood how volatile that market could be, especially in the dawning era of high-output farming. Because farmers were bringing more acres into production and were producing more bushels per acre, and yet were still as vulnerable as ever to crop failures, the food system had increased its potential for both boom *and* bust, and mainly bust. By the 1920s, farmers in Europe and the Americas glutted grain markets so routinely that governments were obliged to intervene. The result was a series of government price supports, production limits, and other intercessions that, while intended as only temporary fixes, became permanent features of the global agricultural system and a permanent reminder of the great flaw in Ricardo's vision of free trade — namely, its inherent corruptibility.

The biggest intervener was the United States. The massive farm bankruptcies of the 1920s had persuaded President Franklin Roosevelt that a free market for food was akin to national suicide. Farmers themselves were simply too prone to overproduction (high prices one year invariably led to overplanting the next); the administration also feared the rising power of the grain trading companies, which used their near monopoly position to buy extremely low and sell high — typically in some much pricier market halfway around the world.

To bring order to this chaos and smooth out these risks, the administration created a policy intended to moderate prices by stabilizing supply. Government would pay farmers to idle their excess acres, in the hope that fewer acres would reduce supplies and shore up prices. Government would also protect farmers from unavoidable price swings (and predatory grain buyers) by guaranteeing a minimum per-bushel price for their grain, in the form of a federal loan. If market prices for grain fell below that target, government would pay the farmer the target price and put the grain in a national reserve, to be released in times of shortage.

These programs, which Congress later bolstered (and which were paralleled in Europe and, later, much of the rest of the world), amounted to explicit rejection of a free food market. In the view of the White House and Congress, and a great many consumers and commentators, food was simply too important to national security and human welfare to be left to the vagaries of an unfettered market. Thus, while policymakers recognized the benefits to be had by trading in food, they knew that these benefits could be assured only if the food trade, and the food market generally, was carefully monitored and managed to avoid volatility. And in the farm program, with its acreage set-asides and price supports, policymakers now believed they had the tools to do just that.

Not everyone shared the government's brimming confidence. Republicans, with some justification, saw the price guarantees as a Democrat ploy to win the farm vote. Food companies complained that intervention artificially inflated the costs of their raw materials. More fundamentally, though intended to control production, the government's price guarantees turned out to encourage farmers to overproduce, because no matter how low the real market price fell, farmers still got paid for every bushel they grew. Whereas a nonsubsidized farmer might read falling prices for corn as a strong signal that the world market for corn was oversupplied and that he should plant less corn, the only signal U.S. farmers were getting now was: grow *more*. And even though government was taking acres out of production, that didn't stop farmers from trying to grow more bushels per acre, which, with the steady advances in seeds, machinery, and chemicals, was precisely what they did — so successfully, in fact, that U.S. grain volume rose faster than government could pull acres out of production. In 1962, the year the government paid farmers to idle sixty-five million acres (an area half the size of California), U.S. wheat stockpiles reached nearly 1.3 billion bushels, or nearly twice our yearly demand.[1]

And there was another, more subtle concern. By the late 1960s, the United States was in deep economic trouble. The costs of the Vietnam War and various Great Society programs were draining the Treasury and spurring inflation. Worse, under increasing competition from low-cost manufacturers like Japan, the United States, once the dominant exporter, could no longer sell as many cars, televisions, and other goods to the world and was now running a negative balance of trade. The only sector where America still led was agriculture — indeed, the United States had such large and regular grain surpluses that policymakers spent much of the postwar era trying to find outlets for it all.*

Now, however, a huge new opportunity was unfolding. World demand for grain had begun to rise, especially in the developing world, where a booming population, rising incomes, and the arrival of Western livestock methods were doubling meat consumption almost every decade.[2] In burgeoning Asia and Latin America, demand for grain was climbing faster than domestic farmers could respond[3] and offering a huge new market to a stymied overproducer like the United States. The only catch was that other exporters wanted these agriculture markets as well: European farmers needed foreign buyers for their own grain surpluses, and even farmers within the developing countries were producing more grain. If the United States was to outbid these rivals, it needed to be not just the world's biggest producer, but also the world's lowest-cost producer, a kind of Wal-Mart of the global grain market.

And this was a problem, because for all its superabundance and surpluses, the United States wasn't the lowest-cost producer: in fact, its grain was often more expensive than grain from other countries — to the point where U.S. exporters had to offer discounts (subsidized by Congress, of course) in order to sell our overpriced grain in foreign markets. And the reason American grain was so expensive, conservative policymakers argued, was the farm program: by propping up millions of smaller, inefficient farmers, the government was effectively dragging down the efficiency of the entire U.S. farm sector. For America to prosper in the new, competitive global market, its agriculture simply could not be left to small players; farming would need to evolve, under the brute force of that market, toward a more efficient model based on a smaller number of larger, more specialized operators.

---

* Much of our early surpluses were used as food aid to keep our allies from joining the Communists. By the 1950s, Washington was lending billions of dollars to developing countries so that they could in turn buy our surplus food. Critics feared such programs turned foreign nations into U.S. dependents, but Washington saw that as a small price to pay to contain Communism. In the opinion of Senator Hubert Humphrey, a chief food-aid proponent, food dependence was actually "good news, because before people can do anything, they have got to eat. And if you are looking for a way to get people to lean on you and to be dependent on you, in terms of their cooperation with you, it seems to me that food dependence would be terrific." Both quotes in Cleaver, "The Contradictions of the Green Revolution," http://www.eco.utexas.edu/facstaff/Cleaver/cleavercontradictions.pdf.

This process of farm market "liberalization" wouldn't be painless. Larger, more efficient farms would need dramatically fewer farm workers. In a 1962 white paper on the subject, one of the leading exponents, the Council on Economic Development, conceded that government would need "to induce excess resources (primarily people) to move rapidly out of agriculture." But this too would ultimately be beneficial, since these ex-farmers would provide a needed labor pool for the new factories.[4] And in any case, in the minds of many export-focused policymakers, economic progress required sacrifices. As Ezra Taft Benson, President Eisenhower's secretary of agriculture, had put it, the modern global economy gave American farmers two choices: "get big or get out." Or, in the words of Earl Butz, President Nixon's outspoken secretary of agriculture — and a primary architect of the new low-cost model — "adapt or die."[5]

Butz's vision was admittedly a rather tough one to sell: farmers, even small farmers, retained a powerful hold on the American imagination and certainly on lawmakers in Congress. But in 1971, Butz and his fellow export advocates got a huge break. That year, Soviet diplomats approached the Nixon administration about buying some of the United States' surplus wheat. Nixon, hopeful that the sale would help spur the U.S. economy, gave his approval — only to realize he'd been duped. The Soviets were reeling from a crop failure that was far larger than U.S. experts had guessed, and they'd been negotiating secretly with private grain companies for vastly more grain than Nixon anticipated. Once the deal was approved,[6] they bought two-thirds[7] of America's wheat reserves. The sale, later dubbed the Great Russian Grain Robbery, pushed up grain prices by a factor of three[8] and shocked the American food economy. The same price spike that made beef so expensive and poultry so popular drove prices for milk, bread, and other basics so high that Nixon came under withering political fire; desperate, he ordered Butz to fix things by increasing wheat production.

Butz was happy to comply. As harsh as the grain crisis was for American consumers, the crisis atmosphere and the high prices it engendered made farmers and Congress far more amenable to a new kind of farm policy, which allowed Butz to disassemble the old farm programs and reshape U.S. farm policy to fit the new global market. First, the practice of idling tens of millions of excess acres was eliminated; farmers were instead encouraged to plant all available acres, or "fence row to fence row," in Butz's immortal phrase, and in the space of a few years, the government reactivated around eighty million acres of previously idle land. Likewise, the nation's grain reserve was terminated: the government would no longer buy up surplus grain or do much of anything to keep supplies tight and prices high. Instead, prices would be allowed to fall to wherever the world market thought they should fall.

Unfortunately, in this new, free market, prices turned out to be a lot lower than anyone, even Butz, had expected. Within several years of the grain crisis, U.S. farmers had planted so many extra acres that markets glutted and the price of a bushel of corn or wheat dropped *below* what it actually cost to produce. As American farmers began going out of business by the tens of thousands and as farm state politicians faced election defeat, Congress lost some of its free-market fervor and again injected government into the business of farming — but with significant differences this time. Whereas previous farm policies had tried to keep prices high by limiting supply, government now simply pays farmers the difference between their production costs and world market prices by way of a per-bushel "deficiency" payment, which, depending on how far those prices fall, can be quite large. In 2005, for example, before the ethanol boom turned grain markets upside down, the world price for corn had dropped to $1.85 a bushel. But because it actually cost American corn farmers nearly $3 to grow each of those bushels, taxpayers covered the difference, which, of course, only rewarded farmers for overproducing in the first place.[9] Indeed, the main reason corn had fallen to $1.85 was that subsidized U.S. farmers were putting too much grain onto the world market.

Impossible as it seems, America's new, ostensibly free-market farm policy is actually less responsive to market signals, and thus more inclined to overproduction, than its predecessor was. Farmers still have every incentive to overplant — they still get paid for every bushel, no matter how much money they lose growing it, and in fact, with the deficiency payments, can now afford to buy more fertilizers and pesticides and better seeds, and thus grow even more bushels. Worse, government now has no way to curb supply, because Washington is no longer idling excess acres or keeping a national grain reserve. In this respect, the "reform" of U.S. farm policy is very much like outfitting your teenage son's car with a turbocharger and then replacing the brake with a bigger insurance policy.

The consequences of Washington's quasi-embrace of a free food market have been both bizarre and transformative, and they explain much of the character of the modern food system. The United States is indeed now the world's lowest-cost producer, but in name only, since our below-cost grain is possible only with massive public subsidies — $20 billion in 2005 alone. Just as Wal-Mart uses its nonfood sales to keep its grocery prices low, the United States uses revenue generated elsewhere in its vast economy to keep its own food costs down. In truth, U.S. food-production costs have actually risen recently, in large part because farmland is becoming so expensive. (And, perversely, one reason farmland is so expensive is that each acre's market price is based not just on its actual productive potential but on whatever subsidy income its owner is entitled to.) This is hardly the kind of free trade or comparative advantage that David Ricardo had in mind. When it comes to food, America's comparative advantage isn't the great quality of its land, or the fitness of its climate, or even the considerable skills of its farmers, but the nation's political inability to reform its farm policy.

Worse, this faux version of free trade is self-perpetuating. Because deficiency payments haven't kept up with the rising costs of land and inputs, profit margins for farmers continue to narrow,[10] and smaller farmers continue to be replaced by larger operations whose greater efficiencies and scale let them survive on those narrower margins. In 1970, the United States had nearly three million farms and an average farm size of 200 acres. By 2005, the number of farms had fallen by more than a third, while the average size had more than doubled, to around 450 acres.[11] And because larger farms cut their costs partly by growing more bushels, the continued shift to larger farming operations has meant . . . even more output, which simply drives down the price of grain, and encourages still larger farms, and still more output — in what has become a permanent cycle of overproduction and falling prices. Between 1996 and 2005, the prices of the most heavily traded agricultural commodities — corn, cotton, rice, soybeans, and wheat — fell by more than 40 percent, one of steepest drops in history.[12]

For the food economy, the consequences of what is in effect institutionalized overproduction are hard to overstate. The rising tide of cheap grain has not only changed the kinds of food we make (like high-fructose corn syrup and chicken nuggets) and the way we market them (supersizing, for example), it has also changed *where* we make our food, especially our meat. Corn and soybeans are now so cheap that instead of having to raise livestock near the crop-producing areas, as was once the practice, livestock companies can now afford to move their animals to wherever regulatory barriers are lowest (Colorado for cattle; Missouri, Texas, Oklahoma, Utah, and North Carolina for hogs) and simply ship the below-cost feed to them. Cheap grain, along with cheap transportation costs, has created what the United Nations' Food and Agriculture Organization calls a new "geography of livestock," in which meat production has been effectively unmoored from its traditional settings — near cropland — and is now free to settle wherever political circumstances are most amenable. So mobile has food production become that in the United States, less than 5 percent of the food consumed in any one community or region is actually produced there.[13]

Above all, below-cost grain has, as its proponents had anticipated, made exports even more important. Despite all the new uses for grain, American grain output still far exceeds domestic consumption. In 2002, before the ethanol boom created a somewhat artificial domestic market for corn, American farmers relied on foreign buyers for more than *half* their grain production; in fact, Vaclav Smil has estimated that between our vast export markets and the rather extraordinary volumes of food we waste every day, Americans actually consume only about a fifth of the grain our farmers produce.[14] Little surprise that American agriculture is still twice as dependent on exports than is any other sector of the U.S. economy[15] — a dependence that continues to shape the way American food companies and food policymakers look at the global food economy and our future place in it. "Americans just can't eat any more than we're already eating," says the U.S. Grains Council's Mike Callahan. "We need to look to the 95 percent of the people who live outside our borders." And nowhere has our attention been fixed more firmly of late than on the emerging economies of Asia.

The breakfast buffet at my hotel in Weifang seems, after yet another long evening of toasting U.S.-China friendships, a wonder of almost medicinal splendor. The white-draped serving tables are crowded with offerings — from plausible facsimiles of pork sausages and other Western dishes to traditional Chinese tidbits: sweetbreads, pumpkin dumplings, porridge, soups, and a welter of items, many of them pickled, that I have to ask Lin to identify. Piling my plate high, I sit down across from Lin, who is smoking what must be his thirtieth cigarette of the morning, and manage to work through about half my breakfast before losing steam. I push my plate away just as our host, a middle-aged portly agricultural official named Tian Lee, appears at the table. He speaks to Lin in Chinese, then glances in my direction and walks away. Lin grimaces. "Mr. Lee was telling me about today's schedule. And he also says you should finish everything on your plate."

Lee's admonishment perfectly captures the reason that the Chinese market is so appealing to Western exporters yet so elusive as well. As recently as the 1980s, China truly was a food system on the verge of a total breakdown. Centuries of relentless subdivision left the nation with hundreds of millions of tiny farms, most less than two acres in size; this fragmented and inefficient system was then subjected to decades of Cold War economic isolation, the suicidal farm policies of Mao Zedong, and rapid population growth. By 1958, while Americans were drowning in grain, China plunged into a three-year famine that would claim at least thirty million lives and set back entire regions by decades: fertile Shandong Province alone lost 7.5 million people, a seventh of its population.[16] Few outsiders knew the extent of the tragedy, but most food experts believed China and the rest of the developing world were fast approaching demographic disequilibrium; India had narrowly escaped famine in the 1950s and 1960s, and Secretary Butz was hardly the only Westerner to imagine a future of permanent and growing grain flows between the United States and grain-deficient Asia. As far back as 1981, when the U.S. Grains Council opened a marketing office in China, Callahan told me, "Our whole premise for being there was that the Chinese would develop a meat industry which would grow to the point where domestic supply was no longer sufficient."

But China had other plans. Reluctant to spend its meager reserves of foreign currency on imported grain, and even less eager to accept food-aid help from its ideological rivals, Beijing embarked on an ambitious strategy to achieve food self-sufficiency that would turn Western export strategies upside down — and ultimately reshape the global food system.

First, Beijing adopted stringent laws to curb population growth, fining families who had more than one child. At the same time, China began building a budget version of Western-style agriculture that substituted labor, which China had in surplus, for machinery, which it couldn't afford, and that made use of Western-style incentives, such as production subsidies. Most important, the state stopped acting as the agricultural market. Where the government had long imposed itself as the only seller of inputs and the only buyer for outputs (and had paid artificially low prices for those outputs in order to keep down food costs in its burgeoning cities),

Beijing now let an actual market develop. Farmers were allowed to sell some of their output to local customers at market prices and to reinvest their modest profits in more or better inputs — new seeds (many of them adapted from successful Western varieties) and chemical fertilizers (which the Chinese use three times as heavily per acre as U.S. farmers do). Then, in the mid-1990s, declaring that socialism and capitalism were "compatible," President Deng Xiaoping removed some barriers between China and the world food market. Tariffs on imports were lowered, as were restrictions on foreign investment. After a half-century absence, China was reentering the global food economy.

Some Western trends haven't taken hold in China. Whereas many American farmers use bank loans to finance each year's planting, Chinese are largely self-financed, out of their own cash flow or family loans. Greenhouses, livestock barns, and other capital assets are often hand-built and crude. And although there has been some land consolidation, most of China's two hundred million farms are still tiny — in part because there is little incentive to expand; farmers here can't afford the big harvester combines that make large-scale farms in the United States and elsewhere so economical. Mechanization of any kind is still rare, and it's common to see farmers walking patiently down each crop row weeding by hand, or spraying for insects with a backpack fogger. "Most farms are really more like big gardens," says Paulette Sandene, a crop expert who monitors Chinese output for the USDA's Foreign Agricultural Service. Such small size, Sandene says, means farmers "can still go out there and flick off bugs by hand, or water by hand, or weed, and they can certainly pick corn by hand."

Whatever China lacks in quality, however, it makes up in quantity. China's small farms are more intensely managed than their Western counterparts: farmers often plant several crops a year, one after the other, and most also keep livestock and even operate fish farms — a diversity that contrasts sharply with the single-crop monoculture model in the United States and which actually generates more calories per acre. Collectively, China's two hundred million family farms produce 20 percent *more* output than do the United States' two million farmers, and on a land base less than three-quarters as big as America's.[17] This isn't free food by any means: China depends heavily on vastly higher inputs of labor and fertilizers. But the results have been stunning nonetheless. Today, China generates more than a fifth of the world's corn and wheat, a third of its rice, an eighth of its fruits, and two-fifths of its vegetables, not to mention a fifth of its chicken and half its pork[18] — all with just 7 percent of the world's arable land and a system that was largely preindustrial fewer than forty years ago. "China is unique," says Fred Gale, another veteran China watcher at the USDA's Foreign Agricultural Service. "They went through decades of war and harebrained agricultural policies, but once they liberalized their economy, they made up a lot of ground very quickly."

This agricultural recovery is even more visible in the cities. After decades of utilitarian cuisine, urban shoppers can choose from the full spectrum of outlets, from classic wet markets, with their fresh vegetables and live animals, to enormous Western-style grocery stores run by Carrefour, Wal-Mart, and Tesco, whose wide aisles are jammed with thousands of new packaged and processed-food items. China is even developing a powerful restaurant culture. While rural residents still hew to the tradition of home cooking, in the booming urban areas, tens of thousands of restaurants have sprung up to cater to the millions of increasingly wealthy, time-pressed urbanites. "When I was young, going out to eat was something you did maybe every five or six years," says Jun Jing, a forty-seven-year-old Chinese anthropologist. "Now, I probably eat out four times a week. Even five years ago, my parents would have never easily said, 'Hey, let's eat out tonight.' Now, they say it every Sunday."

China has, in fact, made up ground much faster than anyone expected — least of all U.S. grain exporters. As early as the 1980s, to the great disappointment of American farmers, China had become a net exporter of corn and soybeans.[19] A decade later, the nation that was supposed to become a new market for American products was actually *competing* with the United States for export customers, including South Korea, until then America's most dependable foreign market. As Callahan told me, "We did not expect China to pinch off our best customer." Nor was it simply raw commodities that were flowing out of China. Chinese producers began to add value to that newly abundant grain by converting it into exportable volumes of meat — poultry, pork, and especially fish. As China's goal of self-sufficiency began to morph into a strategy for export power, traditional exporters, such as the United States and Europe, began not only to look for other markets but to adopt increasingly aggressive tactics to enter them.

In February of 1998, the governments of South Korea, Thailand, Malaysia, Indonesia, Taiwan, and the Philippines — the economic tigers of Southeast

Asia — found themselves on the receiving end of some particularly tough free-market love. For months, these formerly fast-growing countries had been mired in a severe financial crisis, and although the International Monetary Fund, the official banker for the global trade system, had arranged a $120 billion bailout, it was suddenly announced that the money wouldn't be released until the six governments promised to buy substantially more U.S. grain. Given that East Asia already accounted for 40 percent of total American farm exports,[20] such a requirement must have seemed a bit harsh. Washington was roundly accused of using the region's economic woes as leverage to push more of its surpluses into their markets (and indeed, Lon Hatamiya, administrator of the Foreign Agricultural Service, acknowledged that East Asia's financial crisis could well have a "silver lining" for U.S. farmers[21]). But the United States is hardly the first country to use others' economic misfortune as a lever to boost sales. More to the point, such high-pressure tactics offer a window into the increasingly combative nature of the global food trade and raise serious questions about the way the much-touted benefits of the "wonderfully complex interdependency" are actually distributed.

For much of the last century, the global food trade ran on a kind of economic gravity: grain and other products flowed relatively naturally from regions of surplus like the United States to regions of deficit, like Europe and Asia. But by the 1970s and 1980s, these gradients began to change. The European Union, under its own powerful set of farm subsidies, had shifted from a net grain importer to a net exporter.[22] At the same time, the emerging economies of Asia and South America, long anticipated as huge potential buyers, weren't being docile customers. Many of these nations hoped to build up their own agricultural sectors, and they had no interest in competing with below-cost grain from Europe or the United States.[23] Some, such as China, simply closed off their borders to imports. But other, more ostensibly capitalist nations took a more complex tack, shielding their farmers behind barriers in the form of heavy taxes, or tariffs, on imported grain. Although such protectionist measures are common in the West, they were nonetheless heavily criticized by Washington, which was then in the throes of its 1980s free-trade revival. In response to the tariffs, Washington unleashed a barrage of high-pressure tactics designed to knock down Asian trade barriers — tactics that were usually deployed under the banner of free trade but which were widely seen in the developing world as economic blackmail.

Perversely, Washington's new food weapon was actually a direct descendant of the more idealistic policies of the 1960s and 1970s, when Western governments poured large sums into Asia, Latin America, and Africa in the hopes of stimulating economic growth (and slowing the spread of Communism). By the 1970s, this philanthropic spree had attracted Western banks. Flush with new money from the oil boom and urged on with loan guarantees by Western governments and international lending organizations such as the World Bank and the International Monetary Fund (IMF), these banks poured hundreds of billions of dollars into what enthusiastic analysts dubbed "emerging markets."[24]

But by the 1980s, many of these subprime borrowers, plagued by corruption, ineptitude, and political strife, had hit economic hard times. Western banks began to call in their loans, but the debtors had empty pockets. In June 1982, Mexico threatened to default on $80 billion in foreign debt, a third of it with American banks. Fearing the Mexican crisis would spread to other debtor nations, bankers descended on Washington for anxious discussions with the U.S. Treasury, the World Bank, and the International Monetary Fund. The solution that emerged — later dubbed the Washington Consensus — largely formalized the early pushes by Butz and others to create a less restricted, or more liberalized, food trade; in the process, it utterly changed the shape of food power.

The World Bank and the IMF agreed to restructure developing-country debt to make repayment easier.* But in return, debtor nations were required to restructure their dysfunctional economies along free-market lines. Primary targets of this restructuring were debtors' agricultural sectors, which were to be reconfigured as hyperefficient, high-volume machines whose surpluses could be exported and whose earnings would help retire the debts.

To create this new hyperefficient model of farming, debtor nations were required to liberalize their state-run agricultural sectors — for example, by reducing or eliminating "trade distorting" farm subsidies, which, by protecting small, inefficient farmers, delayed the sector's evolution. They

---

* That in itself was radical. The IMF is essentially a publicly funded institution; it draws its support from member governments, most notably the United States, which contributes about a fifth of its budget. By guaranteeing these loans, the IMF was effectively converting private debt into public debt — a trend that would go on to characterize international lending for decades.

were also directed to devalue their currencies so as to make their farm produce cheaper and more attractive to foreign buyers. Moreover, although debtor nations were expected to export more — more beef from Argentina, more soybeans from Brazil — they were also required to open their own markets to more *imports*. More imports of fertilizer and other inputs for their new high-volume farm sector, but also more commodities such as grain — especially if the commodity could be grown more cheaply elsewhere. Last but not least, debtor nations would need to let in more imports of foreign capital — better known as foreign-directed investment, or FDI — because the new ten-thousand-acre farms, meat-processing plants, railroad lines, and other pieces of the modern food economy that these nations were expected to erect required billions of dollars in financing.

Predictably, debtor nations regarded restructuring and the Washington Consensus with both apprehension and anger. Opening their markets to low-cost commodities would essentially leave developing-world farmers to compete with large-scale agribusiness operators from wealthy nations — operators who, it was certainly noted, still benefited from the same subsidies now prohibited in the developing world. Critics were also fearful of the social impacts of such massive changes to such a vital sector: whereas farming is only a small piece of the economy of a developed country (less than 2 percent of all jobs and less than 1 percent of the GDP[25] in the United States comes from agriculture), it can be the source of half or more of a developing country's economic activity. Above all, many developing countries loathed the idea of losing domestic control over so important a resource — food. As their own farmers, unable to compete with foreign counterparts, were squeezed out of the sector, these nations realized they would rely more and more on imported food, putting an end to any aspirations of being food self-sufficient.

Restructuring advocates in the United States and elsewhere discounted these concerns with the familiar arguments of comparative advantage. Given the realities of rising populations and changing food habits, it was no longer sensible or even possible for any nation, especially a poor nation, to produce every calorie its population ate: far better to focus your economic resources on things such as building industry and to buy your food from those who produced it efficiently. Thus, nearly half a century after Western governments had largely rejected the market as a means of managing food supplies, the market was once again the driver for food security

— and this time on a far more global scale. In September 1986, as trade negotiators from 123 nations met in Uruguay to formalize this new freer food system under the General Agreement on Tariffs and Trade (GATT), U.S. agriculture secretary John Block declared the concept of food self-sufficiency to be officially dead. "The idea that developing countries should feed themselves is an anachronism from a bygone era. They could better ensure their food security by relying on U.S. agricultural products, which are available in most cases at lower cost."[26]

That Secretary Block had rather less to say about trade's rewards for those *outside* the developing world — U.S. investors, say, or the makers of farm inputs and machinery (Block himself would later work for John Deere) — did not surprise the many critics of the global food trade. To a growing number of trade skeptics, both in the developing world and within Western activist and advocacy circles, the Washington Consensus was never really about food security, or export earnings, or even debt repayment, but about a deeper effort to rebuild the global economy to suit the interests of the industrial North. In this interpretation, the trend toward a neoliberal food system has been shaped less by the desire to feed a growing planet, or even by the economic strategies of the big, surplus-burdened exporters, than by the business strategies of large transnational food companies whose revenues depend entirely on unimpeded global flows — flows of raw materials from low-cost suppliers, flows of finished goods into consumer markets, and flows of capital in between.

These are, of course, precisely the sentiments one finds on the websites of various and often strident antiglobalist groups, who blame globalism for everything from war to poverty. Yet whatever one thinks of the philosophies or proposed solutions of such groups, it's hard to argue with their contention that large transnational food companies have indeed benefited mightily from, and have been quite influential in the building of, an increasingly global food system.

Although we tend to think of trade as transactions between nations, with the benefits accruing to entire populations within those nations, trade today is better understood as deals between private companies — deals whose benefits may or may not accrue to the nations in which they occur. "In the globalizing agricultural sector, the United States and Brazil do not actually compete with each other for a share of the world soybean market," argues Sophia Murphy, an analyst with the trade-skeptical Institute for Agriculture and Trade Policy. Instead, says Murphy, the United States and Brazil "compete for investment by Cargill or one of the other large grain traders that operate worldwide."

Of course, attracting such investment is beneficial to Brazil, in that it creates jobs and tax revenues. But the reason that these companies invest in Brazil, or Romania, or Poland isn't to help Brazilians, or Romanians, or Poles but to gain things the companies and investors can't find at home: access to new consumer markets, access to cheap inputs, or, ideally, both. This is the rationale behind moves by Tyson, which has spent hundreds of millions of dollars buying up meat companies in Mexico; land and labor in Mexico are cheap, and more to the point, Mexican consumers are getting wealthy enough to, as one Tyson executive puts it, "trade starch-based foods for protein."[27] All in all, it is simply much cheaper to make the food in Mexico with local labor and materials and sell it to local consumers than it is to make it in the United States and ship it to Mexico.

Moreover, in a food economy that is increasingly global, companies invest in foreign countries to reach markets *outside* those countries. Because Brazil enjoys better diplomatic relations with Europe than the United States does, Tyson and other U.S. meat companies that invest in Brazil can export their Brazilian products to the big consumer markets in France and Germany far more freely than companies operating solely in the United States can.[28] For that matter, Tyson's investments in low-cost cattle ranches in Argentina mean it can produce beef there and sell it back into the United States more cheaply than it can produce that beef here.

And, of course, for many food companies, particularly those in a low-margin, high-volume commodity business, developing foreign markets is the *only* way to maintain their necessary growth curve. With steadily declining margins and the need to keep expensive processing plants running at full capacity continuously, American meat producers now overproduce by such a degree that exports are a fundamental necessity — a fifth of all U.S. chicken and an eighth of all its pork is eaten abroad.[29] Lower margin commodities, such as corn and soybeans, are even more trade dependent. Companies such as Cargill or French-based Louis Dreyfus or André depend on selling steadily more corn, soybeans, and other commodities to developing countries to offset the steadily falling margins on those commodities. What's more, much of their profits comes from betting on the price differences between these markets — between, say, the low prices of farmers in Brazil after a bumper crop of soybeans and the high demand in China, which can no longer produce enough. Thus, what a commodity company needs most, other than the prospect of steadily rising global demand, is the freedom to buy wherever a commodity is cheapest and sell it wherever demand is highest.

Proponents defend such arrangements as merely a more global version of the integration and rationalization that made the U.S. and European food industries so efficient and low cost, and it is undeniable that these new transnational arrangements are bringing jobs and lower food costs to places like Brazil. But again, the reason Tyson and Cargill are in Brazil isn't to create Brazilian jobs or lower Brazilian food costs but to generate value for the shareholders of Tyson or Cargill. And even if the new wages and tax dollars do stay in Brazil, much of the shareholder value, or profit, generated by such operations is sent back, or repatriated, to Tyson's headquarters in Arkansas or to the replica of a French chateau in Minneapolis that houses Cargill's executive offices. For all that foreign-directed investment is said to benefit recipient countries, a large share of that money — on average, twenty-seven cents of every dollar invested in Latin America, according to the World Bank — goes back to the investor's home base — usually the United States, Europe, Japan, or, increasingly, China.[30]

Nor is everything that *stays* in the host country always to that country's liking. In the newly global food system, companies look for locations offering the right mix not just of raw materials, consumer demand, and market access but also of political and regulatory convenience. It's worth noting that one of Brazil's main attractions for U.S. and European meat processors, aside from cheaper grain and cheaper labor, is a decided lack of rules governing the handling of sewage.

Because a company's ability to exploit such advantages depends directly on the ease with which the company can move itself, its products, and its capital between countries, one begins to understand why trade skeptics believe the restructuring of the 1980s and 1990s was less about helping developing countries overcome financial problems than it was about reconfiguring their economies for an invasion by Western food companies. "The benign view is that the 'structural arrangements' were designed to 'ready' developing countries for involvement in the global economy," says Doug Hellinger, a longtime skeptic of free trade with the

Washington-based group the Development Gap. "It was sort of, 'We're going to help you get your house in order, help you liberalize your agricultural trade, reduce state involvement, and all other changes that are necessary for global trade.'" The more cynical explanation, Hellinger says, is that the United States and international financial institutions such as the World Bank, although initially intent on addressing debtor nations' financial woes, "began to see that the debt crisis could be used as a lever to control economies." Other trade skeptics, like Eric Holt-Gimenez, executive director with Food First, are less diplomatic and contend that corporate expansion, was the objective for restructuring from the very start. "Debt was always a means to leverage greater access by foreign capital," Holt-Gimenez says. "I don't think anyone ever seriously expected these countries to pay off their debts."

Ultimately, the real question about the global food trade isn't whether economic restructuring was part of a transnational master plan, but whether the results will have positive or negative impacts and whether these impacts are likely to increase or decrease in the coming decades. Due to the speed with which the global population is growing and the fact that many of these newcomers will live in places that cannot easily provide sufficient food by themselves, some food trade — and probably a *lot* of food trade — is not only inevitable but essential. At the same time, given the market's absolute obliviousness to the mounting external costs of its own operations, the idea that we can gain the most benefits from trade by simply leaving the global food economy to develop on its own is an absurd fantasy that is taken seriously only by hard-right think tanks and a few grain company executives.

Many free-trade proponents have, in fact, recognized some of these costs and risks — and they insist that new trade entities, such as the World Trade Organization, and new trade agreements, such as the North American Free Trade Agreement, are now replete with safeguards to protect countries from environmental or social consequences. But such safeguards are enormously problematic. In the first place, global food trade is expanding so rapidly that many of the risks are growing faster than our capacity to manage them. Second, many of the touted safeguards are being shaped, often explicitly, by the same companies whose behavior they would ostensibly govern.

Members of the food industry tirelessly lobby lawmakers and trade officials in the United States and elsewhere in the hopes of gaining favorable treatment in the international agreements that now govern that trade; in many cases, they even help writes those policies. Dan Amstutz, a former Cargill vice president, worked in the office of the U.S. trade representative[31] where he drafted the so-called agreement on agriculture, a subtreaty covering new food-trade rules under GATT,* which has been hugely beneficial to commodities dealers. Allen Johnson, chief agricultural negotiator at the U.S. trade representative's office under George W. Bush, formerly ran the National Oilseed Processors Association, an industry group representing, among others, Cargill, Archer Daniels Midland, Perdue, Purina, and Tyson.[32]

With such close connections between industry and government, one is hardly surprised to find a near perfect convergence between our nation's official rationale for a free food trade and that of the various industries that benefit from that trade — a worrying trend for anyone who believes that global trade, and especially global *food* trade, needs close and impartial scrutiny. During the recent, often rancorous debates on the North American Free Trade Agreement, or NAFTA, which knits the United States, Mexico, and Canada into a single economic bloc, the rhetoric coming out of many commodity companies was strikingly similar to Washington's. "There is a mistaken belief," declared Cargill chairman Whitney MacMillan,[33] "that the greatest need in the developing world is to develop the capacity to grow food for local consumption. That is misguided. Countries should produce what they produce best, and trade." During the height of the NAFTA negotiations, Cargill's employee newsletter put it more bluntly: NAFTA "is important to Cargill because it clears the way for what we do."[34]

A second concern about the consequences of a free food trade concerns the word "free." Most of the thoughtful and reasonably impartial proponents of a global food system — for example, the United Nations Food and Agriculture Organization — contend that the considerable negatives of a freer food trade will be outweighed in the long term by the greater efficiencies of a truly global system — but only if all the players in that sys-

---

* In 2003, Amstutz told reporters he had had "no affiliation with [Cargill] whatsoever" after leaving the company, although records showed he had in fact served on the board of a joint venture backed by Cargill.

tem trade fairly. And that, unfortunately, isn't likely to be the case. For all that countries like the United States and companies like Tyson or Cargill complain about the trade barriers of "foreigners," their own behavior at home would hardly have earned the blessing of someone like David Ricardo.

During the 1980s and early 1990s, as Cargill and commodities traders were championing an unfettered export market, they were also lapping up roughly $4.2 billion in federal subsidies designed to enhance grain exports — that is, to help traders sell American grain in markets where, due to its heavily subsidized prices, it wouldn't otherwise have been able to compete.[35] And although export subsidies have since been phased out, the United States' continued use of deficiency payments allows its farmers to dump their grain on foreign markets at a price that is not only below American farmers' production costs but certainly below that of the foreign farmers, who, of course, are not allowed any subsidies under restructuring. According to the U.S. Congressional Research Service, with deficiency payments and other subsidies, U.S. farmers can export their corn for 27 percent less than it costs them to produce it. There are even higher discounts for wheat (33 percent), milk (39 percent), and sugar (56 percent) — discounts that few foreign farmers have any chance of matching.[36]

To be fair, the United States isn't the only trade cheat. Although American subsidies account for around 22 percent of all U.S. farm income, subsidies make up 32 percent in Europe and more than half in Japan.[37] In addition, the United States has tried for decades to reform its farm policy, if for no other reason than fiscal sanity: between 1995 and 2005, the U.S. Treasury doled out $155 billion for crop and livestock subsidies[38] — more than America's total foreign aid budget.[39] Yet every time the U.S. farm program comes up for another five-year reauthorization, the modern farm lobby, like its predecessor, manages to keep the subsidy system intact. Farmers have no interest in giving up subsidies, especially the larger operations, which garner most of the subsidy payments and thus can afford to lobby Congress most energetically. And farmers aren't unique in their fear of change. If large commercial users of farm commodities such as Tyson or Coca-Cola would love greater access to foreign markets, they also benefit hugely from subsidized grain. According to a 2006 study by Tufts University, below-cost corn and soybeans save Tyson alone $288 million a year[40] compared to what it would have spent to buy corn at its actual production cost.

Politicians too need the U.S. farm program. In 1995, a Republican Congress on a rare tear of fiscal prudence (and aided by a surge in grain prices) actually mustered the will to cut farm payments to about a third of where they'd been a decade before. But the following year, as grain prices fell, Republicans felt the wrath of anxious farmers and effectively rescinded these cuts, which neither party has since had the nerve, or the votes, to reimpose.

Nor have big exporters always been so free when it comes to opening their own doors to imports. Although the United States has substantially reduced many of its import restrictions — per the GATT agreement on agriculture — it continues to restrict or ban outright imports that would undercut its own. For example, American poultry producers have successfully lobbied to block, ostensibly on health grounds, the imports of white chicken meat from Brazil and other low-cost poultry producers. As a result of this rather blatant protectionism, American poultry producers such as Tyson and Foster Farms can charge a small but significant markup for their own white meat, which in turn lets them offer their dark meat to buyers in Russia, a key chicken market, for discounted prices of as much as 60 percent below actual production costs,[41] far below what Russian poultry processors can do.

U.S. poultry-industry insiders acknowledge that the white meat/dark meat arrangement is unfair to both Russian and Brazilian farmers and that it contradicts the United States' stance on free trade. But many sheepishly admit that the existing American poultry industry probably couldn't survive a truly open market, especially one in which Brazil were allowed to send its cheaper chicken to the United States. As Blake Lovette, the former poultry company executive, told me: "[T]he way things are working is that in this country, consumers are willing and able to pay the price for breast meat which lets us produce dark meat more cheaply, and to sell it on the world market for less than the whole birds coming out of Brazil. But if Brazilian breast meat comes up here, it would destroy our market. I see that as probably the biggest threat we have."[42] Alain Revel, a former French agriculture diplomat who has studied U.S. farming, argues that most food production in the United States is so dependent on carefully controlled trade that "a true free market would bankrupt the American agricultural system."[43]

Even as the United States has steadily resisted the full realities of a freer food trade, governments and farm lobbies elsewhere in the world have not had the same successes. Latin America continues to be pried open by trade agreements with Washington, and even the economically powerful East Asian emerging economies, which had managed to avoid heavy borrowing during the 1980s (and had thus been spared the early wave of restructuring), found their advantage crumbling with the 1997 financial crisis, the $120 billion IMF rescue package, and the demands for market opening.* In September 1998, in a move that would be repeated across the region, the IMF sent Indonesian president Suharto a letter demanding that his country abolish all tariffs on imported rice — demands that were soon expanded to sugar, flour, soybeans, and corn.

My tour of China's food system has been moving steadily south, deeper into the densely cultivated coastal corridor between Beijing and Shanghai and farther into the heart of the country's almost mythical new abundance. Outside the car, green fields of wheat and corn flash by, interrupted by orchards of pears and apples and veritable cities of greenhouses that stretch for miles across the landscape like sheets of glittering ice. Every few hours, we reach a new city or town and meet a new contingent of agriculture officials. We hear new statistics on local output, then begin a series of site visits — a seed factory, plant-breeding and research centers — where staff members carefully and proudly outline their roles in boosting China's agricultural productivity. In one particularly bizarre instance, Lin and I are taken to a new duck-processing plant, where an entire employee shift has been brought in, and several thousand ducks done in, just to show me the modern efficiency of the operation.

In Anhui, an Arkansas-size province of sixty-four million people, we roll to a stop at the corner of a large field of winter wheat to find a television crew waiting for us. As the camera rolls, a Mr. Wong, local director of agriculture, outlines the rapid advances of Anhui's wheat sector. Annual output is up to eleven million tons, two-thirds of which is exported. Quality is high and is adapting to changing tastes. "Ten years ago, our wheat lacked the gluten to make good bread," Wong explains. "So our scientists

---

* This was a particularly bitter turn of events for many East Asians, since the financial crisis was widely seen as having been exacerbated by the IMF itself; the fund had pressured the East Asian governments to deregulate their investment markets, which had precipitated a rapid influx of capital and a bubble that eventually shook East Asian financial markets in 1996.

improved the variety and our farmers now get ten percent higher prices."
As Wong pauses, the TV reporter asks Lin to ask me to say something about
China's farm system. I offer a few diplomatic words about "impressive output" and "great advances," while, behind the reporter, in the middle of the
wheat field, a lone farmer with a tank of chemicals strapped to his back
walks slowly down the rows, hand-spraying pesticide on each plant.

For all its very real success in raising food output, China's agricultural
miracle is nowhere near complete — and, in fact, its food economy is at a
perilous tipping point. Population is still rising, and growth rates are now
expected to increase again as Beijing, fearful of supporting a massive army
of retirees, relaxes the one-child policy to beef up the next generation
of taxpayers. Meantime, the same economic liberalization that unleashed
China's farming potential also touched off a domino effect of rising consumption and new food habits. That has meant more vegetable and produce, but also more prepared foods, and considerably more meat, which
means more grain demand. And all this as China is still striving to develop
as a food exporter. Chinese companies are moving aggressively to position
themselves as low-cost suppliers not only of raw materials such as wheat
gluten, soy protein, caffeine, and ascorbic acid[44] but also of fresh produce[45]
and meat, especially poultry and fish. Yet these trends toward further food-
supply consolidation raise numerous concerns about the capacity of the
emerging global food system. As we'll see in later chapters, by raising production volumes of meat and other perishables in a country such as China,
whose food-safety systems aren't adequate for *current* volumes, the food industry there is substantially magnifying the risk of a major food-safety disaster. Such trends also bode poorly for food security. Because China now
controls so much of the market for certain raw materials, and because
global supply chains are so lean and just-in-time, with little slack, a breakdown or other bottleneck in China will quickly disrupt a global food
supply chain that often has few alternative sources.

Even within China itself, the push for more meat has accelerated
China's domestic demand for livestock feed beyond the country's domestic
capacities. Since the early 1990s, China's rapidly rising demand for meat
and soybean oil has outpaced domestic soybean production. Today, a nation that once exported surplus soybeans must import more than twenty
million tons of soybeans a year, a volume that not only exceeds domestic
production and accounts for 40 percent of all world soybean trade but

which, according to the Food and Agriculture Policy Research Institute, will double by 2016.[46] And now, of course, the same trend is emerging in corn. As recently as 2002, China exported 11.67 million tons of corn, nearly as much as the United States.[47] But as more farmers have switched to higher-value produce, and as the Chinese continue to use more corn to fatten livestock and to make high-fructose corn syrup for beverages and ethanol for fuel, China is expected to become a net importer by 2009. "Even the Chinese are saying that they're just on edge," Sandene, the U.S. grain analyst, told me. "If they have a bad year, say, a harvest of only 130 million tons of corn, they would definitely have to go out onto the market." In anticipation of that, Sandene says, Beijing is not only beefing up its import facilities but trying to coordinate a strategy with its leading corn users so that they increase imports only gradually and don't shock markets and push prices to the sky. China, says Sandene, "wants to be gentle on the market."

The question, of course, is whether gentleness is even an option, given the sheer mass of China's food economy, the rapid pace of its population growth, and the transformation of that population's appetites. When China began importing soybeans in the mid-1990s, the new demand not only lined the pockets of American farmers (who now supply a third of China's imports) but touched off a veritable soybean gold rush in Argentina and Brazil, countries with vast untapped agricultural potential and great needs for export earnings. Brazil in particular, with its IMF-imposed restructuring, has positioned itself to exploit China's new appetites. São Paulo encouraged farmers to expand soybean acreage in its massive *cerrado* grasslands and invested heavily in research to develop new soybean varieties that could thrive in local soils and which, with the country's tropical climate, allow two and even three crops a year. Import restrictions were relaxed, resulting in a dramatic increase in the purchases of tractors, seeds, pesticides, and especially the fertilizers necessary to energize the underperforming *cerrado* soils. To pay for it all, Brazil relaxed its investment laws, making it easier for companies like Cargill and Archer Daniels Midland to expand their existing storage and loading facilities[48] and for processors like Danone and Nestlé to buy up local milk and dairy companies. To cement its new relationship with China, Brazil has also welcomed Chinese investors, who are pouring money into farms, port facilities, and other infrastructure. And, in proper restructuring etiquette, Brazil devalued its currency, the *real*, by two-thirds, thereby making its farm products only a third as expensive as those of its main rivals in the United States.

The results have been almost Chinalike. Brazil now leads the world in sugar and coffee exports, while its 175 million cattle — nearly double that of the U.S. herd, and the largest in the world — allowed it to surpass American beef exports in 2004[49] and take an eighth of the entire global beef market.[50] Brazil's booming poultry industry, meanwhile, fed on cheap corn from neighboring Argentina, now exports more than two million tons of chicken, or roughly one in every five birds traded internationally. Within another decade, according to the Food and Agriculture Organization, Brazil's meat exports will be larger than those of the United States, Canada, Argentina, and Australia combined.[51]

More notable is Brazil's soybean boom. Farms are expanding at the rate of around four thousand square miles a year, and Brazilian soybean exports have soared from 8.2 million tons in 1998[52] to 25 million tons in 2006,[53] with much of the increase going to China. Brazil is in fact the world's second-largest soybean exporter, just behind the United States, and is expected to take the lead in the near future. Whereas the United States is limited in its ability to expand soybean production (much of the unused cropland in the United States is only marginally productive, not to mention ecologically fragile), Brazil is currently using just a *fifth* of an arable land base of more than one million square miles. Additionally, Brazil's production costs are lower, making it the preferred provider for China and other importers. All told, America's share of the world soybean market has fallen from two-thirds in 1989 to less than half today[54] — much of that going to upstarts like Brazil.

This shift in agricultural power will only accelerate as Brazil's agricultural system matures while its American rival simply gets older. Today, the United States' primary comparative advantage over Brazil in corn or beans is twofold: we have better farm technology and better roads, rail lines, and other transportation infrastructure, which, when coupled with heavy government subsidies, compensates for our disadvantages (higher land and labor costs) and lets us beat Brazil. But not for long, because Brazil's main disadvantage is poor infrastructure, which is more easily remedied than are the United States' higher land and labor costs. As investment pours into Brazil, the country's infrastructure will improve, and as it does and Brazil becomes a more efficient exporter, the American advantage will fade. "In the United States, we have the infrastructure in place, so there is not a lot of cost to squeeze out," says Chad Hart, a grain trade expert at Iowa State University. "But in Brazil and in Argentina, they have lots of costs left to squeeze." Thus, even as China's huge market opens and as markets like South Korea's are again undersupplied, American corn farmers may lose that business to their competitors to the south.

In fact, at nearly all levels of the food economy, the United States is losing its competitive edge as its production costs rise and as our rivals (often bolstered by investment from U.S. companies) grow more and more efficient. Some agricultural economists, like Steven Blank at the University of California at Davis, suggest that because the United States will surrender more and more of its share of the low-cost commodity markets, it should instead evolve toward the Nestlé model and focus on making and selling higher-value processed foods. But this strategy too seems like a dead end. Foreign processors and manufacturers can already supply many value-added foods so much more cheaply than our own companies can that U.S. retailers simply import these too. In fact, overall imports of food — commodities but also, increasingly, higher-value products such as meats and fresh produce — are growing so steadily[55] that America's food-trade balance actually went negative in 2004; for the first time in decades, in dollar terms, we are importing more food than we export.[56] By 2016, according to one study, the United States will be the world's largest importer of meat.[57] A country that once prided itself on feeding the world will increasingly be fed by it.

This reconfiguration of food power signals the emergence of a new global trade axis, with Brazil and Argentina at one pole and India and China at the other, and the decline in the power of mature countries, such as the United States. Such a power shift is already evident in trade negotiations, yet another arena where America once held largely uncontested sway. For example, because Brazil's massive export earnings and new influx of private capital have allowed it to pay off most of its IMF obligations, that country has far more freedom to pursue a trade policy that is less amicable to Washington. In 2003, Brazil joined with India and eighteen other developing nations, a group known as the G-20, to kill a round of trade talks being held in Cancun after the United States and the European Union had refused to let in more developing-world farm products. Three years later, Brazil and India helped sink the so-called Doha round of trade talks under the WTO, the successor to GATT and another U.S. priority.

China too has been quick to use its new leverage as the dominant market. Just as nineteenth-century England had the attention of governments in the United States and Australia, modern China's potential to absorb farm surpluses has won it leverage in Washington, D.C. "Any time China has come under political pressure on the Hill, like Tiananmen Square, they come in and buy up a bunch of U.S. meat or wheat and placate our politicians," says Kevin Natz, formerly with the U.S. Grains Council. More recently, according to press reports, China used the promise of its massive markets to persuade U.S. officials to override safety concerns about imported food. In 2006, Chinese officials suggested that Beijing might allow the United States to export beef into the Chinese market if the American government allowed China to export cooked chicken into the United States. Despite reported objections by federal safety officials, the U.S. Department of Agriculture began processing the request.

The real impacts of China's emergence as the mother of all markets will go far beyond the competitiveness of our food industry or the balance of our food trade. In the 1970s, the importation of twenty million tons of soybeans would have bankrupted Beijing; today, says one Western analyst, China's skyrocketing nonfood exports have made the nation "rich enough to buy the food off the tables of half the world." Given that many of those tables are in parts of the world where food is already scarce, China's emergence as a power buyer has troubling implications for a future that will have considerably more people but considerably less certainty about how to feed them all. During the medieval period, economically powerful England and western Europe were the highest bidders for grain from Poland and the Baltics — to the substantial detriment of the Polish and Baltic peasants, who lacked similar buying power. In the coming decades, emerging economic powerhouses such as China and, eventually, India may assume the same disruptive role.

What will *not* be the same, however, is the world's capacity to respond. Whereas producers once met rising demand by bringing on new acres or by investing in new technologies or inputs, these strategies are now no longer as assured, or sustainable — and, in fact, may ultimately be undermining our ability to produce food. None of which comes as a surprise. As far back as the 1970s, experts of various stripes had begun asking how the current system of food production could possibly be run on a global scale, with more people, on a more constrained resource base. With the realization of that global system, those concerns have become even more pressing. The greater efficiencies and growth unleashed by this more globalized food economy not only accelerated the negative side effects of that economy but, perversely, make it less likely that we can address those side effects proactively. By allowing producers to shift more of their external costs to less regulated countries, and by allowing nations like China and India to feed their exploding populations with grain from the few remaining under-populated regions, a globalizing food economy essentially lets the world defer the work of reducing those external costs, or of coping with overpopulation, or of confronting any of a number of unsustainable tendencies in the modern food system. And these delays can only make such problems more and more challenging to address.

Perhaps inevitably, it is the Chinese who seem most aware of the way their arrival on the world food market will mark a historic turning point — and raise new risks. The Chinese government has not only worked assiduously to build up its own food production and to build relationships with critical suppliers such as Argentina and Brazil; it has also made careful and strategic assessments of the way this historic shift in global demand will affect both its own security and the security of its trading partners. In a small, dimly lit office at China Agricultural University in Beijing, Tian Weiming, one of the nation's top food-security experts, spends his days trying to predict whether China's changing eating patterns will continue to evolve along the Western patterns and, if so, whether the country and the world will be able to handle the evolution. Traditional forecasts have sought to answer these questions by looking at wealthy Asian countries such as Japan and Taiwan to project China's future growth and behavior. But Tian believes such examples fail utterly to convey the impacts that a country of China's size moving into a postindustrial food economy will have. On a per capita basis, Tian tells me, China will need thirty years to catch up with Taiwan's meat consumption. But even now, he says, China's level of meat consumption is already straining domestic and global markets. "I cannot imagine what the world will be like when Chinese people are as wealthy as Taiwan," says Tian, pausing for moment and getting a pained look on his face. "It will be very different."

6

The End of Hunger

B
Y LATE AUTUMN, the fields on Mango Mutisya's farm, in the high,
semiarid hills of south-central Kenya, lie in perfect readiness for the
seasonal rains. The rust red earth has been carefully tilled; the bags of
seed corn wait in the stick-and-mud granary. Mutisya himself, a thin forty-
two-year-old with a smooth-shaven head, sinewy arms, and hands so cal-
loused they feel like horns on my own soft palms, stands with his wife,
Janet, near their modest mud-brick home, politely answering my questions
about farming while casting surreptitious glances at the puffy white clouds
that march silently across the enormous blue sky.

The Mutisyas are a regular stop whenever an aid organization — in
this case, Catholic Relief Services, or CRS — needs a success story for a
journalist, and it's easy to see why. The couple is educated, gracious, and
patient, and their farm offers a good case study of the ways small-scale, traditional farmers can exploit modern agronomic techniques: everywhere we
look, the red fields are crossed by ditches that Mango dug, at the suggestion
of CRS, to curb soil erosion and raise his productivity. Since the ditches
went in, yields have climbed from barely fourteen bushels to around fifty. A
good harvest now brings around ten thousand shillings, or $130 — money
that has let the Mutisyas put up new outbuildings, save for their children's
education, and buy more land to plant. Their home has a tin roof, instead
of the traditional thatch, and furniture — a table, two battered patio chairs
with vinyl cushions, and a kerosene lantern. They've even been able to afford a few luxuries: meat for dinner every few months, a meal out at the local market, some nicer clothes. As we tour the dusty red fields, Janet, who is
thirty-eight and wearing a red scarf and a new dress, smiles shyly and
points to the ditches. "These helped me buy this dress."

But one look at the placid sky is enough to make clear that the Mutisyas' new prosperity is precarious. Today is November 4, and farmers haven't seen any of the heavy rains that normally come by late October and are essential for the first of Kenya's two growing seasons. Anxiety is already high in the village, where farmers recall only too well the absence of last year's autumn rains. "We didn't harvest anything," says Mango. As food shortages spread, he says, shopkeepers raised their prices, and farmers had to sell their main assets — livestock — to buy maize simply to eat, and even then, Mango says, "We were hungry." I ask him if he thinks the rains will fail this year. He shrugs and looks at the sky. "When it's this hot, it usually means rain," he says as if to reassure me. Janet, rolling her eyes, is far less diplomatic. "*Usually*," she says, "the rains are here by October twenty-seventh or twenty-eighth. *Usually*, we are doing our first weeding by now. That is when we had planned for and it is late."

The Mutisyas' farm offers a crude but accurate metaphor for much of rest of the food economy of Kenya, a nation that, despite the great potential of its agricultural lands and the high hopes of its population — and despite decades of very expensive antihunger programs — is still never more than a season away from disaster. In any given year, four million of Kenya's thirty-one million people go hungry, and in bad years, of which Kenya has seen many recently, that number can easily double. With each new setback — a drought, a plague of plant disease, flooding, border hostilities with neighboring Ethiopia or Uganda — farmers like the Mutisyas will have to wait patiently for the bags of relief corn and soy meal and hope to last the six months till the next planting season. Most will endure, but each year a certain percentage cross some invisible but immutable economic threshold. They sell their land, animals, and whatever household assets are too large to carry and make the long trip to Nairobi, a former regional hub whose enormous slums now teem with more than a million rural refugees and claim one of the highest rates of AIDS in the world.

For all that, Kenya must consider itself lucky. As I was pestering the Mutisyas, Malawi was already so deep in drought that farmers there were digging up wild roots to eat; Ethiopia's meager harvest had been disrupted by civil war; and relief agencies like CRS were either running or preparing to run operations in Somalia, Djibouti, Niger, Chad, South Sudan, Zimbabwe, and other least-developed nations, or LDCs. Such country names appear so frequently in news stories about hunger that it's depressingly easy to misapprehend the human costs in what has become a permanent crisis. Every twelve months across sub-Saharan Africa, malnutrition kills more than ten million people.[1] Hundreds of millions more suffer a collapsing dietary regime and a medieval nightmare of exhaustion, sickness, and ravaged potential. And although sub-Saharan Africa is the poster child for persistent hunger, the affliction is by no means confined to this continent. While China has revamped its food system, India, once the leading light of an agricultural breakthrough known as the Green Revolution, today struggles to handle its more than two hundred million hungry people, including the world's largest cohort of malnourished children. Even in the United States, the wealthiest country in the world, one child in six still suffers from inadequate nutrition.[2]

All told, nine hundred million people — one-seventh of the population — are malnourished, and another one billion suffer chronic and often destructive deficiencies in micronutrients — statistics that, given the fact that food is cheaper and easier to get now than at any time in history, offers the most dramatic proof that the modern food economy is failing catastrophically. In recent years, as world hunger reemerged as a cause célèbre, and as a cause for celebrities, world leaders have committed to cutting the number of food-insecure people in half by 2015 — the so-called Millennium Goals. But many of the numbers are moving in the opposite direction, in part because as food production improves, population increases even faster; each year, the ranks of those who still cannot get enough to eat grow by seven million.[3]

Why hunger persists on such a vast scale and what can be done to alleviate it are two of the thorniest, most controversial questions in the debate over the modern food system. Most hunger specialists now see chronic food insecurity to be symptomatic of the larger problems that plague the LDCs, sub-Saharan Africa in particular. A rampant AIDS epidemic has decimated the region's farm-labor force,[4] while in countries such as Ethiopia and Uganda, decades of war, government ineptitude and corruption, and infrastructural neglect have frozen food production at a preindustrial level of output. Food insecurity can thus be understood as a self-perpetuating cycle that won't be broken until a nation's larger political and economic problems have been solved.

Yet for all that the hunger is in a sense self-inflicted, many of these countries' homegrown problems are clearly being exacerbated by rapid changes well outside their control. Climate change is already having a devastating impact on food production in many LDCs, and in Africa, it may cut yields in half by as early as 2020, according to the Intergovernmental Panel on Climate Change.[5] More fundamentally, the LDCs have not fared especially well in the evolving economic climate. The various technological and commercial revolutions that transformed much of the rest of the global food economy have largely bypassed the poorer countries, while the tough love of neoliberal trade policies has often been *too* tough. Despite early success at bringing Western-style high-yield agriculture to Africa, much of the so-called Green Revolution on that continent has faltered. Crop yields there are still far lower than in industrial countries, with the result that few African farmers can compete in a global grain market now defined by low-cost producers.

More recently, however, as new understandings of the root causes of food insecurity have come to the fore, there is far less certainty that hundreds of millions of poor farmers can, or even should, be fit into a global food economy that is so dependent on large-scale, low-cost production that only the biggest, most heavily capitalized players can keep up. Indeed, so swiftly is the modern food system evolving, and so far have many poor countries fallen behind, that the gap between what the global food economy demands and what the LDCs can hope to deliver now poses an extraordinary divide.

Driving through the rolling landscape east of Nairobi, a visitor can readily understand why Kenyans are so bitterly disappointed by the country's meager progress in the war on hunger. Unlike the Mutisyas' dry surroundings, the landscape here is lush and green, dotted with small farms and vast stretches of tea and fruit operations, and radiating the same vitality that made Kenya the ideal candidate for a Green Revolution during the 1950s and 1960s. Though much of the country is suitable only for grazing, Kenya did have large arable regions with good soils and a wide range of microclimates. The country also had a strong, two-tier farming culture — hundreds of large estates (built by the British to export coffee, tea, and maize, but now owned by Kenya's black elites) and an army of some ten million smallholders working tiny one- and two-acre plots of maize or sorghum and raising goats and chickens. Granted, most of the smallholders used obsolete tools and seeds and got just a few bushels an acre. But Kenya's ambitious president at the time, Jomo Kenyatta, believed such backwardness could be reversed with modern farming technologies, especially chemical fertilizers and new, high-yield crop varieties.

Kenya wasn't the only country banking on new technologies. Across the developing world, euphoria over political independence had been replaced by gnawing fears of economic stagnation, political unrest, and famine, especially in Asia, where India and Pakistan were moving so rapidly toward grain deficits that even with aid shipments from the United States they were losing ground. Although Africa had not reached this critical point, some observers saw it as a matter of time — most famously, Paul Ehrlich, whose book *The Population Bomb* predicted mass famines and hundreds of millions of deaths.

But even as the neo-Malthusians painted their dire forecasts, an alternative, far more encouraging scenario was taking shape. In the late 1940s, as scientists like Thomas Jukes were reengineering meat, botanists and plant breeders set out to develop a stable of high-yield crops specially suited to conditions in the developing world. Varieties of wheat, corn, and rice were bred that could resist fungus and insects — a critical quality in countries that were often hot and muggy. As important, breeders found ways to create plants that could tolerate greater quantities of nitrogen fertilizer. This was essential, because although it was possible to mass produce synthetic nitrogen cheaply, traditional grain crops, especially those in use in the developing world, couldn't use the extra nutrition:* past a certain point, the additional nitrogen simply made the plants shoot up like gangly adolescents, gaining little extra seed mass but growing so tall they toppled, or lodged, before harvest. By contrast, the new varieties of crop plants, many of them bred to be shorter, could take in vastly more fertilizer and convert it into more grain and less inedible stock.

The most famous of these efforts was the research of Norman Borlaug, a plant pathologist who, while working for the Rockefeller Founda-

---

* Disease resistance is critical for farming in the tropics, where blights, rusts, fungi, and other crop maladies are rampant. But the real breakthrough in botany was developing plants' tolerance for nitrogen. In the 1950s, organizations like the World Bank, the U.S. Agency for International Development (U.S. AID), and the Ford Foundation had vigorously promoted increasing fertilizer use in India and other hungry countries — only to find that traditional crop plants couldn't make use of the newly cheap input.

tion, had developed a high-yielding dwarf wheat that revolutionized farming in Mexico and southern Asia. But important work was also under way elsewhere in the developing world. Researchers at the International Rice Research Institute (another Rockefeller Foundation project) developed a sturdy, nitrogen-tolerant rice variety that yielded up to six times as much rice as traditional plants and matured so quickly that two and sometimes three crops could be grown in a single year. And in countries like Kenya, researchers were bent on creating their own versions of the high-yielding varieties of corn, or maize, that had transformed the American Midwest. Although maize isn't native to East Africa, it had been part of the regional culture and agriculture for centuries, and settlers had developed numerous local varieties. African breeders now began to cross their local crops with the higher-yield varieties from Latin America, with impressive results: from 1960 on, Kenya's per-acre maize yields increased by more than 3 percent a year, better than in the United States, and there was every expectation those gains would continue. M. N. Harrison, Kenya's chief maize breeder, went so far as to warn his countrymen to brace themselves for an "agricultural revolution as happened in the USA Corn Belt."[6]

The entire developing world, in fact, was poised for such a revolution. In Mexico, wheat yields almost tripled between 1950 and 1965, allowing a country that had once imported 60 percent of its wheat to become entirely self-sufficient. In 1968, the same year that Ehrlich published his predictions of massive famine, Pakistan and Turkey harvested record wheat crops. The Philippines brought in a record rice harvest, while India's wheat crop was so unexpectedly massive it overwhelmed the nation's primitive storage infrastructure; hundreds of schools were closed and the classrooms used as temporary silos. In a now famous speech in 1968, William Gaud, head of the U.S. Agency for International Development,[7] declared that such "record yields, harvests of unprecedented size and crops now in the ground demonstrate that throughout much the developing world — and particularly in Asia — we are on the verge of an agricultural revolution. . . . I call it the Green Revolution."

The ocean of grain spilling over the developing world by the early 1970s radically altered not only the food supply but the modern debate over hunger as well. Until perhaps one hundred years ago, governments had been more or less content to understand hunger as an unavoidable part of life.

Politicians might try to stop famines (especially when they occurred in strategically useful places such as British India in the 1800s), but in terms of run-of-the-mill hunger, governments tended to let nature take its course. By the early twentieth century, however, events had overtaken this laissez-faire stance. The Great Depression had not only demonstrated the frailty of markets (and, perhaps, of God) but also, by inducing governments to intervene in their own food economies, legitimized the idea of intervention in the food economies of *other* countries. At the same time, there were renewed fears of a population explosion, especially in teeming Asia. Even if outright famine was avoided, Western governments worried that endemic hunger would so destabilize Asian countries that they would be easy prey for Communists.

But even though intervention was now seen as acceptable and necessary, there was no consensus on how best to intervene. On the one hand were the economic pragmatists, who argued that hungry countries shouldn't even try to farm (their methods were so hopelessly inefficient that they had no comparative advantage in agriculture) but should focus instead on industrial development, with Western financing, and use their new earnings to import their food.

However, others, including the Ford and Rockefeller Foundations, held that agriculture did have a role in a nation's food security, but *only* if the sector could become more productive and efficient — a prospect that the new Green Revolution crops now made far more feasible. Rising yields would let poor countries generate exportable surpluses, the earnings from which could help finance schools, factories, roads, and other increments of industrial infrastructure. Higher yields would also bring lower food prices, leaving consumers with more discretionary income for other goods and services, and thereby creating demand for still more industrial development in a virtuous cycle of economic progress. This evolution from agricultural revolution to industrialization was, in a sense, what had occurred in Europe, the United States, and Japan in the eighteenth and nineteenth centuries. And experts believed it could be replicated in poor countries, or developmental states, as they were then called, with hybrid crops, careful government management, and heavy doses of financial aid.

This was certainly the hope in Kenya. The government created a powerful state-run farm apparatus modeled after the American farm program to transform its millions of smallholders into an army of hyperefficient maize producers. New seeds were distributed to farmers at low or no cost, along with instruction and heavily subsidized fertilizers and pesticides. A powerful state grain board was created to act as a friendly buyer, purchasing farmers' crops (at above-market prices), most of which would be sold to urban consumers (at a subsidized discount) with some kept in reserve against bad crop years and speculators.[8] Additionally, to protect the nascent farm system as it modernized, tariffs were enacted to keep out cheap foreign grain.

Such heavily subsidized ventures weren't cheap, especially for a cash-strapped country like Kenya.[9] But these were perfect times to be a poor but promising developmental state. With the Cold War in full swing, Western governments were lending, or simply giving, hundreds of millions of dollars for aid projects in Africa and Asia. Further, it was well understood among development experts that the Green Revolution would be an *expensive* revolution, not least because the new, high-yield crops required more inputs than traditional varieties did.

For example, because the new dwarf varieties were so short, they couldn't compete with weeds for sunlight and so were largely helpless without additional herbicides. And, of course, high-yield plants needed massive amounts of nitrogen and other fertilizers. Thus, a major focus of most aid programs was persuading developing-country farmers to use inputs, especially fertilizer, and indeed, much of the early optimism came not from the fact that crop yields were increasing but from reports of rapidly rising fertilizer shipments to the developing world. In his Green Revolution speech, Gaud actually boasted of the hundreds of millions of dollars Congress was spending to subsidize American fertilizer exports to developing nations.[10]

But the returns certainly seemed to justify the cost. In Asia, soaring farm output not only alleviated famine worries but unleashed the predicted wave of urbanization and industrialization. In Taiwan and South Korea, agriculture's share of the workforce dropped from 75 percent in 1945 to 25 percent in 1970.[11] Africa seemed to have caught the same wave of agriculture-driven industrialization. In Kenya, farm output was growing by 4 percent a year (the same as fast-growing Asia), and the country was producing enough maize to feed itself and to export a substantial volume — which, with the high post-Russian-sale grain prices, earned a substantial income.[12] And just as development experts had predicted, rising farm incomes created a ripple effect throughout African economies. Kenya's urban areas were thriving and businesses were growing, despite the repressive and corrupt Kenyatta regime, and Nairobi emerged as a regional center for culture and education.

But then, almost as soon as it started, Africa's boom was over. By the late 1980s, as Asian yields continued upward, African output faltered; Kenya's per-acre maize yields fell back nearly to their 1960s levels, while the number of farmed acres shrank; other African countries experienced similar problems. The timing couldn't have been worse. Just as Africans were producing fewer bushels, American farmers, still in overdrive from the "fencerow to fencerow" policies of Earl Butz, were glutting markets with their surplus, sending prices plummeting. At the same time, rising oil prices pushed up the cost of fertilizer and pesticides. In Kenya, a desperate government poured more money into fertilizer subsidies and price supports, but that simply meant borrowing even more money from international lenders; eventually the interest payments alone were eating up a quarter of Kenya's entire economic output.[13]

Finally, the World Bank and other lenders stepped in and forced Kenya to restructure its economy. As in Latin America, Kenya (and, eventually, most African debtor states) had to dismantle its state-run farm programs and eliminate most subsidy programs — actions that unsurprisingly did little to improve yields or output. By the mid-1990s, Kenya's output was so low that the country had no choice but to rejoin the ranks of Africa's other Green Revolution has-beens and begin importing maize — a crushing defeat for a nation that had envisioned itself on the leading edge of an agricultural revolution.

As countries like Kenya struggled to survive, development experts fell into a heated debate over what, or who, was to blame for the collapse of Africa's Green Revolution — a debate that has only intensified, and which underlines the massive political and ideological tensions inherent in the fight against hunger. Many Green Revolution advocates, pointing to successes in Asia and Latin America, blame the African failure on poor execution — by corrupt and inept African governments, but also by outside players, especially the major donors, whose antihunger strategies shifted constantly with changing global politics.

Other critics have focused on the revolution's underlying paradigm; its heavy reliance on expensive industrial inputs, they say, was grossly unsuited to the social and physical realities of African agriculture. And indeed, given the deep involvement of the Western input industry (fertilizer, pesticide, and oil companies, among them DuPont, Dow, BASF, and Exxon — all helped distribute the new technologies),[14] it has certainly occurred to some to ask whether the Green Revolution's primary goal wasn't just building food security but building new markets for American farm inputs.*

The truth lies somewhere in between. There is little doubt that African governments grossly mismanaged their farm programs: grain boards routinely manipulated grain prices for their own profit;[15] government seed breeders didn't adequately localize the superseeds developed by international seed programs. But it is also true that the Green Revolution model did impose a set of industrial agricultural practices that didn't fit the realities of African farming and, in many cases, still don't. Most of the high-yield crops, for example, needed lots of water. Asia has sufficient rainfall and river systems to support massive irrigation systems, but that isn't the case in dry sub-Saharan Africa: 85 percent of Kenya's arable land isn't reachable by irrigation and depends entirely on rainfall, which comes only rarely, if at all.† Historically, African farmers coped with water scarcity by planting hardier native crops, such as millet, sorghum, and teff. Maize, which is far less drought tolerant, had been grown only in those areas with high rainfall — the so-called high-potential areas. During the Green Revolution, however, maize production was pushed into unsuitable semiarid areas. According to the World Bank, most of Kenya's maize boom came not because farmers were getting more bushels per acre, but because farmers had planted more acres[16] — an increase that collapsed in the severe drought that struck in 1984.

Even today, African farmers in arid areas insist on growing maize, despite efforts to move them to hardier crops. "It's a huge part of their culture," explains Paul Omanga, a native Kenyan and agronomist who used to work with CRS. "They eat it at least twice a day, and will tell you that it

---

* During the 1960s and 1970s, Exxon had even operated a line of one-stop farm shops in the Philippines offering gasoline, pesticides, fertilizers, and seeds, according to Pat Mooney of ETC. Nor were Green Revolution scientists themselves ignorant of the commercial opportunities inherent in conquering hunger. As one Rockefeller consultant explained in 1942, "When the war is over, there will be millions to feed and farms to be supplied with seed, fertilizer, machinery and livestock" (Kloppenburg, *First the Seed*, 158).

† Researchers were able to breed drought-resistant varieties of wheat and rice, but maize proved far less amenable to such tailoring, and most high-yield varieties required regular rainfall — something that had been iffy even before the recent shift toward drier weather.

doesn't feel like they've eaten a meal unless it includes maize." In fact, when I asked the Mutisyas why they grow maize in such a dry place, Janet just stared at me as if I were an idiot. "Because," she said, finally, "maize is the food that satisfies."

And water is only one of the Green Revolution inputs unsuited to local conditions. The seeds themselves were hugely problematic. In hybridized seeds, the specially bred traits, such as disease resistance and fast growth, aren't stable — that is, they tend to degrade if a farmer recycles the seeds over several generations. To keep yields from falling, says Melinda Smale, an Africa specialist with the International Food Policy Research Institute in Washington, farmers need to buy new seeds every few years, "and you can't do that in an economy where farmers have no cash flow, even when seed is cheap."

But the real Achilles' heel of the Green Revolution was, and is, fertilizer. By conservative estimates, more than a third of the Green Revolution yield increases came directly from using more fertilizer.[17] And yet, as American and European farmers were also discovering, while fertilizers were a necessary ingredient for modern high-yield agriculture, they were not sufficient to ensure its success. Although African farmers saw massive yield increases within the first few years of adopting the new techniques, in a relatively short time, something odd happened — yields fell unless farmers added steadily greater applications of nitrogen and other fertilizers. This effect was so dramatic that over the course of twenty years, a farmer would need to double his nitrogen applications simply to maintain his yields at their initial level.[18]

Why this change occurs isn't entirely clear, but research suggests that under intensive agriculture methods, soils will lose not just macronutrients — nitrogen, phosphorus, and potassium, which can be replaced synthetically — but the carbon-rich organic matter left over by decaying plants and animals. This organic matter is key to good crop yields. The more organic matter in the soil, the more rainwater the soil can absorb and retain, which means water for crops. Organic matter also helps the soil particles stick together, reducing the risk of wind and water erosion. As well, soils rich in organic matter have a greater capacity for additional nutrients — that is, they can absorb more fertilizers, whether natural or synthetic, and convey those nutrients more readily to plants. In short, adding synthetic fertilizers to lands rich in organic matter — such as the American Midwest and certain parts of Africa — could indeed bring massive yield increases. The problem is that soil organic matter, or SOM, can be depleted when farmers raise too many crops without replenishing nutrients with cover crops or manure or other fertilizers. And once SOM begins to fall, the soil's capacity to hold and transport synthetic nutrients also falls, which means that farmers have to add steadily more nitrogen simply to maintain their yields. The loss of SOM also leaves soils hugely vulnerable to wind and water erosion and thus accelerates leaching.

Not all African soils suffered declines in SOM. But whether a farmer had high or low levels of SOM, he still needed lots of fertilizer to grow the better-yielding crops — and unfortunately, just as poor farmers were becoming more dependent on added fertilizer, that fertilizer was becoming less available. In addition to a steep rise in prices due to the oil crisis, the aid paradigm was shifting yet again. Environmentalist groups, worried that farm chemicals were damaging fragile soils, had begun lobbying Western governments to shift their financial support away from farm chemicals and toward "environmentally sustainable" agriculture. At the same time, free-market economists argued that fertilizer subsidies themselves were slowing the development of local fertilizer industries within the poor countries.

These changes were part of a larger retreat by Western governments and lenders, which, frustrated by government corruption and caught up in the tough-love policies of the neoliberalization, slashed aid for developing-country agriculture by nearly half.[19] As a result, fertilizer use in Africa has plummeted[20] — the average farmer uses less than *ten* pounds per acre — less than a tenth of the global average — and a similar fate has befallen other farm inputs. For example, although seeds tailored to African needs did finally become available in the late 1980s, the big donors were no longer funding distribution to farmers. As Jeffery Sachs, a Columbia University development economist, notes, "[B]y the time the [seeds] came to Africa, the model was, 'We don't subsidize an input.'" All of which at least partly explains why, according to official estimates, grain yields have fallen from sixty bushels an acre in the heyday of the Green Revolution to around fifteen bushels today, and experts like Sachs believe that the real average yield may be closer to eight bushels.

Sub-Saharan farmers are almost back to where they were fifty years ago — no inputs, no mechanization, and a preindustrial level of output. The only difference is that farmers are trying to feed a population roughly four times as large. In fact, whereas much of the rest of the world has "beat Malthus," as Sachs puts it, Africa, along with parts of India and Latin America, has been caught up in a deadly Malthusian reprise: as grain output levels off, population growth, spurred by earlier gains in food output, is surging ahead. In Kenya, where population has climbed from eight million in 1960 to forty million today, the country must import nearly half its grain, and even then, nearly half its people are food insecure — double that in 1980.[21] Sub-Saharan Africa as a whole now boasts both one of the highest population growth rates and one of fastest declines in per capita grain supply,[22] and by 2025 will need to triple the amount of grain it buys from foreign suppliers[23] — hardly the future that pioneers like Jomo Kenyatta envisioned at the outset of the Green Revolution.

It is, in many respects, as if that revolution never happened. "When you talk to Kenyan farmers about it, they're very matter-of-fact," says Tom Remington, a Catholic Relief Services agronomist who has been working in Africa since the 1970s. "They don't really see it in macroeconomic terms — that at the height of the Green Revolution, they were receiving a highly subsidized package of inputs that all went away with restructuring. Instead, they tell you, 'Yes, I used to use hybrid seeds and I used to use fertilizer but I don't anymore because they're too expensive.'"

At ten o'clock on a Wednesday morning, inside a walled compound near the Nairobi Airport, the main packing room at the Vegepro Corporation is humming with factorylike precision. At dozens of long stainless steel tables, eight hundred Kenyan workers, mostly women in white smocks and head scarves and green aprons, are sorting piles of green beans into tidy little stacks and carefully packing them into plastic supermarket trays. Within hours, the beans will be sealed, weighed, and labeled, packed on shrink-wrapped pallets, and loaded on planes for the night flight to Europe. "These will fly out tonight on Air France to Charles de Gaulle," says Shaun Brunner, a Vegepro executive, pointing to a stack of green beans in special microwavable plastic bags sitting in a chilled storage area. By the following afternoon, Brunner tells me, the beans "will be in a grocery store display case in France."[24]

On one level, Vegepro is yet another of the hundreds of thousands of suppliers that have sprung into being since the late 1980s to feed the burgeoning global retail-grocery market. Vegepro's planeloads of green beans, baby carrots, baby corn, chilies, and snow peas are specially targeted to hit the European grocery shelves just as summer salad season is peaking, and just as output is tapering from the massive greens farms in Spain and Portugal.

Yet Vegepro is also part of the larger story of the battle against hunger. While few of Vegepro's shrink-wrapped products actually end up on Kenyan tables, the rapid rise in exports of produce and other high-value crops during the last decade marks a significant, if somewhat dubious, shift in the way that Kenya and other developing countries have begun to tackle food security since the collapse of the Green Revolution.

Although Kenya has grown cash crops for export since the late nineteenth century — most notably, coffee, tea, and pineapples — this most recent iteration began in the 1990s, just as the aid community was embracing free trade as the key to food security. To the extent that donors and lenders still supported farming, they targeted the high-value cash crops in demand by food companies — sugar, cocoa, coffee, and palm oil for processors, fresh fruits and vegetables for retailers — as a way for poor countries to join the global food economy, pay off their debts, and generate much-needed income and industry. Many development experts objected, arguing that commodities would be no better for small developing-world farmers than they'd been for small industrial-world farmers. But the momentum was already behind this global version of Welfare to Work, and the results once again demonstrated the complex challenges of pursuing food security in a fast-moving global food economy.

Kenya's coffee industry offers a poignant case in point. After a frost destroyed much of the Brazilian coffee crop in the early 1990s, soaring coffee prices encouraged growers in counties like Kenya to expand rapidly. Within a few years, exports of Kenya's distinctive arabica beans were earning a quarter of a billion dollars a year. Unfortunately for Kenya, the same boom attracted other players, including Vietnam. The Southeast Asian nation couldn't compete with Kenya on quality; Vietnam's climate is suitable only for the inferior robusta bean, famous mainly for its distinctive burnt-rubber flavor. What Vietnam did have, however, was the backing of lenders and of the big coffee companies, such as Nestlé, Procter & Gamble, Kraft, and Sara Lee, which together use almost 40 percent[25] of the world's beans, and which had discovered a new way to make coffee. A process had been developed that stripped out some of the robusta's bad flavor; the remainder of it could be masked by adding flavors, such as vanilla or hazelnut, and the bean could then be used in the newly popular dessert coffees. Because robusta beans are 60 percent cheaper than arabica, the food industry now had a killer proposition: a hot new food product made from a super-low-cost raw material.

With its robusta bean rehabilitated, Vietnam was poised to become the Wal-Mart of the coffee world, a low-cost producer that made up in volume what it gave up on price — and the international investment community was delighted. Money poured in to the Vietnamese coffee industry: $233 million from the Vietnamese government, $16 million from the World Bank, and another $100 million from European governments.[26] Food processors themselves invested heavily. Nestlé, which depends on Vietnam for roughly a quarter of its beans, opened a research center there. The result of so much encouragement was predictable.[27] Between 1990 and 2000, Vietnam coffee production soared from fewer than a million tons to more than sixteen million tons,[28] overtaking Colombia as the world's number two producer and generating hundred of millions of dollars in yearly export income. In 2001, Don Mitchell, principal economist at the World Bank, told the *San Francisco Chronicle* that the bank considered Vietnam to be "a huge success."[29]

In fact, it was quite the opposite — for Vietnam, and any other coffee producer. Coffee production was growing nearly twice as fast as demand, and eventually the market choked.[30] Robusta prices collapsed, and arabica prices were dragged down as well. Between 1997 and 2000, the composite robusta-arabica price fell from two dollars a pound to around forty-eight cents, well below many farmers' production costs. Unfortunately, once a coffee farm is in production, it's no easy thing to stop; because coffee trees require such a large up-front investment, growers have little choice but to keep harvesting for several years to recover at least part of their investment — even if they're losing money in the process. Some producers even raised production to cover their losses, which simply worsened the oversupply and pushed prices down further.

In a perfect world, falling coffee prices would have stimulated consumer demand and kept prices high. In the real world, consumers can drink only so much coffee: like corn or wheat, coffee is what economists call price inelastic. Further, the cost of the bean is such a small part of the retail price (around ten cents per cup)[31] that consumers are rarely aware of any price changes; they certainly don't know if the price is falling, because although manufacturers like Nestlé and retailers like Starbucks eventually pass on to consumers any increase in underlying commodity prices (usually with solemn apologies for "supply difficulties"), they rarely pass on any price cuts. This, after all, is the benefit of adding value. Between 1997 and 2002, the farm price for coffee beans plummeted 80 percent, but the price consumers paid for retail coffee products dropped just 27 percent — a discrepancy that was enormously profitable for food companies (in 2001, Starbucks' profits rose by 41 percent; Nestlé's by 20 percent)[32] but less than helpful for coffee producers.

Such a disconnect between retail and commodity prices is standard in the food business: gains from any improvement, whether in technology or a new marketing channel, accrue primarily to manufacturers and other consumers of the commodity, not the commodity's producers, says Chris Barrett, a Cornell University economist and an expert in food security. "And in the case of export crops," Barrett says, "those consumers necessarily live abroad." Coffee farmers, meanwhile, must produce more and more simply to stay afloat — the tropical version of Cochrane's treadmill.

This effect, known as structural oversupply, isn't new. For any cash crop, such as cocoa, sugar, or palm oil, markets have long been characterized by brief price spikes and interminable periods of low prices (which producers actually make even longer by boosting production to cover falling prices). Typically, prices will stay low until so many producers have failed that supply finally falls and prices finally come back up . . . at which point the cycle restarts. What *is* different now, however, is the way the *developed* world responds to oversupply. In the postwar period, as commodity booms and busts hammered developing countries (and sparked widespread political unrest), the United States and other big importers joined with producers in a system of voluntary production limits, such as the International Coffee Agreement, to stabilize prices (and stave off revolution). But by the late 1980s, as the Communist threat faded and as restructuring moved to the fore, Washington withdrew support for the ICA, and the agreement collapsed, as did agreements for sugar, cocoa, and other tropical commodities, prices of which are all now at or near historic lows.[33]

Today, advocates of restructuring see these shake-outs as critical steps in economic evolution. "It's a continuous process," the World Bank's Mitchell explained in 2001. "It occurs in all countries — the more efficient, lower

cost producers expand their production, and the higher cost, less efficient producers decide that it is no longer what they want to do."[34] But that
free-market justification seems hypocritical, given the World Bank's earlier
financial encouragement for coffee producers. More to the point, such explanations gloss over the trauma this "continuous process" creates among
producers. As commodity prices keep falling, countries that depend on crop
exports for a large portion of their total earnings — which is to say, most
developing countries — find themselves right back in the debt spiral. In
Kenya, coffee earnings have fallen by more than 75 percent,[35] and other exporters have fared worse. In Uganda and Burundi, coffee accounted for
more than half of all export earnings, while Ethiopia depended on coffee
for two-thirds of its export revenues and lost more than $300 million be-
tween 1999 and 2001 alone.[36] By one United Nations estimate, the losses in
export revenues developing nations have suffered due to commodity-price
declines is roughly equivalent to half the aid they received from the industrial nations.[37]

The postcoffee aftermath has been eye-opening. According to figures
from U.S. AID, more than half a million coffee laborers have lost their jobs
worldwide.[38] In Vietnam, coffee plantations were simply abandoned, leaving the exposed soils to erode in the heavy seasonal rains. In some African
countries, unemployed coffee farmers have turned to poaching endangered animals — among them chimpanzees and gorillas — for the thriving
bush-meat market.[39] In South and Central America, many coffee farmers
switched to coca, from which cocaine is made, while many more joined the
exodus northward to America.

Ironically, even before Kenya and other developing nations plunged head-
long into commodities as a way to regain their economic footing, development theorists were changing their stance yet again — away from the commodities focus of the Washington Consensus and toward a more nuanced
and complex approach to food security. Work by Indian economist Amartya
Sen had shown that in most famines, the problem isn't a lack of calories —
famine-stricken nations often had adequate food — but a host of other, interrelated factors. Low wages, for example, prevent the poor from buying
food even when it is available. Bad highways keep food supplies from moving easily from a country's surplus regions into areas of need — a recurring
problem in many African countries, such as Kenya, whose road systems are
among the world's worst.

Focus has also broadened to include the impacts of nontraditional factors, such as AIDS. Not only is the disease itself a massive contributor to hunger (by 2020, according to the FAO, one in four farm workers in ten of the hardest-hit African nations, including Kenya, Botswana, Malawi, and Tanzania, will have died of the disease),[40] but the epidemic itself cannot be slowed until food security improves. Unless AIDS victims are well nourished, they can't tolerate the effective but harsh antiretroviral treatment known as an ART cocktail.

More generally, although food aid traditionally has been used reactively — that is, in response to a crop failure or other specific emergency — experts now push a more proactive approach that uses nutrition to prevent disaster in the first place. For example, deficiencies in certain key nutrients, such as protein, and in vitamins and other critical micronutrients, especially in newborns and young children, can leave populations physically and mentally unable to produce crops or master new farming or other economic skills, making them both more vulnerable to disaster and unable to improve their food security. That's why Barrett and other experts advocate programs to build a nutritional foundation, without which none of the other, more ambitious antihunger and antipoverty ideas can work. "When you look at the sweep of history," Barrett says, "one of the few things that have made a difference in developing countries after improved agriculture has been investments in children's health and nutrition, and especially *early* children's health and nutrition."

Moreover, in the post–Green Revolution collapse, many experts have come to regard large-scale, export-oriented agriculture as an overly blunt instrument in the war on hunger. Although agribusiness has flourished in some developing countries, where land is suitable and infrastructure in place, the model simply does not fit as well in the least-food-secure places, such as India or sub-Saharan Africa. There, farmers are often so small, so focused on feeding themselves, and so poor (the average sub-Saharan farmer has an annual cash income of thirty-five dollars)[41] that they cannot easily upgrade to a more remunerative crop. To be sure, large-scale production and even global trade may be a long-term goal for some small farmers in sub-Saharan Africa. But such a transition won't happen until these farmers can reliably produce marketable surpluses, and surpluses aren't possible until farmers can reliably and sustainably feed themselves. Once this threshold has been reached, argues Barrett, farmers have options. They can sell their surpluses and use that cash to expand their farming operations, adding acres in order to produce more competitively, or investing in different, higher-value crops. Or they can use their new earnings to move out of agriculture altogether — by investing in tools for a new trade or in education for themselves or their children. The point is, improved production leads to economic choices, and it is the choices that make development possible.

In this sense, the small farm has an almost paradoxical function in development strategies: it is a critical first step and cannot simply be dismissed, ignored, or abandoned in favor of "modern" agriculture but instead must be integrated into development plans; indeed, most of the world's poor still live on small farms, and could not quickly move off them even if they wanted to.[42] At the same time, however, in most developing countries, the small farm is increasingly seen as a *means* to an economic end, not necessarily the end itself. We in the West may regard small-scale farming as a noble or romantic pursuit (even if we forget that most small farmers in America survive by off-farm jobs). But in the developing world, says Barrett, few small farmers have any interest in remaining that way: "[T]he only people who would want to keep most rural Kenyans farming the way they do right now are people who don't live there."

So again the question becomes, How best to improve that small-farm productivity? The field is crowded with ideas. Free-marketers like Hernando De Soto of Peru, for example, contend that the key to improved small-farm productivity is property ownership; many small farmers lack title to their lands, which, beyond leaving them vulnerable to expropriation by government and large landowners, means they cannot use their property as collateral to finance expansion or new technology. Other strategies, such as those pushed by the UN's Food and Agriculture Organization (FAO), center on helping farmers enter small-scale enterprise, especially poultry production. Not only do chickens require a tiny investment, they also reproduce quickly, can be sold for cash in times of scarcity,[43] and provide a source of inexpensive protein, including eggs, which can be eaten at home or sold locally.

Columbia University's Jeffrey Sachs is promoting a microeconomic strategy known as the Millennium Project. In dozens of Millennium villages around Africa, poor farmers are receiving new farming packages that include new seeds and fertilizers as well as other resources, such as mosquito netting and clean water, that are carefully chosen to give farmers the best chance of exploiting those new, high-yield farming inputs.[44] The aim, says Sachs, is to allow poor farmers to stop focusing on bare survival and begin moving away from subsistence. "When you have enough food to eat," Sachs explained in a 2006 speech, "you can put aside part of your land and part of your time not to growing maize on nitrogen-depleted soil but actually on growing tree crops. On growing cardamom, on growing spices, on growing fruit trees, on dairy products." Or farmers can begin moving away from farming entirely — taking up metalwork or woodworking or some other enterprise, the income from which can pay for education, or be invested in some new venture, or simply be reinvested in farming.[45]

Once productivity is increased, where should farmers sell their surpluses? Here, too, the field is split. Some advocacy groups contend that the imbalances of a global food economy that is dominated by tightly controlled supply chains and relentless downward price pressure can be rectified by simply *paying* farmers a fair price. If coffee farmers, for example, were actually getting a fair share of the retail price, they wouldn't need to overproduce; this is the essence of systems like the Fair Trade. Many development experts, however, worry that the complexities of fair-trade systems mean they can reach only a small portion of producers. Better to steer clear of the global market altogether — at least at first — and encourage farmers to sell their surpluses locally or perhaps regionally. As Barrett points out, for all the emphasis on export-oriented farming, the bulk of any country's agricultural production is sold *domestically*, and most of the economic linkages between farm growth and general economic development always occur through *regional* trade between rural producers and urban buyers.

The importance of regional markets is one of the lessons from China, and it also helps explain why supermarkets are now growing so rapidly in the developing world, Barrett says: newly urban consumers want, and are willing to pay for, the local foods they can no longer grow themselves. To that end, experts like Catholic Relief Services' Remington and his colleagues are pushing local, small-scale strategies, urging farmers to focus on low-input, medium-value crops such as onions and potatoes, which are in high demand among urban consumers in Nairobi, or commodity crops like chickpeas and navy beans, which are in strong demand in Europe and India, and groundnuts, which have a strong regional market. All these can be produced efficiently at small scale.

No matter how effective the strategy, however, it will not be cheap. Although the long-term goal is to foster self-sustaining business models that don't depend on foreign aid dollars, to reach that point, developing countries will need help paying for the components of this new economy that the market simply won't provide. These components include not only early childhood nutrition and AIDS treatment but also access to bank loans, irrigation systems, farm extension services, and the rest of the infrastructure that Western farmers take for granted but that developing-world farmers lost, if they ever had it, during restructuring. State-run grain boards, for example, were regarded as trade distorting because they allowed governments to control local supply and thus price. However, because these boards also served as large buyers whose economies of scale and access to outside markets could partly compensate for the inefficiencies of smaller farmers, they allowed small farmers to participate in the market to a degree that is no longer possible. In Kenya most farmers lack even the ability to store their grain; silos or storage sheds are too expensive to buy, and the local tinsmiths who once built grain bins have been put out of business by the arrival of plastics. Without storage facilities, farmers must sell their crops as soon as they are harvested, which, because everyone else is also selling then, is precisely when prices are lowest.

Transportation is just as problematic. In many developing countries, and African countries in particular, rail lines are limited or nonexistent and roads are in appalling shape. And although some outside companies have invested heavily to improve local infrastructure in Africa — the roads in Kenya's western highlands, where foreign-backed sugar plantations are expanding, are some of the best in Africa — this private funding of infrastructure cannot meet the needs of an entire country. In fact, such private infrastructure has had the negative effect of persuading African governments and aid agencies that they needn't make such investments. As a result, most of Kenya's roads are falling apart; the main road linking the central and western regions with the port in Mombasa floods so routinely during the rainy season that cargoes simply can't be moved. Asks Remington: "How are you supposed to compete in the global market when you can't even reach the port?"

Some of this new thinking has finally begun to catch on among lenders, aid agencies, and philanthropists. The Bill and Melinda Gates Foundation, for one, is putting financial muscle behind efforts to provide safe drinking water and disease prevention and is also working with the Rocke-feller Foundation to create a new generation of seeds genetically tailored for Africa's needs.[46] And some aid agencies have agreed to promote the res-toration of fertilizer subsidies.

However, some observers warn that unless these new initiatives are carefully planned and executed, they will simply repeat old mistakes. For example, although lack of soil fertility is clearly a large problem for many poor farmers, merely resuming subsidized fertilizer sales won't help farm-ers whose soils are completely exhausted. In some cases, much of the or-ganic matter has been depleted due to overplanting, or to erosion — or was never there to begin with. Barrett argues that work must be done to assure that soils can actually benefit from extra fertility, otherwise, buying more nitrogen may be a poor investment. Adds CRS's Remington, "Simply pour-ing on the fertility isn't going to solve anything."

Development experts are also more than a little anxious about the surge of interest in high-value agriculture such as fresh fruit and vegetables, which has transformed the countryside in many developing nations — and re-vived hopes of an export-led recovery. Demand is especially keen for a pro-ducer like Kenya, which has the climate to grow several turns of green beans and baby corn, as well as the large pool of cheap labor to pick, pack, and process those veggies — a key attraction for cost-obsessed retailers. Outside Nairobi, thousands of acres of coffee trees have been dug up and replaced with massive open-air greenhouses filled with everything from peppers and baby corn to fresh flowers. All told, Kenya's horticultural sector is growing about three times as fast as the global food economy, and is generating nearly $200 million a year, the most of any product, making Kenya the second-largest exporter in all of Africa, after South Africa.[47]

And yet, like so much of the rest of the food economy, these trends have had a mixed effect on Kenyan food security. Paul Omanga, the former CRS crop specialist, says the boom in produce is, as in Asia, pulling land away from staple crops, like maize, while bidding up the price of inputs like fertilizer — an outcome "that hasn't been the best for food security." In ad-dition, devoting an increasing share of Kenyan farmland to products dedi-cated entirely for export (few Kenyans eat green beans, baby corn, or car-rots) is a huge risk: if the produce is rejected by European or American buyers, or barred by a trade dispute, or suffers an outbreak of disease, the country will lose millions of dollars.

Even when business is booming, high-value horticulture is depressingly similar to other commodities. As I follow Vegepro's Brunner around the processing plant, it becomes clear that the company faces many of the same pressures confronting processors everywhere — falling prices and rising demands for quality. Although rising jet-fuel prices boosted the company's costs by 70 percent in a single year, European supermarkets, locked in a constant price war for market share, refused to let exporters like Vegepro raise wholesale prices. "We can't pass anything on," Brunner told me. "In fact, they expect us to cut our costs even further, to be more productive, to get our quality up."

Brunner's complaint is now standard in the produce business, which has become the new stick by which retailers bludgeon one another for market share. But the situation is especially tough for the new generation of long-distance produce exporters operating in countries such as Kenya and Brazil, whose business models were already stretched to begin with by the extra-large shipping costs. As fuel costs continue to rise — and retailers continue to drive down costs — produce exporters are taking desperate measures. In some cases, they are battling back with better technologies, methods, and products. Out on the massive Kakuzi Farms, a sixty-four-hundred-acre commercial operation east of Nairobi,[48] manager Mark Simpkin shows me a new South African breed of avocado whose fruit is anywhere from 13 to 50 percent heavier than the traditional variety. "That means an eight-ton-per-acre yield is suddenly up to as much as sixteen tons," says Simpkin.

But such efficiencies can cut costs only so far, and ultimately, as price pressure continues, more and more of real cost reductions are being borne by Kenyans themselves. Beyond the low wages for farm and processing workers — about three dollars a day — big exporters have also devised production strategies that push their own costs and risks onto the local food economy. For example, because exporters are under such pressure from retailers to deliver specific volumes with clockwork regularity, most supplement their own large commercial farming operations by contracting with smaller local farmers, known as outliers. By contracting for slightly more outlier volume than they expect to sell, exporters effectively buy themselves insurance against crop failure or other surprises that would interrupt output and anger retailers. Similarly, exporters often contract with outliers to deliver produce at the tail end of the marketing season, when the big European buyers are switching to other producer countries and when prices for Kenyan produce are softer and far less certain.

In either case, small local farmers effectively function as a buffer for price risk. Yet, unfortunately, such insurance is economical only if exporters can push down the price they pay outliers. In fact, the diligence with which exporters lower their local production costs helps explain why so much of the money invested in Africa does not stay here: for every four dollars invested by foreign companies, three dollars are repatriated back to those companies' home bases.

For these reasons, many development experts think that while high-value horticulture may be good for Kenya as a whole, the benefits accrue mainly to large operators — not to the small farmers most in need of help. Horticulture "hasn't been the engine pulling the rest of Kenya's agriculture," Remington says. "More, it's two separate systems coexisting side by side, with very little interaction between the two." To the extent that Kenyans participate in this sector, Remington says, "they participate as the cheap labor. Three dollars a day is a lot better than no dollars a day, but we want to connect Kenyans to [the market] as *producers*, not farm laborers."

In fact, while some development specialists still tout horticulture as a way to rebuild agriculture in developing countries, for smallholders, the sector is becoming less and less accessible. Steady price pressure from retailers is leading to massive consolidations at nearly every stage of the supply chain. The surviving companies are larger and have greater buying power and demand even lower prices for the produce they buy, which in turn tends to favor the larger, corporate farms with the scale and the capital necessary to raise yields, cut costs, and respond when market conditions change. "The optimists will tell you that small farmers can move into the fresh produce market by organizing, by becoming more efficient," says Remington, "but I think the market moves much too fast: if an opportunity opens up, investors will come in and capture it long before small farmers could possibly organize themselves."

For example, in recent years, governments and retailers have made increasing demands for produce that is sanitary, bug free, and with a minimum of pesticide residues. And while these new requirements may well be justified (in Kenya, regulators have found workers on some small farms sorting green beans on bare ground, a classic vector for *E. coli* contamination), the costs of compliance are simply beyond what many small farmers can handle. In 2002, after European supermarkets began pushing new quality standards, Kenyan exporters terminated the contracts of some sixteen hundred small farmers and replaced them with larger commercial growers.[49] It's no surprise that the share of Kenya's foreign-bound produce grown by smallholders has fallen from nearly half in 1980 to less than a sixth today.[50]

At Vegepro, Brunner insists that his company is committed to working with their seven hundred smallholders. But he readily admits that the relationship is increasingly challenging. "Supermarkets are pushing us to support the smaller farmers, but they don't give us any extra money for doing it," he told me. "It's getting harder and harder to work with seven hundred individual farmers, when you could be working with one farmer who has three hundred hectares." Most expect it to get harder still. The growing expense of jet fuel is raising freight costs and forcing European buyers to rethink their relationship with countries like Kenya, whose shipping costs are already considerably higher than closer suppliers, like Morocco.

Nothing captures this disparity between large and small producers, and between subsistence farming and cash crops, more dramatically than the recent boom in macadamia nuts. Long touted as the perfect high-value crop for small farmers, the nut actually highlights the failings of smallholding. The new varieties of nut are vastly more productive than traditional ones, but they're also more input-intensive. Their thinner shells, which make them consumer-friendly, also make them more vulnerable to insects, which means farmers must spray more often. Nor are bugs the only threat. At harvest time, the nut trees must be heavily guarded to prevent stealing — a huge problem in a region where hunger is constant, says Kakuzi's Simpkin, who managed a macadamia plantation in Malawi. "They wanted to grow the nuts for export, and they have perfect growing conditions and good soils," he says. "But Malawi is also the sixth-poorest nation in the world and the people are hungry, and nuts couldn't survive the pests or the kids. Those kids could strip a tree in an hour."

In a conference room tucked safely behind the bulletproof glass and the blast-proof doors of the new U.S. embassy in Nairobi, Kevin Smith, the U.S. agricultural attaché to Kenya, is arguing the case for free trade and comparative advantage. We'd been talking about how Kenya is importing more food each year as its grain output declines, and I'd asked whether Kenyans should try to reverse that trend by striving again for grain self-sufficiency. But Smith, a handsome Foreign Service veteran with a booming voice and a North Carolina accent, wasn't buying it. "There's a decline in the belief that a country should be self-sufficient," he reminded me. "We don't believe it. We think food security can be best done by trade." Nations should pursue the crops they grow best and leave the rest to others, Smith told me, adding that Kenya's "best" are in indigenous crops such as cashews or macadamias. "The United States doesn't try to grow cashews; we import them. But we *are* efficient producers of maize, and we believe that if we have that competitive advantage, we should supply other countries."

Agricultural attachés are, of course, in the business of talking up their countries' products, and indeed, the walls of Smith's offices are plastered with posters depicting the American superiority in everything from livestock to feed grains. ("Lactation after lactation," reads a circular for dairy cattle, "U.S. cows prove themselves to be prolific producers and brood cows.") But Smith's advocacy also captures one of the fundamental paradoxes in development debate — trade is rarely fair.

To be sure, agricultural trade has been a powerful engine for progress and has been central in raising economic standards around the world — not least by letting efficient producers generate income exporting their surpluses. Yet these benefits have been slower to accrue to those in the greatest need of economic development, flowing instead to wealthier countries. Beyond the natural advantages of their climate and land bases, and the contrived advantages of their expensive farm programs (which allow production of below-cost grain), countries such as the United States and members of the European Union also enjoy the numerous structural advantages, such as technology, research, and access to cheap financing, that come with economic success. American farmers "have been able to exploit economies of scale to drive down costs as low as possible, and then just go out and search for markets to sell into," says Chad Hart, the ISU agricultural economist. "That means the LDCs need to somehow compete with some very mature sectors that have the advantages of technology and better cost structures. They're really out of their league." And unlike emerging powers such as Brazil or China, which are rapidly gaining their own structural advantages, LDCs simply do not have the money to upgrade.

Developing countries do possess exploitable comparative advantages — Kenya *can* produce baby corn and green beans competitively, at least until jet-fuel prices rise much higher. Yet such advantages aren't equally distributed across the country but are generally held by a specific region or economic sector; in this case, large producers and exporters operating in the best growing areas. Small farmers, especially in arid and semiarid areas, are simply unable to participate in the global food economy. And if shielding these minor-league farmers behind heavy tariffs and other protectionist measures hasn't been entirely successful, exposing them to an unbridled market hasn't been an entirely positive experience either. Because Kenya has few barriers to imported maize, its farmers compete with producers in more developed countries, whose low-cost output is simply unbeatable. Nearly half of all Kenyan farmers lack the productivity to grow maize as cheaply as their counterparts in South Africa or in faraway Brazil, who are taking more and more of the Kenyan maize market as Kenya's own production falls behind its massive population growth.

This is a standard pattern in developing countries in the post-restructuring food economy. In Mexico, for example, corn is the traditional basis of Mexico's several million subsistence farmers, who consume roughly half of what they grow and who, until fairly recently, could sell their surpluses in local markets for a price that was kept high by bans on cheap imported corn. After the passage of the North American Free Trade Agreement (NAFTA) in 1995, however, that protectionist ban was phased out, opening Mexico to a rising volume of corn from farmers in Canada and the United States. And although larger Mexican corn farms are nearly as competitive as their U.S. and Canadian counterparts, on nearly 80 percent of Mexico's corn acreage, farmers lack both the good soils and the technology to produce competitive yields. Unable to compete with below-cost imports, nearly two-thirds of Mexico's domestic corn production has either shut down or been reduced since NAFTA's passage.[51]

Certainly, cheap imported food is a boon for consumers in developing countries. Between 1995 and 2005, Mexican consumers enjoyed a 70 percent drop in the price of corn, as well as cheaper meat. Likewise, cheap maize has been a blessing for urbanites in Nairobi, but also for rural households that once grew their own corn but that now, with rising input costs, have shifted from being net producers of grain to net consumers. In fact, because these small farmers can afford to buy grain more cheaply than they can produce it themselves, the arrival of imported grain allows those farmers who want to leave the sector or shift to different crops to keep themselves fed in the process. In this sense, any barriers to cheap imported grain are in effect barriers to economic progress — which is certainly the line offered by trade liberalizers in Washington. As President George W. Bush told Congress early in his first term, "I want America to feed the world. I want our great nation that's a land of great, efficient producers to make sure people don't go hungry. And it starts with having an administration committed to knocking down barriers to trade, and we are."[52]

But the opening of local food systems to free-market forces poses enormous risks as well. By removing grain boards and grain reserves, for example, developing countries have exposed themselves to some of the downsides of a free-market economy. In 2002, officials with the International Monetary Fund advised the government of Malawi to sell off a large portion of its strategic grain reserves in order to pay off an outstanding loan* — just as the country was moving into a massive maize shortage that sent prices skyrocketing and caused several hundred starvation deaths.[53]

To be fair, Malawi's tragedy resulted from a combination of events — not the least being a profoundly corrupt government that reportedly conspired with grain dealers to drive up grain prices. But even when economies are running smoothly and without corruption, the argument that food self-sufficiency is "an anachronism from a bygone era" — and that a developing country is better off importing cheap grain, focusing its energies on more remunerative industry, and simply *buying* its food — is valid only as long as the country can depend on cheap imported grain. And even though cheap grain has been a safe bet for the last half a century, as countries such as the United States generated far more grain and other commodities than the market knew what to do with, these conditions may not always prevail. In late 2006, as dozens of new ethanol plants came online in

---

* IMF officials later claimed that they had instructed the government to sell only a small portion of the reserves, and most observers agree that the IMF was genuinely misled by overly optimistic harvest forecasts. Further, Malawi's government is widely regarded as among the most corrupt in Africa: its National Food Reserve Agency, which manages the grain reserve, is routinely accused of having sold public grain to private traders who then sold it back to consumers once the shortages were apparent and prices soared. But the debacle couldn't have come at a worse time: not only was the harvest much lower than forecasted, but the big donors, suspicious about the state's grain reserves management, delayed sending aid, and once aid was sent, transportation bottlenecks slowed its distribution.

the American Midwest, prices for corn more than doubled, driving up food costs not only in the United States but in Mexico, which is now increasingly reliant on American grain, and where tens of thousands of angry consumers took to the streets to protest the quadrupling of tortilla prices.

Many market analysts contend that the ethanol spike is only temporary because U.S. corn farmers will respond by planting more acres. But as will become clear in coming chapters, even if the ethanol spike is short-lived, many forecasts suggest that longer-term prices will rise — both because population will continue to rise and because the rapid growth in crop yields that characterized the early modern food economy have tapered off and will be much harder to maintain going forward. If such forecasts bear out, then countries that have abandoned the production of primary staples on the promise of cheaper imports may find themselves once again in a position of steep food insecurity — lacking adequate domestic production and increasingly unable to afford imports. Such are the risks of the new food economy.

A few hours after leaving the Mutisyas' farm, our caravan of white Land Cruisers pulls into Malatani, a farm village on the banks of the Athi River. Here too farmers have been deploying new techniques and new technologies. The fields have been cut with deep trenches to curb erosion, and yields have more than doubled since. Inside a crude nursery, villagers are tending hundreds of seedlings — a new variety of sweet potato rich in vitamin A, to help cure local micronutrient deficiencies, as well as mango, passion fruit, and papaya, which will be grown for cash crops and sold in the local markets. Nearby, a handsome young man, his face glistening with sweat, carries a pail of water up from the river and pours it tenderly at the base of a newly planted mango, one of a row of mangoes that the village hopes will be a first step into a more modern food economy. Each plant costs about seven cents and requires three years of more or less constant care before it's able to bear fruit. But if the villagers can keep them alive that long, each piece of fruit will bring about fifteen cents in nearby Embu, and even more if the fragile fruit can be trucked into Nairobi before it rots and before all mangoes from all the other enterprising villages hit the market and drive down prices.

In a sense, the food economy of much of the developing world is waiting to see what happens next. After decades of corruption and ineptitude, a steady progression of natural and medical disasters, and continually shifting aid strategies, the food systems of sub-Saharan Africa and other poor regions are on the knife's edge. In many cases, local officials and aid workers have identified critical factors in food insecurity and have launched targeted solutions that are bringing small but significant successes. The United Nations' World Food Program now feeds some two hundred thousand Kenyan children in the Nairobi slums alone — often in school lunch programs — which means many of those children may stay in school long enough to learn the beginnings of a trade. CRS has had successes showing farmers how to build rain basins, so they don't have to walk miles for drinking water, and setting up seed fairs to supply them with newer, more locally suitable crop varieties.

At the same time, the food economies are so fragile that even a small disruption — a drought, a flood, a border conflict — is sufficient to push the system into collapse. In 2006, just months after my visit, drought destroyed yet another harvest and killed tens of thousands of cattle and other livestock. More than 2.5 million Kenyans survived on handouts of Unimix, a soybean-corn mix distributed by the United States, while uncounted thousands of others abandoned their farms and moved to the slums of Nairobi or other Kenyan cities.

And even during years when no single calamity befalls Kenya, its food economy feels more and more pressure. Unable to fully participate yet in the rapidly evolving global food economy, with its large-scale, capital-intensive operations, but lacking the political power to protect its struggling agriculture sector so its farmers can gain twenty-first-century capacities, Kenya finds itself moving deeper and deeper into a nation-size poverty trap. Population continues to grow, which pushes more people into less and less suitable lands, where their maize and other nontraditional crops are even more prone to failure. As settlements spread and roads deteriorate, relief supplies take even longer to reach those in need. Daily nourishment levels fall, both in overall caloric intake and, crucially, meat; whereas meat consumption has climbed by a third in Latin America and nearly doubled in Asia since the 1970s, it has actually fallen in sub-Saharan Africa,[54] leading to a new surge in stunting. All told, most of the gains in food security made during the 1960s and 1970s are being lost: infant mortality in sub-Saharan Africa and India is soaring, while life expectancy is dropping like a stone. In Kenya, life expectancy rose from forty in

1970 to nearly sixty by the mid-1990s, but it has since fallen back to nearly forty and drops a year with each year.[55]

For decades, the operating assumption of the aid community was that no matter how dysfunctional a country's food system might be, it would eventually respond to the right combination of policies and technologies and join the global food system. Such an outcome may still be possible for a country like Kenya. But we also now understand that food insecurity comes not simply from bad government, fickle aid strategies, and postcolonialism, but also from the pressures of a burgeoning population coming up against natural constraints such as poor soils, scarce water, and a changing climate. In this, it is possible to imagine the crisis in Kenya not as a vestige of our food history but as a vision of our food future.

Here in the village of Malatani, as elsewhere, Kenyans seem to regard such possibilities with the ambivalence of soldiers who have fought too many battles to believe that the war will end anytime soon. A farmer named Jacob Mutua, dapper in a pink shirt, pale trousers, and a Real Madrid cap, tells me all the crops he plans to grow this year: maize, of course, but also beans, pigeon peas, and cotton, for insurance. The rains haven't fallen here either, and I ask him how long he thinks it will be before they fall and he can start planting. Jacob looks at the sky and shrugs. "We don't feel it will come," he tells me. "But we are still planting. What else can we do?" I don't have an answer for him. We stand in awkward silence for a moment and then I ask him what he will do if indeed the rains don't come. He shrugs again, glances at his neighbors, and looks back at me. "If it doesn't rain, then the hunger comes. Then we wait for the hunger."

7

We Are What We Eat

O N FEBRUARY 15, 2004, Dr. Stewart Ritchie, a veterinarian in British Columbia's Fraser Valley, got a troubling call from a local egg farmer. Ritchie, a tall, soft-spoken man with thirty years' experience looking after commercial flocks, had just spent the previous week helping this same farmer handle a low-grade virus in one of his layer barns and now assumed the man was calling to say his birds were well. But this was not the case. The layers in the first barn had indeed recovered, but in a second barn, the nine thousand birds were oddly quiet and off their feed. Worse, where the normal mortality rate was four birds a day, the farmer was now losing that many an hour. "Stew," the farmer said, "something serious is going on here." Ritchie drove back out to the farm, but there was little he could do: within two days, Ritchie recalls, mortality had soared from one hundred birds to two hundreds birds "to too numerous to count." As the alarm went out to health authorities, the question wasn't whether the valley was dealing with an outbreak of avian influenza, but how bad it would get.[1]

The early signs weren't encouraging. Although the virus in the first barn was a mild, or low-pathogenic, strain, the bug had rapidly mutated into a high-path strain by the time it reached the second barn, and it now threatened not only the valley's eighty-million-dollar poultry industry but its human inhabitants as well. For although this virus was identified as an H7N3 subtype, and not the H5N1 that had killed dozens of people in Southeast Asia, an H7 is fully capable of becoming zoonotic — that is, jumping from birds to humans. In 2003, an H7 had infected a hundred people and killed one in the Netherlands. Because zoonotic potential is entirely unpredictable, Canadian health officials opted to depopulate the farmer's eighteen thousand chickens and isolate the farm within a three-mile biocontainment zone.[2]

What happened next has become a case study in the vulnerability of the modern food system. Efforts to euthanize the sick birds went almost comically wrong. Workers pumped the barns full of carbon monoxide, which didn't kill the birds but did blow virus particles out of the barns and into the surrounding air. A second strategy — electrocuting the birds in huge portable stunners — bogged down because the stunners were designed for old, or spent, birds, which are mostly skin and bones, whereas the infected birds were large and fat; their executions generated huge plumes of greasy, virus-laden smoke, feathers, and other poultry particles. All the while, the containment boundary was violated repeatedly by locals, delivery trucks, and media vehicles. Within three weeks, the virus had reached three other chicken farms and eventually spread to forty-two farms, requiring the culling of nineteen million birds.

Then, on March 16, came word that a health worker with flulike symptoms had tested positive for H7N3: the virus had gone zoonotic. By April 4, fifteen people were sick, and public health officials, stunned by the speed with which the virus was spreading, wondered if they were watching the onset of another Netherlands outbreak, or worse. As BC epidemiologist Aleina Tweed reminded me when we spoke later, "the 1919 Spanish flu began as a low-path avian flu."[3]

This is worth keeping in mind. Although the media have been filled with what-if flu scenarios, a reprise of the 1919 outbreak would far exceed anything Hollywood has imagined by several orders of magnitude. Based on the small-scale epidemics in Asia, and taking into account the ease with which illnesses are transmitted in modern cities and the utter inadequacy of our medical system, forecasts for a full-blown worldwide pandemic suggest that a high-path virus could kill as many as seventy million people.[4] There would be trillions of dollars in economic damage as billions of workers stay away from work, and massive political instability as supplies of flu vaccines, antivirals, and even chlorine to purify municipal water systems are quickly exhausted, and governments buckle under the gruesome but essential chore of "corpse management."[5]

In the end, for reasons researchers still don't fully understand, the British Columbia outbreak never went fully zoonotic; although the H7 virus did jump into the human population, it didn't become highly pathogenic to humans nor gain the ability to jump easily from human to human

— the two requisites for a killer pandemic. All human patients recovered, and attention quickly shifted to the massive tasks of disinfecting hundreds of poultry barns, disposing of forty thousand *tons* of chicken carcasses, and enacting new emergency procedures to ensure that future outbreaks would be contained far more quickly. Despite the palpable sense of relief among health officials, however, many of those who had participated in the outbreak had a newfound anxiety about avian influenza. Given the virus's talent for rapid mutation, and given the character of the modern poultry business, with its large flocks of densely concentrated, highly susceptible birds, many experts regard the prospect of a lethal human AI outbreak as a question of "when," not "if." "All it takes is a single mutation and you suddenly have a very pathogenic virus," Victoria Bowes, an avian pathologist with the BC ministry of agriculture and lands, told me. And on the modern chicken farm, says Bowes, "we provide them with barns full of virus incubators."[6]

Of all the concerns about changes in our food economy, none gets our attention so quickly, or brings home the paradox of modern food so sharply, as that of food-borne disease. Despite dramatic advances in food production, preservation, and packaging, food-borne diseases continue to strike some seventy-six million Americans — one in four — each year,[7] and although the vast majority suffer little more than an upset stomach or diarrhea, 325,000 require hospitalization, and of these, anywhere from 5,000 to 9,000[8] die.

Granted, even these stark numbers represent a drastic improvement over a century ago, when food-borne pathogens killed tens of thousands of people a year; in terms of actual risk, I'm much more likely to die in a car crash than from food poisoning. Yet the belief that our food supply is "among the safest in the world," and certainly safer than it used to be, repeated endlessly by food companies and FDA officials, requires more caveats and qualifications by the month. For although the frequency of overall food-borne illness is falling, certain pathogens, such as listeria and salmonella, have become more prevalent, more pathogenic, and more resistant to antibiotics. Even more alarming are the so-called emergent pathogens — microbes that until recently existed only in mild forms or didn't trouble the human food chain at all. Three of the most dangerous bugs circulating today — *Salmonella enteritidis,* campylobacter, and the deadly *Escherichia coli* O157:H7 — weren't significantly present in the food system before 1979, while the most infamous emergent of all, AI H5N1, is barely a decade old.

Why is our food-borne epidemiology changing so dramatically? One factor, clearly, is that we're much better at detecting problems: with high-speed testing, computer modeling, and trace-back systems, seemingly isolated cases can now be recognized as part of a larger outbreak. But as observers such as Pollan and Schlosser have noted, there are also undeniable parallels between these shifts in food-borne disease — the kinds of pathogens circulating, the patterns of outbreaks, and the difficulty in treatment — and the emergence of a food system geared toward high volume, low costs, and rapid, worldwide distribution. The rise of cheap global sourcing, for example, means that formerly isolated pathogens can now move between countries and regions with greater ease. High-speed distribution means contaminated food can be in consumers' homes, and stomachs, long before the contamination is detected. Even improved convenience, especially in the ability to buy food nearly anywhere and at any time, has increased the opportunities for microbial attack. In many cases, the very innovations that let us feed so many so well can also nourish an epidemic — and ensure that its impacts will be devastating.

Perhaps most worrying, although the increasing risk of food-borne disease is a perversely democratic phenomenon (even advanced food economies like the United States and Europe aren't immune), the risk is escalating most rapidly in the developing world. In Latin America, Africa, and, especially, Asia, the scramble to produce and market cheap protein — ironically, a key step in improving national diets — has not only given pathogens everything they need to enter the food system but also has made it nearly impossible for companies and governments to prevent these intrusions from becoming outbreaks and even pandemics. In fact, for all the concern about terrorists poisoning the food system, it now seems more likely that our food system will attack itself. As David Nabarro, director of the United Nations' avian flu program, put it at a 2006 flu conference: "Why do we spend so much defending ourselves from terrorism or natural disasters, but so little defending ourselves from animal disease?" The "major threat to human life," Nabarro argued, wasn't al Qaeda or hurricanes, but "bugs in the animal kingdom."[9]

One of the most striking things in the war on bugs isn't that the battle is so challenging but that we ever thought we could win in the first place. In a world of single-serving packages and refrigerated supply chains, it's easy to forget just how *normal* food contamination is — how susceptible food is to bacterial colonization, and how formidable our microbial rivals are. Beyond their sheer numerical superiority (meat will generally not exhibit an off-odor or a slimy surface until bacterial concentrations exceed ten million per square centimeter),[10] food-borne pathogens are remarkably well-equipped. They are highly potent (a single droplet of blood from a raw chicken contaminated by campylobacter is sufficient to induce fever, cramping, and abdominal pain) and resilient (salmonella can endure freezers and 85-proof Scotch).[11] And while most are eventually killed by high temperatures, in some environments, such as raw milk, the normal bacterial load can be so high (with everything from *E. coli* to staphylococcus) that the heat needed for complete sterilization would destroy the food itself.

Above all, food-borne pathogens, like all microbes, are quintessential adapters: they can modify their genetics, and thus their physical structure and behavior, to defend themselves against antibiotics and to exploit new opportunities. And when it comes to creating new opportunities for microbes to exploit, the system of industrial food production has few rivals. Nearly everything about the way we make food today — from how we use land and manage animals to the ways we process and distribute our finished products — has presented our adaptive and entrepreneurial micro-rivals with a steady progression of opportunities.

Thus, until fairly recently, most efforts at defending the food supply from pathogenic attack have been the regulatory equivalent of shooting in the dark. Most of our early food-safety laws focused not on pathogens (which people knew very little about) but on additives. Not until the 1950s, with the emergence of the fields of microbiology and epidemiology, could governments truly address microbial threats, and it wasn't until the 1960s that Congress beefed up inspection in factories and processing plants. Even then, there was no technology to easily detect bacteria; inspectors relied on a method known derisively as "poke and sniff."

By the late 1970s, however, the rising number of food-illness outbreaks made it clear that such safeguards were inadequate. The sheer volume of perishable foods moving through the supply chain made "poke and sniff" impossible. As well, the new high-speed mechanized processing in meat plants, so critical for the high-volume, low-margin business, significantly raised the risk of contamination: mechanical handling routinely punctured animal intestines, coating carcasses and equipment with bacteria-loaded feces. And once contamination had entered the food supply, the industry's increasingly centralized structure assured that pathogen of the widest distribution. Hamburger, for example, which was once ground locally, is now made in huge batches, using meat trimmings from multiple carcasses purchased from multiple suppliers. Batches are constantly intermingled (one processor may sell a batch to another processor, who may then add it to a larger batch before making patties, sauces, or other items), so that finished products typically contain meat from dozens or even hundreds of animals. One DNA analysis by researchers at Colorado State University found that the average four-ounce burger patty contains tissue from fifty-five separate cows; some patties had tissue from more than a thousand animals.[12]

But it wasn't just the frequency of outbreaks that was changing; the pathogens themselves were evolving in ways that mirrored the evolving food system. In the 1980s, investigators discovered that *Salmonella enteritidis,* the rarest and most toxic of the salmonella coterie, had somehow jumped from its historical reservoir in rabbits and horses to the ovaries of commercial chickens. Here, the pathogen began to seed itself inside the yolks of otherwise normal-looking eggs — a brilliant adaptation that simultaneously exploited the peculiarities of mass production while undermining our own traditional strategy for disease prevention. Because the chickens themselves continued to lay eggs normally, farmers could not easily spot and cull diseased birds or eggs.

Even more troubling were changes in the *E. coli* bacteria. Until the late 1970s, *E. coli* was one of hundreds of relatively harmless bacteria that thrived in the guts of cows and other ruminants and that sometimes jumped to the human food supply by way of fecal contamination but posed little threat to human health. At some point in the late twentieth century, however, the *E. coli* bacteria acquired several new and dangerous traits. First, the bacteria interacted with another germ, shigella, known for its high human toxicity, and from this coupling it acquired the genetic codes to manufacture so-called shiga toxins. These nasty compounds work by shutting down protein synthesis in the victim's intestinal wall; as protein production stops, the wall perforates and the toxins enter the bloodstream, where they begin killing off red blood cells and, in about 5 percent of cases, destroy the kidneys.

Ordinarily, *E. coli*'s newfound toxicity wouldn't have been relevant to humans, because human stomach acid kills the *E. coli* long before it reaches the intestines. But several decades in the evolving food system had produced a second *E. coli* adaptation. Because cattle were increasingly fed corn, and because corn contains far more sugar than does grass or hay, the bovine gut had been gradually pushed to a sweeter, and more acidic, state, and this had forced *E. coli* to become steadily more acid-resistant. Eventually, a new strain emerged — O157:H7 — that could withstand the acid shock of the human stomach and reach the intestine intact, where the shiga toxins could work their deviltry.

Precisely when these adaptations occurred isn't known, but by 1982, when an outbreak of O157:H7 sickened forty-seven McDonald's customers,[13] the bug's new weaponry was both fully formed and more lethal than anything investigators had ever seen. Whereas many food-borne pathogens need massive numbers of individual bacteria to overwhelm the immune system and cause serious illness, O157:H7 needs fewer than fifty bacteria per hamburger patty.[14] And, like salmonella, O157:H7 is hard to detect: cattle intestines lack receptors for the shiga toxins, so cows show no symptoms that would alert ranchers, feedlot workers, or the government meat inspectors who work in every meat-processing plant in the country. In effect, O157:H7 could enter the meat supply chain undetected and unhindered. "We had always believed that if you kept the livestock from getting sick, the food was safe," Lester Crawford, a veterinarian who ran the USDA's meat inspection department in the 1970s and later headed the Food and Drug Administration, told me. "The phrase we used was 'healthy livestock, healthy people.' But here was a case where livestock were thriving and people were getting not just sick, but *violently* sick."

For all the anxiety, it would be years before regulators mounted a real counterattack. Beyond the fact that salmonella and *E. coli* are hard to detect, federal health regulations simply didn't treat such food-borne pathogens as illegal. Whereas toxic additives were classified as adulterants and prohibited, microbes were still officially defined as naturally occurring substances, and thus beyond government authority. This omission was not entirely accidental. The meat industry had no desire to see pathogens reclassified as adulterants. Doing so, companies argued, would force ranchers and slaughterhouses to make changes that would be both costly and unnecessary. In the industry's view, the most effective (and certainly the cheapest) "kill step" for meat-borne pathogens wasn't in the slaughterhouse but in consumer kitchens, through proper handling and cooking.

For decades, industry's arguments (amplified by a steady flow of campaign contributions) dissuaded Congress from taking a tougher stance on pathogens and prevented federal agencies, such as the U.S. Department of Agriculture, ostensibly the overseer of the meat industry, from exercising much oversight. Although USDA researchers and officials were keenly aware of the rising pathogenic tide, the department was hobbled by its traditional role as a promoter of the nation's farm products — a role that had led to a cozy relationship between agency and industry, and especially the meat sector. Many of the USDA's top officials were, and are, drawn from the agricultural industry, and farm and industry groups heavily lobby Congress, the governing body that sets the USDA's authority. As a result, the USDA behaves less like the industry's regulator and more like its marketing arm.* Thus, the thousands of federal and state inspectors stationed in each and every meat-, poultry-, and egg-processing facility were authorized, trained, and equipped to check only for *visible* signs of disease or adulteration.

Such a laissez-faire approach to meat safety changed abruptly in December 1992, when a Seattle doctor noticed a spike in the cases of children with bloody diarrhea. Within two months, the outbreak of *E. coli* O157:H7, eventually traced to Jack in the Box restaurants, had sickened more than six hundred people and killed four, pushing food-borne illness into the public consciousness in a way that even industry lobbyists could not quash. Beyond the young age of most of the victims and the horror of the injuries (many suffered permanent kidney and other organ damage; one nine-year-old girl was comatose for seven weeks and endured three strokes and more than ten thousand seizures),[15] the complicity of government regulators and the company itself was astonishing. Jack in the Box and its parent company, Foodmaker, were found to have routinely undercooked their burgers, despite employees' complaints — because, as one internal memo later revealed, "If patties are cooked longer . . . they tend to become tough."[16] The Jack in the Box outbreak, says Bill Marler, the Seattle attorney who repre-

---

* As late as 1994, the newly appointed head of the USDA's Food Safety and Inspection Service could arrive at his office on his first day to find his speed-dialer programmed for the National Cattlemen's Beef Association and the American Meat Institute; http://www.competitivemarkets.com/news_and_events/newsletters/2003/webocmNewsJuly03.pdf.

sented many of the victims, "was the kind of event that no one could simply sweep under the rug. For the food industry, it was their 9/11."

Within a year, the USDA reclassified *E. coli* as an adulterant (despite heavy industry opposition) and began an overhaul of the entire system of meat safety. Whereas inspectors had traditionally tried to identify and contain contamination *after* it had happened, companies were now required to prevent contamination before it occurred. Under this new approach, known as Hazard Analysis and Critical Control Point, or HACCP, companies must identify the points in their manufacturing process where contamination is most likely and take steps — with new technologies, better procedures, or both — to prevent it. Routine testing, using microscopic analysis, not "poke and sniff" methods, measure how well each critical point is being controlled. Repeated failures to meet federal standards for pathogens can lead the USDA to pull its inspectors, and the all-important "inspected and passed by USDA" stamp, effectively shutting the facility down. These measures (coupled with intensive pressure from retailers, who apparently realized that dying customers were bad for business) had a tremendous impact on meat safety. In addition to the mandatory HACCP programs, most processors have invested heavily in state-of-the-art technologies, among them a method called steam pasteurization, where carcasses are blasted with bacteria-searing steam and then vacuumed of all stray foreign matter.

For all the very real progress, profound gaps remain in the system for protecting both meat and other products. Although HACCP has improved conditions in slaughterhouses, those facilities are simply one link in a much larger supply chain, and most pathogens enter that chain long before they enter the slaughterhouse. Because of high-corn diets and the realities of feedlot confinement (cattle spend their lives walking in and on manure), half of all feedlot cattle harbor the O157:H7 strain of *E. coli*, and that fraction climbs to as high as four-fifths in summer, when warm weather speeds bacterial growth and when, not coincidentally, most outbreaks occur.[17] So many of the steps in the supply chain are vulnerable (one study found the pathogen on nearly one in ten cattle trucks)[18] that many experts speak of *E. coli* as having effectively "colonized" the United States' beef supply.

In fact, steps could be taken to attack this colonization: *E. coli* can be significantly reduced early in the supply chain by feeding cattle more grass or hay and less corn just before shipping them to slaughter.[19] But feedlot owners have few incentives to make such expensive changes. Although slaughterhouses are now required to control *E. coli*, ranchers and feedlots aren't legally obligated to keep pathogens out of the animals they send to those slaughterhouses. And since neither *E. coli* O157:H7 nor salmonella noticeably diminishes the quality of the meat, there is little economic penalty for failing to reduce pathogens. Further, because large meatpacking companies buy from dozens or even hundreds of feedlots (Cargill Meat Solutions, the number two beef packer in the United States, slaughters nearly thirty thousand cattle a week, partly from its own massive feedlots, but also from field purchases by forty-five Cargill buyers throughout North America),[20] tracing a pathogen back to a particular feedlot or animal is almost impossible. Given such realities, one former FDA official told me, "there is no advantage for a producer to try to get rid of *E. coli*. It doesn't hurt the cattle, or the quality of the beef, and it's impossible to trace. You could ask the owner of the largest feedlot in the U.S. and he probably could not articulate what *E. coli* is or what he ought to do about it."

For these reasons, few experts expect that pathogens will be eradicated from the beef herd or kept out of the supply chain, which is why most of the industry's efforts focus on stopping the pathogen at the supply chain's narrowest channel: just as the animal enters the plant. But while this arrangement has helped lower the incidence of *E. coli* in meat, the threat is by no means gone. Studies by the USDA's Animal Research Service, for example, found that even after multiple antimicrobial procedures, including chemical washes, steam washes, and some physical trimming of suspect meat, nearly 2 percent of cattle carcasses tested still had *E. coli* O157:H7[21] — a dramatic improvement from pre-1990s levels, but more than enough, apparently, to pose a continued threat. In October of 2007, the New Jersey–based Topps Meat, the nation's largest maker of frozen beef patties and, one would assume, a company with some food-safety expertise, was forced to shut down after an *E. coli* O157:H7 contamination in its burgers sickened thirty-eight people in eight states and forced the recall of nearly twenty-two million pounds of hamburger. (USDA inspectors would later report that Topps employees had mixed meat from one day's batch into a batch the following day.) The Topps recall was the sixteenth of 2007, a disturbing reversal of the meat industry's improving record on *E. coli*.[22]

There are myriad other concerns that protocols like HACCP don't even touch on, such as the safety of imported foods, which now make up an eighth of our food supply.[23] Although much of the attention in the recent Chinese food scandals has focused on the abysmal conditions of the Chinese food system (refrigeration is not mandatory for fresh produce and most meat products) and the deviousness of Chinese exporters,[24] the fact remains that the United States cannot police its own borders; the FDA currently examines less than 2 percent of the food shipments entering the country,[25] and even when the shipments are examined, each cargo gets an average of thirty seconds' scrutiny.[26]

Domestic producers, meanwhile, maintain practices that raise outbreak risks: livestock and poultry producers still add protein to their animals' diets by feeding them animal products, such as blood or offal, left over from the slaughtering process. Although it is no longer legal to put these byproducts directly into cattle feed[27] — a measure to halt bovine spongiform encephalopathy, or BSE — it is legal to put cow blood and offal into chicken feed. Further, it is legal to collect the litter from chicken houses, prized as a low-cost source of protein and calories (mainly from the feathers and spilled corn), and feed that to cattle — raising the possibility that the BSE prion from a contaminated cow could find its way back into the cattle feed stream after a detour through the digestive tract of a chicken.

Feed isn't the only problematic input. Decades of heavy subtherapeutic antibiotic use by livestock producers, which now accounts for nearly half of all antibiotics used worldwide,[28] has produced numerous new strains of bacteria that are immune to entire classes of antibiotics. Such resistance means that livestock and poultry producers must constantly upgrade to different antibiotics — a demand curve that even some pharmaceutical companies aren't sure how long they can meet. More to the point, it means that some of the most common and inexpensive antibiotics can no longer treat humans infected by these resistant food-borne pathogens. Already, according to the U.S. Centers for Disease Control, many forms of salmonella are immune to medicines like ciprofloxacin, once regarded as the heavy artillery of antibiotics, and medical experts now worry that escalating resistance may be pushing humanity into what Mary Gilchrist at University Hygienic Laboratory in Iowa calls "the 'post antibiotic era' . . . a period where there would be no effective antibiotics available for treating many life-threatening infections in humans. If this proves true, deaths due to infection will once again become a very real threat to substantial numbers of children and young adults as well as the sick and the elderly."[29]

In response, lawmakers in the United States are considering legislation to limit use of growth-promoting antibiotics[30] (as many European governments already do); meat producers, reacting to consumer fears, are using fewer antibiotics or giving consumers the option of buying antibiotic-free meat.[31] But like everything else in the interconnected food system, these changes have had unanticipated effects on food pathogens. Because commercial poultry is still produced in massive, densely packed barns on bacterial-rife litter, the decline in antibiotic use has fostered a rebound in some diseases among commercial flocks, which has actually raised the threat for contamination. Sick birds, it turns out, don't process as cleanly as healthy ones do. They eat less, which simultaneously weakens their internal organs, including their intestines, while increasing the pathogen load inside those intestines.[32] And because sick birds are typically stunted, they're harder for the mechanical eviscerators to handle and are four times more likely to suffer ruptured intestines. (As one researcher dryly put it, stunted chickens "present poorly to mechanized plant equipment that is set for the 'average' bird.") When these guts burst, the bacterial-loaded contents contaminate not only the processing equipment, workers, and inspectors but the meat itself. This may help explain why meat from sick birds was roughly twice as likely to test positive for E. coli and salmonella. It may also explain why more than half of all raw chicken meat is contaminated with Campylobacter jejuni,[33] a germ that causes two million annual human illnesses (some of which progress to the acute neurological disorder Guillain-Barré syndrome) and is increasingly ciprofloxacin resistant.

Perhaps more fundamentally, massive gaps remain in the legal framework for food safety. While E. coli is now officially regulated as an adulterant, other pathogens, like salmonella and listeria, aren't — in no small part because the food industry has persuaded policymakers that the threat these pathogens pose to consumers is lower than that posed by E. coli and merits less onerous regulation. Thus, despite the fact that salmonella sickens well over a million Americans, six hundred of them fatally, every year[34] and is the most common cause of food-borne deaths,[35] the pathogen is still considered naturally occurring under the law.

The results of such a regulatory failure are fairly discouraging. Like E. coli, the Salmonella genus is nearly ubiquitous in the meat supply chain: according to a Colorado State University study, the pathogen is routinely found in soil, water, and feeding areas of feedlots, on a majority of cattle transport trailers, on nearly all cattle at the preslaughter level,[36] and, according to the Food and Inspection Service, is found at the grocery store in all classes of regulated meat products.[37] Researchers and regulators now believe that salmonella, like *E. coli,* can be controlled primarily during processing, but because salmonella isn't regulated as strictly as *E. coli* is, and thus involves fewer penalties, food companies haven't been as rigorous in battling the pathogen.

This explains why the Food Safety and Inspection Service recently found that rates of salmonella, after trending down in the mid-1990s, are moving back upward, especially in chicken products. According to the FSIS, the percentage of whole broilers contaminated with salmonella dropped from 20 percent in 1994 to 13.5 percent in 2004, but then rose to 16.3 percent in 2006. The problem is far more prevalent in ground poultry products: rates of contamination in ground chicken, for example, fell from 44.5 percent in 1994 to 25.5 percent in 2004, but have since risen to 32.4 percent.[38] Given such trends, Dan Engeljohn, the FSIS's deputy assistant administrator for policy, explained in a 2006 speech, that the FSIS "no longer could accept the performance by the industry," and would "take steps immediately to address this particular issue."[39]

What steps regulators can take, however, isn't clear. Beyond the sheer numerical impossibility of the task (poultry inspectors today spend, on average, 1.5 seconds with each bird),[40] regulators lack legal authority to mount an effective antisalmonella effort because, again, Congress refuses to classify salmonella as an adulterant, like *E. coli.* The absurdity of this was dramatically illustrated in 2001, when a U.S. appeals court in Texas ruled that the USDA could not close a Texas meatpacking plant despite repeated discoveries of salmonella contamination. Lawyers for the now defunct Supreme Beef Company (whose customers had included public schools) didn't deny that their client's hamburger was contaminated (indeed, one test showed the bug in 47 percent of the meat). Rather, they persuaded a federal judge that adulteration technically couldn't have occurred because (a) salmonella hadn't been added to the product in the plant but instead had been in beef trimmings purchased from suppliers,[41] and (b) the likelihood of actual harm to consumers was small because the pathogens could in fact be killed by proper cooking. In the wake of *Supreme Beef,* federal efforts to force meat companies to eliminate pathogens other than *E. coli* ground to a halt: the Bush administration decided not to challenge the ruling, which meant that companies failing to control salmonella, although they might face greater government scrutiny, wouldn't be shut down.

But *Supreme Beef* points to larger unresolved questions about the direction of food safety. Despite evidence of continued pathogenic presence in the food supply, much of the official rhetoric and regulation is still driven by the notion that consumers, not producers, are ultimately responsible for killing pathogens. On the surface, this idea seems quite reasonable, given that meat *does* have a kill step and that consumers *can* eliminate pathogens through proper cooking and handling. But there are two flaws in this argument. First, in the modern food economy, consumers prepare fewer and fewer of their own meals in their own kitchens and instead rely on others, such as makers of prepared foods or employees at restaurants, to exercise that responsibility on their behalf. Second, it's not clear that consumers are even aware that precautions need to be taken. Surveys show that consumers still routinely undercook hamburger, in part because, as Marler puts it, "consumers assume their food is safe, because otherwise, why would the government let stores sell it?"

To Marler, *Supreme Beef* and the nation's general reluctance to toughen federal antipathogen laws simply represent the meat industry's continued success at "having it both ways. On the one hand, if bacteria from the gut of a cow or the feces of a cow get into a hamburger, they want to be able to claim that it is 'naturally occurring,' and to continue to shift the burden on the consumer. On the other hand, they want to continue telling the consumer that their product is healthful and good and you should eat a whole bunch of it."

Granted, Marler, whose litigation has cost the food industry nearly a quarter of a billion dollars over the last decade, may not have the most objective perspective here. But his cynicism isn't unique. Discussing *Supreme Beef* in a *Frontline* interview, Dan Glickman, who was USDA secretary when the suit was filed, complained that the Texas court had effectively excused meat producers from any pathogenic responsibility. "What the judge basically said was, 'Well, if there is salmonella in this meat, consumers [should] cook it out. That's all you need to do. . . . Wash your hands, cook it out, and smile."[42]

Just after nine on a frigid January morning in 2007, nearly six months after the disastrous *E. coli* outbreaks in bagged spinach, some two hundred vegetable farmers and shippers, tailed by a small army of reporters and a few trial attorneys, filed into an open-air auditorium on the Monterey County Fairgrounds. The occasion was a hearing before the California Department of Food and Agriculture, and the lone item on the agenda, not surprisingly, was food safety. Monterey County is in the heart of the Salinas Valley, a vast and fertile plain once billed as the nation's salad bowl but now known as the new ground zero for *E. coli*. Since 1995, more than half of all major produce outbreaks of *E. coli* O157:H7, and nearly all lettuce and spinach outbreaks,[43] including those in 2006, were traced to farms or packing sheds in Salinas Valley. The result, as Jim Bogart, president of the Grower-Shipper Association of Central California, told the audience at the hearing, is "a crisis of confidence in our industry that extends throughout the consumer marketplace." To restore that confidence, the produce industry launched a zero-tolerance program to eliminate pathogens through voluntary measures by producers themselves — an idea that most of those in the auditorium seemed to support. "If we are talking about how farmers should farm," declared Tom Nassif, head of the trade group Western Growers, "they know better than any regulators."[44]

Nassif's perspective is not exactly the mainstream view. Although it's clear that vegetable producers are genuinely and desperately interested in halting outbreaks (spinach growers alone suffered $200 million in lost sales in 2006), there is a growing consensus outside the industry that the crisis may already be beyond growers' ability to fix. Despite the media's focus on the feral pigs as the killer vector in the spinach outbreak, for example, researchers have identified dozens of other "nodes of risk" where pathogens could have breached the industry's safety systems. And, as with the meat business, many of those nodes were created by the very technologies and business practices that allow the industry to deliver ever greater volumes of produce year-round at declining costs. "These guys are in supply-chain mode," says Trevor Suslow, a University of California at Davis microbiologist and a leading expert in food-safety investigations. "And when you're in that mode, when your objective is to fill orders, you tend to stretch your system — in terms of capacity and throughput, but also in terms of what you can really handle while paying attention to all the details of quality and safety."

Suslow is intimately familiar with supply-chain mode. Before joining UC-Davis, he worked in the produce business himself, developing specialty vegetables for the then new market in "branded" and "fresh-cut" produce. More to the point, the gist of his critique is something even industry insiders no longer dispute — namely, that the upward curve of fresh produce outbreaks began in the 1980s, just as grocery retailers began pushing suppliers for larger volumes, more varieties, and year-round coverage.

To satisfy this huge new market, growers had to reengineer the produce business: they not only created a massive reciprocal network of farms in different growing zones — from the Salinas Valley to Arizona to Mexico and even to South America — that could generate a continuous volume twelve months out of the year, but they significantly expanded production volume within those zones, and nowhere more dramatically than in the Salinas Valley.

Some of that new volume has come from better farming methods: planting crops more densely in the field, for example, or harvesting more crops per season (each crop of baby spinach is ready in just twenty-six days); and by using mechanization: leafy greens that were once laboriously hand-picked are now carefully mowed by huge harvesters, which in turn allowed farmers to plant ever larger seedbeds. But much of the new output has come the old-fashioned way: by adding new acres. And while some of these new acres would be found elsewhere, like Arizona or Mexico or Chile, there was considerable economic pressure to look for new acres in existing produce-growing regions, such as the Salinas Valley, where growers could exploit an existing processing infrastructure and an existing labor pool.

And it is here, says Suslow, that the problems may have begun. Because the Salinas Valley is already crowded with farms (and under new pressure from encroaching urban development), expansion meant growers had to leave the traditional farming areas on the flat valley floors and move into the surrounding foothills, which, unfortunately, were already occupied — by cattle and dairy operations and by wildlife habitat — with the result that "suddenly, you have produce fields surrounded by cattle and feral pigs." It's probably not a coincidence that the spinach field identified as the source of the E. coli O157:H7 strain found in bags of contaminated Dole spinach was not only next to a Salinas Valley cattle ranch but was itself a cattle pasture only a few years earlier.[45]

In the mainstream press, explanations for this collision of farms, ranches, and wildlife follow what Rob Atwill, another UC-Davis researcher, calls the Typhoid Mary theory: namely, that a specific carrier — a feral pig, say, or a stray cow, or even a bird — picks up the *E. coli* from cow manure on a nearby ranch, and then "runs through the field and craps." And in fact, the fresh produce industry very much hopes this is how the contamination occurred, because it means that the problem can be prevented, or at least minimized, through efforts to keep critters out of the fields — building stronger fences around them, say, or by setting animal traps or offering cash bounties to hunters who shoot feral pigs — all of which the industry has begun doing. (Fresh Express, the nation's top producer of bagged salads, requires its growers not only to fence their fields but to set rodent traps and operate noise-making, pest-startling carbide cannons.)

Unfortunately, there are any number of other ways contamination can occur, most of which are far harder, if not impossible, to control. For example, Atwill has been tracking a theory known as winter migration, which suggests that during the heavy winter rains, the pathogen washes from its natural reservoirs (such as cattle pastures) down the hillsides and into the valley below. "If you look at the terrain," says Atwill, who often spends entire days looking for clues in fields and pastures, "it's clearly demarcated: you have the foothills, and then the plains, and linking them are all these historic streams that have either been left intact or else canalized to run next to the farm fields."

Once a pathogen enters this massive drainage system, it has access to the agriculture complex through thousands of fixed points, such as irrigation wells, canals, and ditches — a massive pathogenic opportunity that modern growing methods are actually making even worse. Because the new large seedbeds are too wide for traditional drip or furrow irrigation methods, Salinas Valley farmers use overhead sprinklers, which means that if *E. coli* infiltrates the irrigation system, farmers are simply spraying pathogens right on the leaves. What's more, once *E. coli* O157:H7 enters the watershed, it can access farm areas not only via *known* entry points, like irrigation systems, but through the nearly infinite number of nonpoint sources, like riverbanks or flood areas. Researchers have found *E. coli* O157:H7 at edges of rivers and streams, hiding in clay, or on rocks in large, slime-covered microbial colonies. During floods, overflowing rivers can carry these pathogens onto roads and into adjacent fields.[46] Because of such nonpoint risks, many big distributors now reject any produce grown in a recently flooded field.

Yet even these measures get at only part of the risk, because although

E. coli O157:H7 is clearly using water systems and flooding to migrate, the bacteria don't need water to move. While many bacteria populations die off once soils dry out, *E. coli* O157:H7 will tolerate weeks and even months of dryness. Such drought tolerance means that *E. coli* O157:H7 can not only survive in dry dirt, but can aerosolize, or become airborne on dust particles, which significantly increases the pathogen's range — and could seriously undermine the industry's efforts to protect the supply chain. Atwill says dust laden with *E. coli* O157:H7 can travel nearly anywhere: it could be carried by wind from a dry streambed, for example; it could be kicked into the air by a cow stomping its own dried manure; it could be thrown from a dirt road by a passing vehicle. "All you need is a delivery truck driving down the dirt road next to a pasture, stirring up the dust and letting that dust cloud move downwind and deposit itself in the middle of a field of leafy greens."

If the winter-migration theory is correct or, more likely, if the cause turns out to be some combination of winter migration and Typhoid Mary or some other vector still to be conjectured, prospects for fixing the problem suddenly become far less certain. Fencing against dust isn't practical, nor is securing every inch of the hundreds of miles of streambeds and canal banks that run through the Salinas watershed.

Worse, once a pathogen enters a field, the industry has no assured means to keep the bug from reaching consumers. Unlike their counterparts in the meat business, fresh produce processors have no kill step — processors cannot subject greens to steam washes nor reasonably expect consumers to boil their salad mixes. And while such risks have always been present in fresh produce, they've actually been magnified by many of the same improvements the industry undertook in order to increase its output and lower its costs.

The huge mowing machines that growers rely on leave cuts on the spinach and salad greens — wounds that not only provide the *E. coli* O157:H7 with a place to attach itself but also give it a source of nutrient-rich leaf juice. And as soon as *E. coli* begins to feed and replicate on the leaf, it is given the perfect opportunity to spread its progeny: the freshly harvested greens are mixed with hundreds of pounds of other greens and thousands of gallons of water in the processing plant. This procedure not only provides *E. coli* with lots of moisture, which it craves, but it also gives the germs virtually unlimited opportunities to touch, and potentially contaminate, other freshly cut leaves.

Processors work hard to interrupt this pathogenic cascade: spinach and other greens are sent down huge water-filled flumes in order to physically dislodge the bacteria, then dunked in chlorinated water, then kept in refrigerated warehouses, trucks, and display cases until the consumer buys it. But these measures are proving insufficient against the newer strains of *E. coli*. Once tucked inside the cut edge of the plant, the pathogen is formidably difficult to shake free. And although chlorine kills *E. coli* as it floats in the water, the chemical can't always kill those pathogens still lodged on plant surfaces. This is especially true with the low chlorine concentrations that processors often use in order not to affect the odor or flavor of the greens, according to Robert Mandrell, a microbial expert with the USDA's extension service in California and one of the top federal experts on *E. coli* and produce. All told, Mandrell said, the washing process is "hit and miss." For that matter, even the industry acknowledges that washing "removes between 90 and 99 percent of all pathogens."[47]

Refrigeration is also proving less and less effective. Older forms of *E. coli* suffered in cold temperatures, but newer strains can tolerate the "cold chain" quite well. "In school, you learned that *E. coli* wouldn't grow in temperatures below 45 degrees," Suslow recalls. "Well, it can clearly grow — it just grows slowly." Slowly, that is, until the pathogen encounters what food microbiologists call "temperature abuse" — a temporary gap in the chain of refrigerated storage or transport between the processor and the consumer. Perhaps the bagged salad sits too long on a loading ramp; maybe a consumer leaves it too long in a warm car or on a kitchen counter. Whatever the case, once temperatures reach an optimal level for the pathogen, the bug starts reproducing rapidly. And because warm temperatures accelerate the process of leaf decay, the decomposing greens begin to emit nitrogen, another key nutrient for *E. coli* — all of which, Mandrell says, helps explain why even a short bout of temperature abuse can yield a tenfold increase in the pathogen load.

Recognizing these vulnerabilities, the industry is responding with better technologies. Earthbound Farms, the largest producer of organic salads and spinach — and the source of one of the 2006 outbreaks — now uses extremely sensitive testing methods to examine produce samples. Greens are tested before they're washed and after they're packed, and they're held in storage until the test results are known. "We've completely reinvented our food safety protocol," a company spokeswoman told me. "No one else is doing anything like this." Yet it is possible that *E. coli* is eluding even these sophisticated methods. Suslow, Atwill, and other researchers now suspect that in some cases, the initial contamination in the farm field occurs at such low bacterial concentrations and is so unevenly distributed that the pathogens may simply not be detected during routine sampling, even with the latest testing technology. Only after conditions have begun to improve for the pathogen — when the leaf is cut, for example, or immersed in water — do its numbers begin to multiply. What all this has led some researchers to conclude, in what Suslow calls "speculation with a reasonable foundation," is that tiny doses of *E. coli,* under the right conditions, could be reaching infectious concentrations just as the product is reaching the consumer.

What's more, for all the industry's efforts to prevent such a scenario from ever occurring, most of the trends now transforming that industry actually make such a scenario more likely. The emphasis on rapid delivery, for example, means produce is usually picked, packed, and on its way to a customer in seventy-two hours or less, and often in less than thirty-six hours. The massive consolidations that came about as the industry cut costs lower and lower meant that smaller, less efficient farms had to give way to larger, integrated operations in which huge factory farms feed their output into enormous centralized packing houses. (Four-fifths of the U.S. lettuce crop now comes from farms of five hundred acres or larger,[48] and a single packer, Fresh Express, handles 40 percent of the prepackaged salad market.[49]) And while this growing consolidation and integration has helped drive down the retail costs of your fresh salad, it has also made the task of keeping pathogens out of that salad considerably harder. "When there's a problem," says Suslow, "it's being disseminated and shared rapidly, among many locations, to a degree you wouldn't have seen even five years ago." In this respect, Suslow says, the modern salad is "a lot like the modern hamburger, but without final opportunity for a 'kill step' at the end."

Given these realities, one begins to understand both why outbreaks keep occurring and why the regulatory response has been so tentative. For all the legislative hearings and the threats of tough new laws, state and federal agencies, especially the Food and Drug Administration, which has authority over fresh produce, have been slow to offer any dramatic proposals for fixing the system. For starters, the FDA lacks the funds and personnel to carry out the proper oversight: budget shortfalls have forced the agency to cut hundreds of field inspectors,[50] and the current scandal over the FDA's failure to stop tainted Chinese imports has further distracted from the agency's domestic duties.

More to the point, regulators recognize that there are few feasible policy fixes. The only sure protection from preharvest pathogens would be to require that all produce be grown in greenhouses — an enormously expensive solution that no politician would even consider. Yet the alternative — mandating changes in traditional farming practices — is no more attractive because, as we've seen, no one is exactly sure how the pathogens are entering the supply chain and, thus, which practices should be encouraged or prohibited. "A lot of people are calling for regulation, but what is it, exactly, that you would have the farmers do?" asks Jim Prevor, a former produce farmer and industry columnist. "If you could assure a farmer that if he built a fence 10 feet high and five feet deep that he would never have another problem, he'd do it tomorrow. But no one knows that. A lot of people ask, 'If new federal regulations had been in place last year, would the outbreaks have happened? And the answer is, yes, because the companies that got caught up in this were already top performers, in the upper ten percent in terms of safety."

Instead federal agencies have been more or less content to let the industry develop its own fixes, in the hopes that market pressure will force the necessary changes, just as buyer pressure helped push meat processors to take on *E. coli*. Many in the U.S. industry point admiringly to Britain, where heavy pressure from grocery retailers, whose agents routinely visit farms, transformed the produce supply chain. In the United States, by contrast, retailers don't enforce their own standards but generally insist only that suppliers be legally licensed — a move that pushes responsibility for food safety onto whatever government agency issues the license while allowing the retailer to claim it is in total compliance with all regulations.

Soon after the big produce disease outbreaks in 2006, retailers and industry groups, in consultation with the FDA, began negotiating voluntary safety standards that, in theory, could bring a system similar to Great Britain's to the United States. Under a new marketing agreement, suppliers failing to meet the guidelines would lose access to big grocery-retail outlets. Yet the flaw in this solution may be its expense. Each improvement in safety, from better testing and stronger fences to rodent traps, sonic cannons, and new irrigation systems — all add costs to what is already a low-margin business. Farmers might be willing to make such investments if they could be assured that spending more money would get them a higher selling price. But safety isn't an added value that consumers feel they should have to pay more for, and so there is no way for retailers to easily pass on the extra costs to consumers, which makes the retailers much less willing to accept higher prices from their suppliers. Add to that the fact that, statistically, most outbreaks are never successfully traced back to a particular farm or even officially confirmed, and, says Marler, one can understand why some growers are willing to gamble and take fewer precautions. Most food companies know "that the chance of them ever getting caught in a food-borne illness outbreak is frankly, quite small," says Marler. "In most cases, the victims of the [outbreak] never figure out what caused it." Statistically, says Marler, "spending more money on food safety isn't necessarily going to help your bottom line."

In the meantime, as we wait for government and industry to come up with new standards, the produce business continues to operate in Suslow's "supply chain mode." More acres are planted. Growing seasons are stretched in order to get one more crop in before winter — even if it means working in the rain and mud. "You have to be running all the time, with something being planted, harvested, or packed, year round," says Suslow. "Under a different scenario, a company might say, 'We need to wait until this field dries out a bit, because we don't want to be taking this heavy equipment into a rain-soaked field, because it makes everything more difficult — makes harvest more difficult, it kicks soil onto the product." But because the current scenario is one in which companies must maximize output, Suslow says, they are more and more likely to press ahead in less than optimal conditions, thereby "making all of the things they have in place to meet their quality and food safety expectations more difficult." Sometimes, Suslow told me, "you just want to say, 'I can't believe you're harvesting right now.'"

On a poultry farm in the busy, smog-choked outskirts of Hefei, a city half-way around the world from the Salinas Valley, the collision between food production and food pathogens enters the epidemiological equivalent of the major leagues. Our host, a compact, neatly dressed man who gives only his surname, Wu, raises a breed of chicken known for its blackish meat that is said to possess medical powers and that Wu sells for a high price to "rich city women worried about circulation problems." It's a tiny operation by

Western standards — twenty-six thousand birds housed in seven crude, Quonsetlike barns — but such a small scale means Wu can cut costs by handling many of the daily chores himself, from feeding and cleaning to the routine injection of antibiotics for "disease control." The small scale may also explain Wu's low-budget approach to avian flu: whereas poultry farmers in North America or Europe would make me put on a protective suit and footwear (or, more likely, wouldn't let me visit at all), Wu simply directs me to walk through a pan of chalky white antibacterial powder at the farm's gated entrance. The fact that such a barrier is unlikely to stop a pathogen like H5N1, which can travel through the air, doesn't appear to be too great a concern to Wu. He is far more focused today on the fact that demand for chicken is rising so fast that he is building an eighth barn to keep up. "Yes," says Wu of the meat business. "It should be a great future."

It is, of course, precisely the ambition of men like Wu in places like Hefei that causes so much anxiety among experts in avian influenza. For all the ready benefits of rapidly expanding meat production in Asia, Africa, and elsewhere in the developing world, the pace and character of this belated livestock revolution are generating a rising tide of pathogens whose potential for havoc makes *E. coli* and salmonella seem almost insignificant. And of these emergent and reemergent diseases, none has the potential of high-pathogenic avian influenza, or HPAI. Beyond the sheer nastiness of the disease itself (60 percent of those infected with the current HPAI variant, H5N1, die excruciating deaths), its expansion into the human food supply appears inexorable, driven not only by the pathogen's great adaptability but also by the fact that its points of attack are on two food items — rice and poultry — that the world, and especially developing countries like China, couldn't live without.

Avian flu has been lurking in the margins of the food industry for decades. Although the virus's native reservoir is wild waterfowl, like other pathogens have, it has entered the human food chain at points where farm activities have pushed into wild habitat. In this case, the intersection occurred in the massive wetlands that Asian farmers use as rice paddies. After each harvest, farmers bring their domestic ducks into the paddies to eat the spilled grain in the water. Unfortunately, these wetlands are also feeding grounds for migratory waterfowl, some of which carry the virus. It is excreted in the water, passed to the domestic ducks, and carried home to the farms. From here, many of the ducks are sold at local wet markets, especially at the Chinese New Year, when hundreds of millions of Chinese families buy and cook ducks.

By itself, the duck vector would pose little threat to humans. Ducks have become fairly resistant to the virus; they can carry and transmit it to one another without becoming gravely sick. In other words, the AI virus is becoming so well adapted to its duck host that it can live, reproduce, and spread sustainably while remaining in a mild, low-pathogenic, nonfatal form. Which means most infected ducks don't develop severe respiratory symptoms or, more to the point, give off, or shed, so much of the virus that nearby humans become infected.

Chickens, however, are another story. Although chickens lack the right genetics to serve as a long-term host for the virus (the virus tends to kill them off too quickly), they can survive long enough to act as a bridge to a new host — ideally, another duck population. And in fact, because chickens are vastly more numerous than ducks in China, and are more heavily traded, chickens now provide the virus with a much larger number of potential bridges, and thus greater chances of reaching a new duck population and a new stable host. "If we look at the virus as a community seeking geographic expansion," says Jan Slingenbergh, a flu expert with the FAO, "then for the Chinese virus to find its way into the ducks of the Red River delta or Mekong Delta or the central plains of Thailand, it helps a lot if, in between those places, you have lots of hugely susceptible chickens. Even if they're not a real host, the spillover and rapid reproduction helps the virus invade a much wider pool of domestic ducks." This dynamic explains both how the virus colonized Asia, by continually cycling between domestic duck and chicken populations, and how it spread out of Asia: once the virus returned to the rice paddies, it passed back to the populations of wild birds, many of which migrated hundreds or even thousands of miles and helped spread the virus to Central Asia, Europe, and Africa.

The chicken vector has another consequence. Once the virus turns into a high-pathogenic version (and researchers are still not sure why this happens), the odds for human infection go up. As the virus switches into a more deadly form, it pushes the chicken's immune system into overdrive, causing massive inflammation in the bird's virus-filled lungs. As the lungs fill with fluid, the gasping bird expels hundreds of millions of virus particles into the air. Here, the virus can easily infect other chickens, setting in motion a firestorm that can burn through a flock of ten thousand or even one hundred thousand in a week. More to the point, the massive release of viruses now increases the chance that at least some of these viruses will posses the genetic mutations necessary to cross the zoonotic barrier and infect a human.

Granted, simply crossing that barrier isn't sufficient to touch off a pandemic. Due to the large genetic differences between humans and birds, most avian flu viruses that do spill into the human population have no staying power: they either remain so low pathogenic that human immune systems quickly kill them (as with the British Columbia outbreak), or, in those rare cases where they turn lethal, the viruses never "learn" how to transmit themselves efficiently from person to person. The infamous H5N1 strain, for example, kills most of its human hosts, but it hasn't gained the ability to pass efficiently or sustainably between humans and so cannot become a pandemic — at least, not in its current form. However, once an avian virus (or any other zoonotic virus, like, say, HIV, which jumped to humans from chimpanzees decades ago) does manage to become both lethal and transmissible in humans, the results are almost always devastating. During the 1919 flu pandemic (caused, researchers now know, by another mutated avian virus, H1N1), the pathogen was not only highly contagious but blindingly quick: a victim might feel perfectly healthy in the morning, report symptoms in the afternoon, and be dead by nightfall.

What isn't clear is whether H5N1, or H7N3, or any of the dozens of other avian strains now circulating around the globe will acquire the genetic mutations needed to be both lethal *and* efficiently transmitted in humans. Researchers know that avian viruses are constantly mutating. Even when happily settled in an ideal host, like a duck, the virus must continually reassort its genetic material. Researchers also know that viruses increase their chances by drawing genetic material from an unusually wide pool: unlike humans and other macroorganisms, which can mate only with related species, some viruses linked to one species (birds, say) can swap genetic material with viruses linked to another (pigs, for example, or humans). This ability not only allows viruses to create novel progeny with impressive new powers (the ability to go zoonotic, for example) but also adds a level of complexity to viral genetic selection that overpowers most forecasting models and makes it very hard for researchers to predict when such progeny might show up.

This complex genetic reassortment is one reason that avian flu experts grow more anxious during our own flu season: if someone sick with human flu encounters a bird with avian flu, the resulting hybrid might be a virus that is not simply lethal, like H5N1, but also has acquired enough human genes to become the next pandemic. Such a replacement is, in effect, what happened in the 1919 pandemic, and again during the smaller, but still quite lethal, flu outbreaks in 1957 and 1968. Studies show that the genetic shift needed to create such a bug needn't be extensive. According to Terrence Tumpey, a CDC expert who has studied samples of the 1919 Spanish flu virus, just two relatively minor changes in the genetic code switched that bird virus into a human pathogen that killed as many as one hundred million people. But because each virus is different, researchers have no way of knowing how many mutations a modern variant like H5N1 might need to turn pandemic, or how long that process might take. As Michael Osterholm, director of the University of Minnesota Center for Infectious Disease Research, told an interviewer in early 2007, the missing mutations "could be very complicated and could never happen, or they could be just one or two mutations that could happen at any time. The trillion-dollar question is, which one of those assumptions is closer to the truth?"

Nor is there any certainty as to how bad the next avian flu pandemic might be. Some predictions detail a pandemic as large as the Spanish flu, which, given today's population densities and global travel and transportation systems, would be staggering in terms of deaths and economic damage: the World Bank estimates seventy million deaths worldwide,[51] and some studies have suggested trillions of dollars in economic damage. Even a mild pandemic, on par with the 1968 flu, would kill 1.4 million people and cost $330 billion.[52] Other experts, like Slingenbergh, theorize that although H5N1 will continue to devastate poultry flocks, the virus may take years or decades to develop the efficient, sustained transmission necessary to cause a human pandemic, and even then, it will probably be considerably smaller than the Spanish flu. The reason: the human flu virus is more stable than it was in 1919, and thus less in need of genetic replenishment from any avian cousin, which means that a new variant probably won't be significantly different — or more pathogenic — than the current flu variant.

On at least three points, however, flu experts are in fairly close agreement. First, given the avian virus's vast capacity for genetic reassortment, there are absolutely no assurances that H5N1 or H7N3 or any of the other avian viruses currently circulating *won't* mutate into something far more dangerous. Second, given the ease with which the virus has moved from Asia to Europe and Africa, there is little doubt that some variant will wind up in the United States, despite the heavy biosecurity of U.S. poultry companies and despite national defensive efforts that now include tracking infected bird flocks with spy satellites. "It is only a matter of time before we discover H5N1 in the Americas," says Michael Leavitt, U.S. secretary of health and human services, who has advised citizens to stock up on tuna and powdered milk.[53] "The migration patterns of birds make its appearance here almost inevitable." Health officials, said Leavitt, "are in a race. We are in a race against a fast-moving virulent virus with the potential to cause a pandemic."[54] Adds Secretary of Agriculture Michael Johanns: "There's no way you can protect the United States by building a big cage around it and preventing wild birds from flying in and out."[55]

Third, to the extent that human activity can quicken that race and increase the risk of an outbreak — for example, by providing the virus ever more opportunities for mixing or bridging — few activities are as worrying as the rapid expansion of meat production, particularly in the developing world. Whereas meat operations in Western countries historically have been tied to grain-producing rural areas, meat production in developing countries has been an urban phenomenon. Because countries like China or Indonesia or Vietnam lacked surplus domestic grain production, they relied heavily on imported feed for their burgeoning pork and poultry operations. (It was in fact the arrival of cheap, imported feed — often by way of restructuring requirements — that made the mass production of meat a possibility in the developing world.) Livestock and poultry farmers tended to build their operations near big cities, close to both port terminals (for grain and soybeans) and wealthy urban customers. The result has been a massive concentration of meat production, especially swine and poultry, in and around the big urban port cities like Shanghai, Bangkok, Hong Kong, Jakarta, and others centers in South and Southeast Asia.

In the context of avian flu, this rapid buildup of urban poultry production has two primary implications. First, the trend toward commercial poultry flocks that are closely confined and genetically uniform makes it far easier for the pathogen to generate and then be shed in massive quantities. Second, it has created millions of opportunities for bird-human mixing. Although much of Asia's poultry production is handled by large commercial operations, which can afford high-tech biosecurity measures, nearly half of poultry production in China, Indonesia, Vietnam, and neighboring countries still occurs in backyard flocks and so-called small-enterprise operations of between thirty to a thousand birds.[56] Unfortunately, these small-scale producers play a leading role in the colonization of avian flu. Moving frequently between farms and local markets, farmers and traders can easily transmit infection from flock to flock and market to market — often well before the birds show any visible signs of infection.

Asian governments, desperate to protect their economies and export earnings, have moved aggressively to contain the virus, which is reckoned to have sapped regional farmers of more than $10 billion since 2003. Thailand, which has lost $1.5 billion in export revenues since the outbreak resurfaced in 2004, has taken the regional lead and has fielded an army of some eight hundred thousand paraveterinarians who go door to door looking for sick birds and people. Vietnam has embarked on programs of vaccination and intensive monitoring for additional outbreaks. In Indonesia, which has suffered the highest number of human cases, authorities have banned backyard flocks in densely populated Jakarta. And through the region, authorities have sought to halt outbreaks by culling more than two hundred million birds since 2003.[57]

But such efforts have not been as concerted or as consistent as might be hoped. Large commercial operations are obligated, both legally and to protect their reputations as exporters, to halt shipments and cull their flocks in the event of an outbreak, but smallholders typically can ill afford such measures and will often avoid culling. In Vietnam, efforts to rein in the traffic in poultry has simply forced the business underground, with farmers now smuggling birds from their backyard flocks into city markets — in some cases, in the panniers of motorcycles.[58] "Vietnam did a bang-up job on vaccination, but now we have to see how well the other initiatives that they should have been putting in place over the last two and a half years will work," says Andre Ziegler, a professor of Veterinary Population Medicine at the University of Minnesota. "And if what we are seeing already in the south of Vietnam — the large number of duck outbreaks — is any indication, we're going to have trouble."

But the gravest fears center on China, where the disease started in 1996 and where efforts to halt its spread are most lacking. Critics say that the federal ministry of agriculture, excoriated for years for letting the flu crisis get out of control, now routinely underreports outbreaks so as to appear to be in charge of the situation. And even when governments do take action, their capabilities are severely limited. Although Chinese researchers have begun aggressively manufacturing new vaccines, some of them quite effective, the government lacks the capacity to monitor the more than ten billion vaccine applications that would be required each year while simultaneously tracking the way the virus is changing in the field. Chinese flu researchers, Slingenbergh says, "are very good scientists, but they are facing a task they cannot hope to handle."

Viral research isn't the only food-safety element that is lagging in China. By conservative estimates, it will take the Asian giant ten years and $100 billion to bring its food system up to safety standards that would be acceptable in the West — in a sense, the cheapness of Chinese food is a reflection of the billions of dollars that have yet to be spent on improved safety.[59] Yet few outsiders can imagine Beijing spending that kind of cash or, as significantly, willingly reining in the rapid growth of its food exports until it can better control food quality and safety. Instead, Beijing appears to have decided that the best defense is a good offense: the Chinese have mounted a major PR campaign to polish China's food's tarnished image[60] and have redoubled lobbying efforts in Washington to block restrictions on Chinese imports.[61] In one case, Chinese officials even charged that the FDA's insistence on testing all Chinese seafood amounted to an unfair restraint of trade.

Ultimately, China's real challenge isn't just its lack of medical or scientific resources or the intransigence of its political leaders, but the sheer momentum of its food economy. Between rising population and the surge in per capita meat consumption, demand for cheap protein will continue to soar over the next several decades, putting additional pressure on a meat industry that is already incriminated in China's avian flu crisis. In this context, recent plans by Beijing to buy more U.S. grain and use it to produce more chicken, some of which is to be exported back to the United States, is especially worrying — and hard to square with an administration that claims to be "in a race against a fast-moving virulent virus with the potential to cause a pandemic." Quite aside from the health risks posed by *any* food product from the woefully dysfunctional Chinese food economy, any deals that encourage China's already overburdened meat system will only exacerbate the H5N1 problem significantly. Although most of the exported meat will come from large-scale "biosecure" poultry operations, these exports will result in a domestic shortfall that, for the foreseeable future, will be filled at least in part by small-scale producers.[62]

Today, according to FAO figures, more than five hundred million Chinese raise poultry, and although that number is gradually falling, smallholders will make up a large share of domestic production for decades. "Rural development is a long, slow process," says Slingenbergh. "We know more or less how many smallholders will be producing poultry in 2030, and we also know that there will still be a lot of virus circulating, and if you combine that with all the commercial and industrial development in a country that will be crowded as hell, the issue becomes very simple: in terms of mitigating the risk of avian flu, it doesn't add up." In this context, says Slingenbergh, the current efforts to supply more U.S. feed to China "so China can take the current absurd level of poultry production and go much higher still" need to be reconsidered. "What we need to be asking is," says Slingenbergh, "is China *really* the place to be producing twelve million tons of poultry a year?"

8

In the Long Run

A T THE VAN DE GRAAF RANCH, a large cattle feedlot that occupies a square mile of hillside just outside Sunnyside, Washington, faith in capitalism is at a low ebb. It's early March, a time when forty-nine-year-old Rod Van de Graaf would normally be loading some of the twenty thousand head of beef cattle that have fattened up here all winter and shipping them off to slaughtering plants. But this year has been anything but normal. A maintenance shutdown at the big Tyson packing plant in nearby Pasco meant Van de Graaf had to hold, and thus feed, his cattle longer than normal. Such a delay would be costly in any circumstances, but this year, it's been catastrophic. With the booming ethanol industry in the Midwest, feed corn that was selling for $1.80 a bushel last year costs $4.57 today — bad news for a man whose herd eats a ton of corn every five minutes. "The grain brokers kept telling us that corn prices couldn't keep rising, that they had to settle back a bit," says Van de Graaf, who is tall and slender with a gray mustache and a ruddy, weathered complexion. "They were all dead wrong." So wrong, in fact, that Van de Graaf's break-even costs are now about ten cents a pound *above* the price buyers are paying, which, on a 1,350-pound steer, means a loss of nearly $140 a head. "You lose $30 or $40 or even $70, that's one thing," says Van de Graaf. "But $140 is a disaster. How are you supposed to make *that* back?"

The question is largely rhetorical, since neither Van de Graaf nor any of the big CAFO operators expect to recover these losses anytime soon. The heavy subsidy that the federal government now pays ethanol refiners has not only encouraged the rapid buildup of new refineries and, thus, boosted demand for corn, but it's also allowing those refiners to bid up the price of corn well above what traditional users are accustomed to paying. That's been a sweet deal for farmers, who haven't seen good prices since 1995. But it's sending painful ripples down a supply chain built on the promise of nonstop cheap grain. Food and beverage companies and especially meat and dairy processors have all seen their profits narrow: every ten-cent increase in corn costs Tyson $17 million in profit.[1] Even consumers, long accustomed to food prices that do nothing but fall, are paying more at the grocery store — prompting a chorus of warnings by industry officials that America's new energy policy is putting American food security at risk. "Energy security is an admirable goal but [the] benefits of ethanol must be weighed against its consequences," admonished the American Meat Institute's Patrick Boyle.[2] Added Tyson CEO Richard Bond, ethanol is forcing America to choose between "corn for feed or corn for fuel."

For all the complaints about biofuels, the hullabaloo over "food versus fuel" is obscuring a larger and much more important question about supply and demand in the modern food economy. Ethanol refineries do indeed consume gobs of corn — nearly 30 percent of the U.S. corn crop, up from just 10 percent in 2002. But rising grain prices are also being driven by other, more familiar factors, including crop failures in Australia and, most notably, the rapid increase in meat consumption in China and the rest of the developing world.

Further, while there are many reasons to be skeptical about the U.S. ethanol program (for example, the fact that corn is among the least efficient crops to make into fuel), the spectacle of the meat industry criticizing anyone else for misappropriating corn or endangering food security is more than a little ironic. In terms of scale, ethanol is dwarfed by the livestock industry, which vacuums up more feed than all other end users combined (more than a third of the two billion tons of grain produced worldwide in 2006 was fed to animals)[3] and will continue to be the dominant end user for decades to come. For although the corn-based ethanol craze may fade (as corn becomes too expensive, for example, or, preferably, as we get better at making biofuels from nonfood crops), the livestock revolution can anticipate no such endgame. Because most of the three billion or so extra people anticipated by midcentury will be born in the developing world, and because the diets of most developing countries are still catching up with richer Western diets, demand for grain-intensive foods like meat will actually grow far more swiftly than the population does. While population will probably peak by 2070 at 9.5 billion (from the current 6.5 billion), global meat demand by that point will be somewhere between twice and three times its current level.[4]

Such projections are all the more disconcerting because at this juncture, we have very little idea as to where we're going to get all the grain needed to make all that meat. Global grain reserves today are at their lowest levels in thirty years, despite near record harvests,[5] and although farmers everywhere are planting new acres as fast as they can (American farmers, for example, haven't seeded this much corn since World War II), forecasters still expect that rising demand for grain — from ethanol refineries, but also from the global livestock sector[6] — will push global grain prices up by as much as 50 percent above historic averages through 2017.[7] And if that's a disappointing outlook for the Tysons and Coca-Colas of the world, it could be a disaster for poorer countries, many of which increasingly rely on imported grain — and are already struggling to maintain food security. In Mexico last year, the spike in corn prices brought tens of thousands of angry consumers into the streets — a scene that Lester Brown, president of the Earth Policy Institute, worries will become common as rising global food prices leave food-insecure governments "with a much thinner margin between stability and instability."

Indeed, while many mainstream food agencies continue to forecast a future of expanding grain supplies, higher yields, declining food insecurity, and steadily rising per capita intake of meat,[8] such rosy predictions are now routinely contradicted by reports from those on the front lines of the food economy. In July of 2007, the United Nations' World Food Program announced that soaring grain prices meant it would be able to feed considerably fewer than the ninety million hungry people it had assisted each year since 2002.[9] After decades of all but drowning in excess food, producers and consumers alike are waking up to the possibility of a food economy that is once again defined by scarcity. "We face the tightest agriculture markets in decades and, in some cases, on record," the UN's Josette Sheeran told the *Financial Times* in mid-2007. "We are no longer in a surplus world."[10]

Of course, the world has heard Malthusian forecasts like these before, and each time it has averted catastrophe through a combination of market forces and technological advances. But the world has also begun to confront the fact that its food economy will be harder to fix this time. A system so focused on cost reduction and rising volume that it makes a billion of us fat, lets another billion go hungry, and all but invites food-borne pathogens to become global epidemics is now running into other problems as well. Arable land is growing scarcer. Inputs like pesticides and synthetic nitrogen fertilizers are increasingly expensive. Soil degradation and erosion from hyperintensive farming is costing millions of acres of farmland a year. Water supplies are being rapidly depleted in parts of the world, even as the rising price of petroleum — the lifeblood of industrial agriculture — is calling into question the entire agribusiness model. More recently, forecasters have begun to tabulate just how damaging even a minor change in global climate will be to a food system built on the assumption of stable temperatures and consistent rainfall. In this context, the question for many resource specialists isn't simply whether we'll be able to feed 9.5 billion people by 2070, but how long we can continue to meet the demands of the 6.5 billion alive today.

Optimists like Norman Borlaug, godfather of the Green Revolution, insist that the answers lie in transgenic crops and other new farm technologies; in fact, most of the upbeat forecasts for future food supplies assume a rather massive degree of technological innovation. But even optimists acknowledge that if these predicted breakthroughs fail to materialize, or don't come soon enough, the entire food economy could gradually slip back into a state of demographic disequilibrium where productivity is once again in a race with population growth and where the most heavily populated countries compete for access to large surpluses of grain and soybeans, just as the big industrialized nations now compete for oil. On this level, the ethanol debate is better understood as the prologue to a much larger argument — about the sustainability of current food systems and practices and, more specifically, about whether the dramatic improvements in diet — and the spectacular rise in meat consumption in particular — that occurred during the last century can be maintained during the next.

At the heart of any discussion about the sustainability of the modern food system is what might be called the protein paradox. Since the end of World War II, our capacity to produce protein, generally in the form of meat and dairy products, has outpaced even our capacity to produce people. With cheaper grain, better breeding, and larger and more efficient livestock operations, the food industry can make a four-ounce serving of meat almost as inexpensively as it does a can of soda or a loaf of bread — a capability that has improved the lives of billions of consumers around the world but has also come at substantial cost.

Some of these costs are quite familiar: America's per capita meat supply is around nine ounces a day, nearly four times the federal recommendation intake for protein[11] and a likely factor in U.S. obesity rates. But some of the costs are more subtle. Meat's increasing cheapness has not only allowed more people to eat it more often but has also effectively embedded meat deeply in the food economy and given us fewer options for alternatives. Just as a century of declining gasoline prices encouraged the rapid expansion of a gasoline-powered fleet, and thus made it even harder to shift to alternative fuels or technologies, the rationalization of meat production has helped the world achieve a level of meat consumption, and has helped foster expectations *about* meat consumption, that will be increasingly challenging to maintain and very hard to change. And yet, nearly every credible forecast shows that if we're to have any chance of meeting future food demand in a sustainable fashion, lowering our meat consumption will be absolutely essential.

To understand this dilemma, consider the cows that move through Van de Graaf's massive feedlots. By raising these animals in pens on rations computer-designed to produce rapid weight gain and marbling, Van de Graaf can turn a six-month-old five-hundred-pound feeder calf into a 1,350-pound, slaughter-ready steer in about four months. By comparison, a steer raised entirely on grass (and all cattle were once raised that way) takes two years to reach a slaughter weight of barely 1,100 pounds.[12] The fact that most U.S. beef cattle spend most of their lives in CAFOs, eating below-cost corn, is one of the main reasons that the U.S. beef industry has been able to more than double its beef herd since 1950 while reducing pasture and grazing acreage by more than a fifth.[13] It is also why the price of beef has fallen by nearly half since 1960,[14] and why beef consumption, despite the meat's declining popularity in the United States, is rising dramatically worldwide (even in poorer countries, like China, where beef eating was almost unknown before 1980), and is expected to climb by 25 percent over the next fifteen years.[15]

The problem is that, for all the marvelous efficiency of the CAFO, these operations still can't compensate for the *inefficiency* of the cows themselves; even in the best-run feedlot, the modern cow needs at least seven pounds of feed to put on a pound of live weight — nearly twice that of pigs, and more than triple that of chickens.[16] Worse, because so much more of a cow's weight is inedible — 60 percent is bone, organ, and hide — than is the case with smaller livestock, beef's true conversion rate is actually far lower: it takes a full *twenty* pounds of grain to make a single pound of beef (compared to 4.5 and 7.3 for chicken and pigs, respectively).[17] What this means is that every additional ton of beef consumed represents another *twenty* tons of feed added to world demand — a rather impressive bit of caloric leverage that helps explain why 90 percent of the grain that Americans consume is eaten in the form of meat or dairy.

The Hummerlike inefficiency of the beef cow never really mattered when corn and other feed grains were cheap. But in a world where grain prices are rising and where the cheapness of grain is increasingly understood to be the artificial result of subsidies and other off-the-books factors, beef has gradually lost ground to the more efficient pork and poultry. Already, these "white" meats account for most of the growth in meat demand in developing countries. For decades, in fact, this steady global shift away from red meat and toward white was a source of great optimism because it offered a vision of a hyperefficient global meat sector capable of generating more and more protein for less and less feed.[18] But such optimism is fading: although pork and poultry are still gaining global market share, their better feed efficiencies are being steadily offset by two countervailing trends. First, the growth in the sheer number of pounds of white meat being consumed is already overwhelming the gains in per-pound efficiency. Second, the once phenomenal improvements in feed efficiency have flattened out as breeders push up against biologic limits. Consider the ultraefficient chicken. Although poultry's per-pound conversion rate improved from 4.5 pounds in 1925 to 2 pounds in 1985, it has since gained only marginally — to 1.95 pounds — despite significant advances in breeding technologies. "We might see 1.6 pounds or 1.5," poultry analyst Paul Aho told me. "But on feed conversion rates, the low-hanging fruit has already been picked."* Improvements in pork efficiency have also flattened.

As gains in feed efficiencies taper off, and as global meat consumption continues to rise, a great many experts are trying to understand where the new global average for per capita meat consumption will settle and what

---

* Beef-industry officials argue that cattle, too, can be bred for better feed efficiency. However, because cattle take so long to reach sexual maturity, breeders would need decades to achieve meaningful gains. As one specialist told me, "Even if you were doing a constant selection, turning your population over as fast as possible, you'd need thirty years to get a significant improvement in feed efficiency, at a bare minimum. And even then, you'd be lucky if you got it down to five to one."

that number will mean in the long term. And the conclusion is fairly stark: under any model for a future food system that is both sustainable and equitable, the meat-rich diets of the West, and especially of the United States, simply don't work on a global scale. If the American level of meat consumption — about 217 pounds per person per year[19] — were suddenly replicated worldwide, our total global grain harvest could support just 2.6 billion people[20] — or less than 40 percent of the existing population, and barely a quarter of the ten billion expected by 2070.

Of course, no one, not even the most optimistic beef-industry lobbyist, expects an American level of meat consumption to become the global norm. But even if we use a more modest Western level of meat consumption, such as that of Italy, where per capita meat consumption is about 80 percent that of the United States',[21] world grain supplies would still be adequate for just five billion people. In fact, according to the Earth Policy Institute's Brown, it's only when the world adopts an Indian level of meat consumption — that is, around *twelve* pounds of meat a year (which is possible only because Indians consume 90 percent of their grain directly, in the form of bread and other products — and tens of millions of Indians still don't get enough calories of any kind) — that current global grain supplies would be adequate to feed 9.5 billion people.

Granted, Brown's country comparisons don't take into account the mountains of additional grain that optimists believe the world is still capable of generating (although, as we'll see, such projections are mighty uncertain). But the comparisons are useful in emphasizing the meat gap between the developed and developing worlds and, thus, the likely acceleration in overall meat demand as poorer countries catch up with richer ones. Between 1960 and 2002, per capita meat consumption in developing countries more than doubled, from a tiny 22 pounds to 56 pounds, and is on track to hit 74 pounds by 2030.[22] That's still less than the gargantuan 220 pounds that each consumer in industrialized countries is expected to be eating by 2030.[23] But owing to the massive size of the developing world's population and its rapid rate of growth, even a relatively small gain in per capita meat consumption there translates into a staggering increase in total global demand. Thus, while the worldwide average for per capita meat consumption is expected to rise just 25 percent by 2030 (to about 99 pounds), total meat demand will climb by more than *70* percent — from 229 million tons to 376 million tons.[24] And by midcentury, global meat demand will be at 465 million tons,[25] or well over *double* the current level, which in turn will require us to produce an additional one *billion* tons[26] of feed grains — an increase that will challenge much of what we know not simply about science and farming but about population, food security, and, ultimately, progress. "The world has had many years' experience feeding an additional 70 million or more people every year," notes Brown, who used to work as a foreign demand forecaster at the USDA. But "it has no experience with some 5 billion people wanting to move up the food chain at the same time."[27]

A few miles east of Rod Van de Graaf's feedlot, you can almost see a meat-happy world moving up the food chain. Because corn prices are so high, many of the farmers here who normally grow hay are instead plowing their fields into corn — a living demonstration of the market response that, according to optimists like former U.S. agriculture secretary Mike Johanns, will soon rebalance overheated grain markets and put the global food system back on the pathway to smooth expansion.

But for Van de Graaf and other market participants, the story has a slightly different spin. The shift from hay to corn will indeed mean more corn, but it has also driven up the costs of hay, which Van de Graaf feeds his younger animals, to heights not seen since the late 1980s. And hay isn't the only crop getting short shrift. In an output-obsessed country like the United States, where most farmland is already in use, the great corn expansion is occurring almost entirely at the expense of other crops. Across the Midwest, farmers who might have planted soybeans have instead gone to corn, which is making soybeans so expensive that they are, in turn, "bidding away" acres from sorghum, and wheat, and peanuts, and even cotton, and so on in an agronomic chain reaction that is driving up the costs of nearly anything grown in the ground and pointedly illustrating the enormous challenge facing a food economy as it struggles to meet rising demand.

After centuries of rapid expansion and population growth, the old-fashioned method of increasing output — plowing up new acres — is becoming less viable by the year. Most of the world's obviously arable cropland is under cultivation, and much of what remains, mainly in places like sub-Saharan Africa and South America, is occupied by forest or grassland. In heavily populated regions such as Africa, South Asia, and even North America, the existing agricultural base is actually shrinking as urban and industrial areas expand; in California's Central Valley, source of a quarter of the nation's produce, fifteen thousand acres of farmland disappear each year to residential and commercial development,[28] and nationally, every new American baby or immigrant translates into another 1.7 acres of lost farmland.[29] For this reason, most forecasts suggest that of the one billion tons of extra grain needed by 2030, four-fifths must come not by planting extra acres, but from intensification — that is, getting more food from existing acres.[30] By 2030, average yields must climb from the current rate of 1.1 tons of grain per acre to 1.5 tons, according to the FAO.

And here is the problem: while the FAO and other optimists *assume* such increases will occur and, indeed, factor them into their forecasts, neither the FAO nor anyone else can say precisely *how* such yield increases will come about — or even whether these increases are truly possible. After decades of steadily more tons per acre, yields are going up by just 1.3 percent a year, barely half the rate of thirty years ago[31] and much more slowly than demand is growing.[32] And within certain crops, growth rates are even worse. Rice yields are growing only a third as fast as they did in the 1970s and 1980s.[33] Even in the United States, corn yields are rising by just 2 percent a year — a rate that won't come close to meeting future needs. And while no one expected the explosive yield growth of the Green Revolution to continue indefinitely, the rate of decline has been surprisingly swift and has created what the FAO acknowledges is a yield gap between forecasts for future supplies and forecasts for future demand.

Why are yields slowing? Conventional explanations tend to see the problem in terms of inadequate inputs such as good seeds, fertilizers, irrigation, and other material or infrastructural deficiencies, especially in the developing world. Such theories suggest that if input levels in poor countries can be brought up to those in Western nations — for example, if we can restore fertilizer subsidies in countries like Kenya — the world should be able to close, or at least significantly narrow, the global yield gap.* In this view, solving the food-security challenge is mainly a case of steadily narrowing the yield gap (by restoring inputs, developing better-yielding crops, and carefully husbanding our remaining land) long enough to get human-

---

* According to an FAO study, to meet future food needs, the world as a whole will need to increase its fertilizer use by 1 percent a year, and by nearly 3 percent a year in nutrient-deficient regions like sub-Saharan Africa; see http://www.fao.org/docrep/005/Y4252E/y4252e06b .htm#TopOfPage.

ity past both the population peak and the "meat peak" — at which point food demand will plateau and the food crisis will, in a sense, be concluded.

There are, however, other explanations for the yield gap, the implications of which are far less rosy. Plant breeders, for example, are running into the law of diminishing returns in the plants themselves: past gains have come in part by increasing a plant's ratio of edible matter (seeds) to inedible matter (stocks and leaves) — and at some point, this ratio simply can't be further expanded without hurting the plant's ability to grow.[34]

Slowing yields may also be related to the physical limits of the land. By one FAO estimate, the soil of nearly a third of all arable land is so acid that it can't support high-yielding crops.[35] Similarly, half a billion people now live (and farm) on lands so hilly and erosion prone that further intensification won't be possible without a considerable cost,[36] according to a World Bank study, and, globally, erosion is so severe that in the words of one expert, by 2050, the world may be trying to feed "twice as many people with half as much topsoil."[37] (In China alone, erosion and contamination have reduced annual grain production by six million tons, which, as Vaclav Smil points out in *Feeding the World,* is essentially the margin of additional supply the country must produce each year simply to keep up with rising demand.) And as we saw in previous chapters, natural soil quality is so low in many parts of the world that even adding more fertilizers won't raise crop yields sufficiently to pay for the extra fertilizer, much less boost net food output.[38]

To be sure, in much of the world, existing farm soils are healthy enough to respond to additional inputs. In sub-Saharan Africa especially, experts say that increasing fertilizer applications by 50 percent over the next fifteen years would bring huge increases in yield. But here we meet another problem: fertilizer costs are now climbing so fast that such application increases can no longer be assumed. Since the corn-ethanol boom, demand for nitrogen (and corn is the most nitrogen hungry of all commercial crops) has more than doubled the price for ammonia fertilizer, to $500 a ton.[39] Phosphate, another key nutrient, is also becoming more costly despite the fact that in the United States, the world's biggest producer and exporter of phosphate,[40] factories are running at full capacity.

Fertilizer-industry officials contend that these shortages are temporary and that today's high prices will soon encourage more production and bring more supply onto the market. But given the emerging constraints on the raw materials from which fertilizers are made, this scenario is less than assured. Synthetic nitrogen is made from natural gas, the price of which has more than tripled since 2002,[41] and which, like its geochemical sibling oil, is expected to get even pricier in the not-so-distant future. Supplies of natural gas are already tight; even in the United States, one of the largest natural gas producers in the world, drilling companies are finding fewer and fewer new gas fields each year. Demand, meanwhile, is skyrocketing, in large part because natural gas is the preferred fuel for generating electricity in power plants. (The thirty-three thousand cubic feet of natural gas needed to make a ton of nitrogen fertilizer could be used instead to generate 9,671 kilowatt hours of electricity[42] — enough to run the average U.S. home for ten and a half months.[43]) In other words, fertilizer companies (and thus farmers) are now directly competing for natural gas with utility companies and their customers — and losing. Natural gas prices have already climbed so high that many U.S. fertilizer companies have off-shored their factories to countries with cheaper natural gas — with the result that the United States has lost about a third of its nitrogen production capacity in the past decade and must now import more than half its nitrogen fertilizer.[44]

Given that the current model of industrial food production is simply impossible without ready access to synthetic nitrogen (40 percent of the population survives on the extra calories produced from synthetic nitrogen, a fraction that will rise to 60 percent[45] by 2050), this increasing dependence on imported fertilizer is worrisome not simply for developing countries, but for *any* country with large farm sectors. China, for example, expects to be using 27 percent more nitrogen by 2011, much of it from foreign sources,[46] while the United States, which already uses one-eighth of the world's synthetic nitrogen, will see its imports rise as well.

In a perfectly peaceful world, it shouldn't matter where a country gets its nitrogen; indeed, if a country's comparative advantage is large reserves of natural gas, then everyone else should gladly buy fertilizer from them. But as we saw in chapter 5, this world is not perfect, nor are its resources distributed in the most democratic of patterns. The countries with the largest untapped surpluses of natural gas, and thus the largest potential as power brokers in a future global nitrogen market, are Iran and Russia — two states not known for their stability, global philanthropy, or concern for the welfare of U.S. consumers or farmers. Russians and Iranians have talked openly of forming a natural gas cartel to control prices, much as OPEC now does with oil — a worrying prospect for any country dependent on fertilizer imports for food security. In fact, considering that the United States imports more than half its nitrogen, this country's food security is now nearly as uncertain as its national energy security. Such developments are not lost on ethanol critics, who point out that for all that ethanol has been touted as a means to cut U.S. dependence on "foreign oil," ethanol will actually *increase* national dependence on nitrogen from many of those same foreigners.

The irony of the looming fertilizer shortage is that when it does materialize, it will occur simultaneously with a fertilizer *surplus* — not in world markets, where it could be used, but in the soils, where it poses a substantial risk to nearly everything it touches. Although high-yield plants need ever greater volumes of fertilizer, much of the fertilizers that farmers apply to their fields never reaches the plants but instead accumulates in the soils, with serious consequences for environmental and public health.

Some of the problem stems from simple overuse. In Asia, heavy fertilizer subsidies and lack of expertise have led to routine overfertilization. Even in advanced agriculture economies such as Europe and the United States, excessive fertilization is common — in part because overfertilization has long served as a form of crop insurance: farmers would rather invest in a little extra nitrogen than risk a subpar harvest. Making matters worse, at the same time as farmers have been adding too much fertilizer, new farming methods have accelerated the tendency for those excess fertilizers to leave the soil. Because farmers now rely mainly on synthetic fertilizers, they're less likely to bother with the traditional cover crops that were once sown between cash crops — with the result that many fields are now left bare between autumn harvest and spring planting. Unfortunately, when uncovered soils are exposed to weather, soil nitrogen (both synthetic and naturally fixed) quickly converts into nitrate, a highly mobile chemical compound that is easily leached from the soils by rain. By conservative estimates, of the 230 pounds of synthetic nitrogen applied to the typical acre of U.S. corn, as much as 50 pounds will leave the soils and enter the surrounding environment.[47] Throw in the hundreds of millions of tons of nitrogen-rich manure that accumulates in our huge livestock feedlots and often leaks into surrounding water sources, and our modern agricultural system becomes an unrivaled source of liberated nitrogen whose effects are pervasive and devastating.

In rivers and lakes, this wayward nitrogen essentially fertilizes whatever is in its path, such as milfoil, which clogs waterways, and various algae,
which coat rocks, beaches, and piers in green muck. Worse, when these organisms die, they set off a chain reaction known as eutrophication, which
sucks the oxygen from ponds and lakes and even coastal waters and creates
massive fish-killing dead zones. According to a 2003 report by the United
Nations Environment Program, the number of dead zones worldwide is
nearly one hundred and fifty — more than double the number in 1990.
Excess nitrogen is also linked to a number of human health risks,
among them miscarriages and cancer — which is why the city of Des
Moines, Iowa, now spends $300,000 a year filtering agricultural nitrates
from the local water supply,[48] and why federal and state environmental
agencies now regard farming as one of the biggest polluters of our water
systems.

Nitrogen's most lasting impact, however, isn't in the water. By binding
with oxygen, the migrating nitrogen becomes nitrous oxide, a major pol-
lutant that causes smog, depletes the ozone layer, and is a greenhouse gas
three hundred times more potent than carbon dioxide.[49] As much as 70 percent of all human-generated nitrous oxide comes from the farming sector.[50]
All told, since the beginning of industrial agriculture, the volume of excess
nitrogen circulating between soil, water, and air — and wreaking havoc in
each of these zones — has more than doubled and is sure to keep rising as
the farmers of the world try to grow more grain and raise more meat.

The noxious effect of excess fertilizers is only one of industrial agriculture's
less desirable yields. The synthetic weed killers and bug sprays, collectively
known as pesticides, that most high-yield crops cannot survive without
also routinely follow fertilizers out of the soils and into water supplies,
where they cause their own particular kinds of mischief. Atrazine, one of
the most widely used herbicides in the United States,[51] is linked to heart
and lung congestion, muscle spasms, degeneration of the retina, and cancer
(not to mention the wholesale extinction of amphibians), and yet despite
long efforts by federal and state regulators, it remains the second most frequently detected herbicide in drinking-water wells.[52]

Riskier still are modern insecticides and fungicides. Many are based
on complex molecules known as organophosphates, which disrupt the
pests' central nervous systems by causing nerve cells to fire nonstop. Sold
under a variety of names — among them Malathion, Supracide, Monitor,

Cygon, and Sniper — organophosphates and a related compound class, carbamates, account for roughly one-third of global insecticide sales and are heavily relied on by growers of alfalfa, almonds, carrots, grapes, apples, strawberries, peaches, walnuts, and, above all, corn and cotton, which together account for half of all organophosphate use.[53]

The problem is that organophosphates' neural-disruptive abilities aren't confined to pests. The chemicals, which pass easily through human skin, eyes, and mucous membranes, can cause cardiac arrhythmia; intense contractions of the stomach, intestines, and bladder; seizures; mental impairment; depressed heart and lung function; loss of muscle coordination; and coma; such side effects explain why the German military tested organophosphates as a human nerve agent in the 1920s and why, nearly a century later, farm work is still regarded as among the most dangerous occupations.[54] In 2002, U.S. health agencies reported some 97,000 cases of organophosphate exposure, more than half of them in children under six.[55] Further, while organophosphates and carbamates are gradually being phased out in the United States, they're heavily used in many other countries, including some countries that export to the United States.

Beyond their immediate health impacts, however, pesticides have other more insidious effects on the food system — effects that illustrate some larger, if less visible, challenges to the sustainability of modern food production. For example, insecticides kill not only organisms that we don't like but also many that we do: the beneficial insects, such as bees and others that naturally prey on pests, as well as the nearly countless microbes that live in the soils and help circulate nutrients and water, and whose demise typically accelerates a decline in soil productivity.

Pesticides also tend to require fairly constant upgrades and replacements — sometimes because they are found to be excessively dangerous, but more often because these high-powered chemicals simply stop working. Just as food-borne pathogens can develop resistance to our antibiotics, so too insects, fungi, and weeds become tolerant of our strongest chemicals. Such resistance, which has been noted since the 1940s, requires chemical companies to develop (and farmers to buy) generation after generation of new farm chemistry. What emerges is yet another variant of the technology treadmill — one that enriches input companies but degrades soil health and, thus, harms the farm productivity that the chemicals were supposed to protect in the first place.

Chemical companies contend that with proper application, farmers can avoid pesticide resistance. But Joe Lewis, an insect expert formerly with the U.S. Agricultural Research Service, argues that resistance is unavoidable: all natural systems, whether human, animal, or plant, have a built-in tendency to adapt to outside interventions. No matter how powerful or targeted the pesticide, Lewis says, its very lethalness will eventually provoke natural "countermoves" by weeds or insects that "'neutralize' their effectiveness"[56] — leaving farmers either to suffer greater losses to weeds and bugs or to buy new pesticides.

Breaking this expensive and ultimately unsustainable cycle of chemical dependence is, at this late date, extraordinarily difficult. Just as the industrialization of meat production made antibiotics a virtual necessity, our commercial crops and farming methods have reached a point where we are essentially unable to farm without massive and continuous chemical interventions. Giving up organophosphates, for example, would cut wheat farmers' revenues by 10 percent[57] and cost the California produce industry half a billion dollars a year.[58] Nor is this chemical addiction just a Western problem: as developing nations adopt industrial agricultural practices, they tend to replace their native or traditional crops with one of the four supercrops — wheat, corn, rice, or soybeans — that are increasingly dominating world production. And although these high-yield crops often outperform their native predecessors, their arrival in developing countries has had the unintended effect of exposing the supercrops to new populations of native bugs, fungi, viruses, and other pests to which these crops have had no time to develop natural resistance. In such cases, developing-world farmers have little choice but to apply pesticides — and thus initiate their own chemical treadmill.

The consequences of this reciprocal relationship between crop and chemical, says Lewis, are manifold. Not only must farmers apply more pesticides, but even with these additional chemicals, crop losses due to pests continue to rise and have increased by 20 percent since 1960.[59] What's more, as the world continues to rely on a smaller and smaller number of crop varieties and to bring those varieties to a larger number of regions, the risk of some truly massive outbreak of plant disease escalates. In the 1800s, more than a million Irish died when a fungus, or blight, wiped out the potato crop. Plant breeders have since successfully developed crop varieties that resist various attacks, but as the number of crop varieties shrinks, the probability of an emergent pest or other plant-attacking pathogen destroying a significant share of the food supply grows substantially. Such an event may in fact already be under way. In 1999, plant pathologists in Uganda noted the appearance of a fungus known as stem rust that was destroying even wheat varieties specially bred to fend off rusts. Once in a field, this new fungus can wipe out three-quarters of the crop, and its spores spread easily by wind. Since its emergence, the so-called UG99 rust has migrated into Kenya, Ethiopia, and Yemen and is now moving north and east, threatening India, China, and, eventually, even North America with a wheat-crop failure that could cost billions of dollars while adding significant pressure to global grain supplies.[60]

In economics, such unanticipated side effects are known as externalities — that is, costs that don't show up in the retail price of the product or practice, but that eventually must be paid by someone — and that ultimately may make the practice or product unsustainable. All sectors have external costs (tailpipe emissions are an external cost of gasoline), but in industrial agriculture, external costs have become a defining characteristic. The increased use of nitrogen fertilizers, for example, helped bring yields up and food prices down, but it also contributed to a surfeit of calories, runoff, weed-clogged waterways, coastal dead zones, and a decline in commercial fisheries — all of which carry enormous price tags. Yet because these costs aren't included in the retail price of the food we buy, they tend to be invisible to consumers and policymakers and thus are rarely counted when we evaluate the efficiency of our food system or praise our capacity to generate so much food so cheaply. "It's not that we really produce cheap food," says Steve Suppan, director of research at the Institute for Agricultural and Trade Policy, a left-of-center think tank that focuses on sustainability.[61] "It's just that by externalizing a lot of the cost, we've made it appear to be cheap."

The proposition that modern food production carries numerous hidden costs isn't new. By the 1950s, the problems of pesticide resistance and chemical runoff were raising alarms among some agricultural researchers and journalists, including Rachel Carson, who later wrote *Silent Spring*. And while mainstream farm scientists, with plenty of encouragement from the input industry, were able to downplay the seriousness of these impacts for decades, by the 1980s, "externalities" had become a buzzword too prominent to quash. The silence was officially shattered in 1989, when the National Academy of Sciences published a detailed and occasionally scathing report on the costs of modern farming.[62] The 465-page tome described not only the ecological impacts of agribusiness, such as chemical runoff, erosion, and pesticide residues in food, but the financial repercussions as well. Paying for the consequences of surface-water contamination, for example, was calculated to cost the United States $16 billion every year, and contaminated water was only the beginning.[63]

In the decades since the academy report, hundreds of studies have shown modern food production to be an astonishingly costly business when all such externalities are counted. The true costs of livestock, for example, include everything from the massive amounts of liberated nitrogen and methane (another potent greenhouse gas) to the millions upon millions of acres of land that have been degraded since the livestock revolution began. In Latin America, nearly three-quarters of the former forest base has been cut down to make pastures,[64] while in China, overgrazing transforms fourteen hundred square miles of grasslands into desert each year.[65]

Ultimately, in any finite system, even the most externalized of costs will eventually emerge and require payment. The loss of commercial fishing due to expanding dead zones, for example, means billions of dollars in lost fishing revenues. But more recently, the external costs of food production are registering not simply in economic terms but in the form of reduced food output and thus lower food security. In Asia, much of the decline in rice yield growth is now attributed to degradation from overuse of fertilizers and pesticides, and similar connections are being seen elsewhere. If such declines continue, and there is little reason to think they won't, it will confirm what critics like John Ikerd, agricultural economist at the University of Missouri, have long argued — namely, that the celebrated high output of industrial agriculture is by definition "short run," because it is "supported by the extraction and exploitation of both the natural and human resources upon which its long run productivity inevitably depends." Industrial farming, Ikerd contends, like much of the rest of the industrial model, "is using up nature and it is using up society and when these natural and human resources are gone, there will be no means of sustaining the economy."[66] In other words, precisely at the moment in history when we need to shift our system of food production into overdrive, our agricultural engine is in the process of breaking down.

Exponents of the industrial model have long countered that such anx-

ieties about depleted resources are themselves functions of outdated thinking about the limits of technology and productivity. In the 1920s, agronomists gloomily predicted that our flat grain yields were permanent — only
to see new breeding techniques double those yields in a decade and touch
off the greatest explosion in food productivity in history. Nearly a century
later, modern breeders, armed with new tools of transgenic technology, are
poised for what they say will be an even more dramatic breakthrough — in
yields, but also in traits such as insect resistance, nitrogen efficiency, and
other qualities tailored to a resource-constrained world.

Given the rapid advances in genetic and plant sciences of the last decade, such promises bear careful attention, as we'll see in the next chapter. In
the meantime, however, suffice it to say that if we truly are to keep the current production system not only operating but operating at a steadily
greater output, it will take more than boosted yields or better nitrogen uptake. Because for the foreseeable future, the food system won't simply be
dealing with familiar problems such as rising fertilizer prices and dwindling farm acreage and pesticide runoff, but with a trinity of more recently
discerned limits — energy, climate, and water, the veritable Big Three of
yield constraints — that will ultimately force a complete rethinking of the
way we make food.

The most visible driver transforming the food economy will probably be
the rising price of oil. Since September of 2001, crude-oil prices have jumped
from around $26 a barrel to well over $90 a barrel in 2007, and even the
most hopeful analyses suggest prices will remain high at least through 2010,
when new oil fields now under development are brought online. Optimists
promise that the volumes of new oil soon to enter the market will more
than replenish world stocks and push prices back down — as has happened
with every previous price spike. Oil pessimists, however, warn that this
spike is different and quite likely permanent — both because oil demand is
now being fueled by massive emerging Asia and, more important, because
oil supplies are being depleted: each year, the oil industry is finding fewer
and fewer new barrels of oil to replace all the barrels they sell — currently
eighty-five million a day. At some point soon (and some pessimists claim it
has already happened), world oil production will hit a *peak* and begin to
decline; Samsam Bakhtiari, a former senior executive at the National Ira-
nian Oil Company, expects the peak in the next year or so and sees world

oil output dropping to fifty-five million barrels by 2020,[67] at which time to-day's price will seem like a bargain.

In truth, we don't even need to wait for a peak; because the oil market is already stretched so tightly between supply and demand, any disruption in the oil world — continued unrest on the Turkey-Iraq border, for example; or worsening relations between Washington and Tehran; or a terrorist attack in Saudi Arabia, the largest oil supplier on the planet — could easily send prices toward the sky. "The unrelenting pressure of increased demand has left the market a coiled spring," oil market analyst John Kilduff told Bloomberg.com. "We're only a headline of significance away from $100 oil."[68] Matt Simmons, an oil supply expert in Houston (and an adviser to the current White House), goes further, arguing that with the right combination of political events, oil prices could rise within the year to well above $200 a barrel.[69]

Two-hundred-dollar oil would be a disaster for the industrial economy as a whole, but it would be particularly devastating for the agricultural sector, which is based entirely on cheap oil. Our tractors, harvesters, and irrigation pumps rely on oil, as do all the trucks, trains, and cargo ships that supply farmers with their fertilizers and pesticides (themselves made from oil's geologic cousin, natural gas) and transport all the farmers' outputs to market. Food processing and packaging are also incredibly energy intensive: each new layer of added value represents a substantial increase in energy use. The energy needed to make a pound of breakfast cereal from wheat, for example, is about thirty-two times the amount needed to make a pound of flour from the same wheat,[70] and in many cases, companies use even more energy packaging the food than making the food itself.[71] By one estimate, it takes 2,200 calories of hydrocarbon energy (from oil, natural gas, or coal) to produce a can of soda that contains just 200 calories of food energy — a degree of energy intensity that helps explains why food production accounts for nearly a fifth of the United States' total energy bill.[72]

But the ramifications of pricier oil go beyond the energy content of our food products. Cheap oil (and for most of the twentieth century, oil has been well under twenty dollars a barrel) not only allowed us to replace our horses with tractors and all our manures with fertilizer, it also let us extend our reach. With cheap oil and, thus, cheap transportation, farmers were no longer limited to producing the things they could sell locally; they could sell anywhere on the planet. Without cheap oil, it is hard to imagine how the United States would have evolved so rapidly from a nation of tiny, multicrop farmers into a tightly coordinated community of massive, hyperefficient regional agribusiness specialists, each focusing on whatever crop or animal they can grow for a few fractions of a cent less per bushel or pound.*

Cheap oil allowed the idea of comparative advantage to move from theory to lucrative reality. With expanding fleets of ships and, eventually, planes, countries and regions could truly leverage their natural advantages: farmers could not only make their grain, meat, and produce as cheaply as anyone else on the planet, but with cheap and increasingly rapid transportation, they also could deliver those goods to distant buyers almost as quickly as local competitors could.

But the tradeoff for this faster pace of food transportation has been high, because as speed increases linearly, fuel use climbs *geometrically:* going twice as fast requires significantly more than twice the energy. Thus, as the average speed of food deliveries increased over the last century, as we moved from slow ships to faster ships, and from ships to airplanes, our energy use has soared. In other words, our sprawling just-in-time global food economy, the very foundation of year-round produce and seafood, has effectively locked us into massive oil consumption — consumption that was feasible when oil was cheap but may be unsustainable in a world of two-hundred-dollar oil.

Ultimately, losing our fresh pineapple and salmon may be the least of our worries. Given how important oil is in the making and moving of food, and given how critical cheap oil has been to the rapid expansion of food supplies in the last century, one has to ask what an oil peak will mean for our ability to sustain current food production — or to boost output to the level we'll need by midcentury. Daniel Davis, one of the gloomier of the oil pessimists, lays out a particularly bleak peak-oil scenario: by 2020, the same year that Iran's Bakhtiari sees world oil production falling to 55 million barrels a day, our global population will be 7.5 billion, according to the United Nations. Yet the last time the world was producing 55 million barrels a day — 1985 — our population was just 4.75 billion.[73] Granted, our farms will be

---

* With cheap oil, fishing fleets could reach fisheries that had previously been inaccessible (and which are now, as a result, increasingly depleted), http://list.web.net/archives/food-news/2006-January/000012.html.

much more productive in 2020 than they were in 1985 — specifically, farmers will surely be able to generate more food for each barrel of oil used. But Davis's exercise does offer a useful baseline to consider the looming challenge of feeding more people on a resource base that, whether the resource in question is arable acres, soil, water, or some other input, is shrinking by the year.

Even if there were no shortage of oil (and as oil optimists point out, serviceable alternatives can be made from coal, tar sands, and, of course, crops), the oil problem goes beyond a question of mere scarcity. Oil, like coal and other fossil fuels, has its own set of externalities, not least the emission of pollutants like carbon dioxide, or CO_2, the most pervasive of all climate-altering gases. Since the start of the Industrial Revolution, atmospheric concentrations of CO_2 have risen from 270 parts per million (ppm) to more than 370 ppm, and are on track to hit 550 ppm by the middle of this century, at which point, according to most forecasting models, the changes in climate will begin to have devastating and potentially irreversible effects on the planet's various biological and ecological systems, including those on which food production now depends.

The good news is that climate's potential food consequences are finally gaining recognition among some policymakers. The bad news is that even the most motivated policymakers can do very little to soften the impact of climate change in the short term and perhaps in the long term. Even if the world were to embark on a radical program of emission reduction (a dubious scenario, given how rapidly industrializing Asia expects to increase its consumption of oil and coal, and given the reluctance of mature economies, like the United States', to curb their own CO_2 emissions), the impacts from climate change would still be significant. This is especially true for regions like Africa, which are already suffering severe climate problems and whose agriculture systems are already precariously balanced. Worse, because it will take decades for surplus CO_2 to leave the atmosphere, even if we did somehow shift from fossil fuels to clean energy sources tomorrow (while simultaneously addressing the other greenhouse gas sources, such as livestock), atmospheric concentrations of greenhouse gases would keep rising for decades, which means that some degree of climate change — and thus some degree of damage to our food-production capacity — is now inevitable.

How damaging? The world's food systems — from the crops farmers grow to the patterns of planting and harvesting to the models of management and financing — have evolved under particular climate regimes, so even a modest shift in climate could have massive consequences on yields and revenues. Most obviously, crops bred for a particular level of rainfall or a particular range of temperatures will see dramatic changes in yields in hotter, drier weather. But hotter, drier summers are just *one* of the effects that a changing climate will impose on world food production. Most climate models envision more frequent extreme weather events of all kinds — droughts, of course, but also severe rainstorms, hailstorms, and flash floods, which can be just as damaging to yields. Higher temperatures boost pest populations and allow insects, fungi, weeds, and other pests to migrate into farming regions that were previously uninfested, leading to substantial crop damage. Higher temperatures also stimulate soil bacteria and fungi, which accelerates the decay of soil organic matter and thus reduces the soil's capacity to store and transport nutrients and water.[74] Such soils will not only erode more easily, they will also need more fertilizers to maintain yields; yet because they have less organic matter to retain those fertilizers, they will simply surrender more of that added nitrogen into groundwater.

Such impacts will be particularly devastating in developing counties, where agriculture is still a major economic engine and where governments lack the financial and political capacity to take adaptive actions, such as switching to new, more suitable crops. Projections by Robert Mendelsohn of Yale University, an expert on climate and agriculture, and his colleagues suggest that eight countries in sub-Saharan Africa — Zambia, Niger, Chad, Burkina Faso, Togo, Botswana, Guinea-Bissau, and Gambia — could lose nearly three-quarters of their agricultural output, while the African continent as a whole could see its total food output fall by as much as $194 billion.[75] Overall food security will also suffer: one report predicts that by 2080, climatic shifts will have increased the population of malnourished people by fifty-five million, nearly all of them in Africa.[76]

Further, although the discussion of climate and food security has focused mainly on poor countries, it's clear that even moderate climate change could hurt food production in many wealthy, food-surplus countries. Much of the agricultural prowess of the United States stems directly from its climate — reliable rain and moderate temperatures; this is particularly true for the production of corn and soybeans, two crops that account for a huge share of U.S. acreage, total output, and export earnings, yet whose modern, high-yield varieties are relatively intolerant of swings in climate. In fact, for all the importance of fertilizers and crop subsidies and even cheap oil, were it not for the climate's stability during the twentieth century, the United States could not have become a great grain power. And unfortunately, says Frederick Kirschenmann, a sustainability expert at the Leopold Center in Ames, Iowa, there is every reason to believe that the twentieth century was a climatic anomaly. By most accounts, temperatures and rainfall in the last hundred years have been far more stable than they were during earlier periods,[77] and certainly more stable than we can expect in any of the futures described in various credible climate forecasts — not the best of news, argues Kirschenmann, "for highly specialized, genetically uniform, mono-cropping systems [that] now dominate the agricultural landscape." Given how damaging a *single* extreme weather event can be for yields — the typical drought in the U.S. Midwest, for example, can lower corn yields by 30 percent in any given year[78] — the prospect of a future in which extreme weather is routine is alarming indeed for a country that now supplies a huge share of the global grain production[79]and which would clearly be called on to help feed a climate-starved world.

Kirschenmann, who has spent the last three decades working on alternative food systems, sees climate change as a fundamentally new kind of crisis — one that cannot be addressed simply by substituting one kind of fertilizer or fuel for another but which will require a much deeper rethinking of the way we produce food. Whereas traditional externalities like chemical runoff and even energy scarcity can indeed be addressed through a combination of new technologies, better farming methods, and simple conservation, climate poses a problem that will push innovation to the edge, and perhaps beyond. "When oil gets too expensive, we know we can move back to the traditional agriculture that we had before," Kirschenmann told me. "But with climate, once we reach that point of 550 parts per million, then we've crossed a threshold that we can't come back from very easily. Climate is the one we really need to be paying attention to."

Unfortunately, by the time the climate problem becomes sufficiently visible to policymakers, climate will share center stage with a problem of even greater consequence for food security: water. On average, every ton of grain we grow requires a thousand tons of water,[80] which is why agriculture now accounts for roughly three-quarters of all freshwater use.[81] Some regions, like the American Midwest, can count on ample rainfall for their crops. But in much of the rest of the world, farming is utterly dependent on irrigation. Half of the developing world's grain crop is grown on irrigated acres,[82] and the success of the Green Revolution in India and Southeast Asia[83] wouldn't have been possible had the area of irrigated land not doubled since 1960. This critical link between output and irrigation is why most forecasts for future food supplies assume that overall water use will continue to climb: the FAO, for example, envisions an increase in irrigated acreage of 20 percent by 2030.[84] But as with other crucial inputs, such increases in water use are by no means certain; many studies indicate not only that a 20 percent increase isn't simply impossible but that even *current* usage is now so far above sustainable levels that it is already hurting global farm output.[85] Further, because water, unlike energy or fertilizers, has no alternative, this emerging scarcity poses a constraint on food supplies that in some ways is more final than that of oil or climate.

The problem is twofold. On the demand side, agricultural users are becoming more numerous and increasingly must compete for water with industrial and residential users — a conflict already playing out in urbanizing regions from coastal China to the western United States, and especially California, whose farms suck up four-fifths of the state's water.[86] On the production side, the picture is more complicated, because not all agricultural water is created equal. Farms can be supplied by two kinds of water: water that comes directly in the form of rain, and water that is stored in rivers, lakes, glaciers, reservoirs, and underground aquifers. The difference between these two forms is critical. Rain-fed, or "green," water, in hydrological parlance, is what is known as a free good — it falls from the sky and doesn't need expensive reservoirs, dams, irrigation ditches, or wells; rain-fed crops also tend to be more water-efficient than those fed from stored water, because stored water — also known as "blue" water — has to be transported, when a huge amount is lost via leakage and evaporation. (Moreover, groundwater often contains traces of dissolved minerals, including salt; when such water is used for irrigation, these minerals accumulate in farm soils and can destroy yields.[87]) Because of these differences, green-water agriculture can generate up to five times the calories for the same ton of water as blue-water agriculture can.[88]

The downside to green water is that it's finite: once you've put the last rain-fed acre into production, you can't expand your green-water resource. Instead, you must supplement your rain-fed capacity with blue water, which is what the great push for irrigation systems in the last century was all about. And while tapping our blue-water resources was a big reason we could feed so many additional people, it's also why we're in such dire straits now. Whereas green-water supplies are effectively self-regulating — once you use up all the rainfall, you can't use any more — blue-water resources can be exploited faster than they can be replenished. Rivers can be pumped so extensively they no longer flow year-round, or at all. Aquifers can be drawn down faster than they can be recharged by rainfall. (And some aquifers, known as fossil aquifers, are unable to recharge at all; once emptied, they remain that way.) In India, a country whose food self-sufficiency was essentially paid for with overdrawn blue water, aquifers have been so heavily tapped that water tables are falling by up to twenty feet a year. In North Africa, water is being withdrawn from aquifers as much as five times faster than it can recharge, forcing farmers to drill their irrigation wells to depths of nearly a mile. Even in the rain-rich United States, the huge Ogallala Aquifer, which supplies one in five irrigated acres nationwide, is being overdrawn at a rate of 170 million tons (3.1 trillion gallons) a year and is gradually forcing many farmers to either shift to new "dry land" crops or abandon agriculture altogether.[89]

Perhaps the most serious overuse occurs in eastern China. In the so-called 3-H region, a massive area covering the Huang, Hai, and Huai river basins that contains 40 percent of the nation's population, half its grain production, but just a tenth of its water resources, water use now exceeds the sustainable flow by more than six hundred million tons a year, according to a 2001 report by the World Bank.[90] Overdrafting is so severe that water tables have fallen by up to three hundred feet, ground levels are subsiding (Beijing's elevation has actually dropped by several meters), and in coastal areas, freshwater wells are now sucking in seawater. All told, reports one researcher, China is now thought to be feeding perhaps two hundred million people — around a sixth of its population — with water withdrawals that cannot be sustained.[91]

Certainly, the United States, China, India, and other big water consumers could ease their water shortages by reducing water waste. In some areas, poorly designed irrigation systems lose three-quarters of their water. Such losses can be curbed dramatically through better technologies, such as drip irrigation, and by policy changes. For example, if governments stop subsidizing water sales to farmers, higher water prices will naturally stimulate conservation efforts.*

But there are limits to such efforts. Some blue-water resources are now too depleted to ever recover. As one World Bank study notes, even if China pursues aggressive countermeasures, "restoring aquifers to groundwater levels encountered even ten years ago is by-and-large no longer possible and all that can be done . . . is to halt further decline in water levels and pressure."[92] And given that most countries aren't trying to just sustain *existing* levels of food production but need to massively *increase* food output, at some point, water conservation will no longer suffice: to make more food, farmers will need *more* water. And that water, by most accounts, may simply not be there. According to the World Bank, even if China adopts a rigorous water-management system — with higher price incentives, better system efficiency, recycling of waste water, and massive water transfers (as much as 270 million tons a year) from the wetter southern part of the country to the drier north — the 3-H region will still face a water deficit of about 600 million tons (11 trillion gallons), or about two-thirds of the total yearly flow of the Huang River.[93] And China is only one of a growing number of countries that will soon be slipping into severe water deficits.

In conventional economic terms, such looming water scarcity will be massively expensive: China's shortages are expected to cost more than $8 billion a year in lost farming income.[94] But even this large figure misses some of the more important impacts of water shortages. Declining water supplies are one of the main reasons for China's falling grain production — both as China's farmland loses its productive capacity and as farmers abandon grain for higher-value, less water-intensive hothouse crops. In other words, water shortages are a principal driver in China's growing grain insecurity and its rising demand for imported grain: every ton of grain the Chinese buy on the world market is another one thousand tons of water Chinese farmers don't have to pump from their overdrawn aquifers or that anxious Chinese government officials do not have to divert from other regions.

This so-called virtual water that is contained in grain and other food

---

* This market-based approach has provoked massive opposition in the U.S. Southwest, where farming interests have considerable political power, but it is already bearing fruit in the more autocratic China.

has always been an implicit factor in the grain trade; when countries like the United States export surplus grain, they are also exporting their surplus water. But in the future, food economists say, virtual water will become a far more explicit and defining driver in the global food economy, as countries in Asia and Africa, which will no longer have the water to feed their growing populations, will have to rely instead on what is effectively becoming an international water market. By turning water into a globally traded commodity, the global food system is essentially being called upon to rebalance a world water market that is in serious disequilibrium. From this standpoint, the many million tons of grain that are exported globally each year, along with all the traded meat and produce, represent a redistribution of some 980 billion tons of water[95] — a massive reallocation of a resource that is going on largely without public notice.

How long such world water trade can continue, however, is quite uncertain. Alexander Zehnder, with the Swiss Federal Institutes of Technology in Zurich, says that while the number of big water exporters — among them the United States, Europe, Brazil, Argentina, and Australia — will remain largely unchanged during the next two decades, the number of water importers will rise dramatically as populations climb and water tables fall.[96] And given the already recognized gap between total food demand and total available water, it's clear that there is only so much rebalancing that the new global water market can do. According to the World Water Commission, even assuming irrigated agriculture is made as efficient as possible, "humanity will still need at least 17 percent more fresh water to meet all of its food needs than is currently available."[97] Indeed, just to produce the extra grain the world is forecasted to need by 2050 will require us to somehow come up with as much as a *trillion* tons of additional water — a challenge that may simply exceed our technical, political, and physical capacities.

As Brown puts it, "[W]hen demographers project what future populations will be, they use standard assumptions, like sex ratios, or the expected number of children per woman, and they come to the conclusion that our population, which is 6.5 billion now, will be 9 billion by 2050. But they never ask the real question, 'is there enough water to support 9 billion people?'"

Brown's point is worth some serious thought. Many mainstream forecasters continue to operate on the assumption that past patterns of productivity increases will continue into the future, and that the necessary increases in inputs, yields, and innovation will emerge in time to avoid calamity. Thus, while institutions like the FAO recognize the massive challenges ahead, their projections still show per capita global food supplies growing inexorably from the current 2,800 calories to 3,000 calories by 2030, and per capita meat consumption rising by 25 percent,[98] as if the sheer momentum of past growth will carry humanity into the future.

But more and more, such assumptions are being publicly challenged — by activists, of course, but increasingly by scientists, policymakers, and even some business leaders, who have come to question the sustainability of many basic production practices. Commissions and panels now issue regular reports on the impact that nitrogen pollution or rising energy prices or worsening water scarcity is having on global food systems and food security. Advocates such as Lester Brown, who spent decades being dismissed as a latter-day Malthus, are finally gaining some space in mainstream discussions. (Brown's famously gloomy 1995 tract *Who Will Feed China?* was closely read by Chinese government officials and a great many regular Chinese citizens.)

Resource experts have begun creating tools to analyze agriculture's impact on the environment. Among the more interesting is a concept known as the "ecological footprint," in which everything needed for a particular activity or product (such as energy, chemicals, or waste disposal) is distilled into a single value indicating how much cost that activity or product incurs against the planet's biological capacity to sustain life. Advocacy groups such as the World Wildlife Fund contend that our footprint now exceeds the planet's biocapacity by about 25 percent.[99] And, for the record, the largest piece of that footprint, according to WWF and others, comes from producing food, especially meat — a finding that is hardly surprising, given the massive amounts of land, energy, water, and chemicals required to raise livestock.

What has yet to emerge is a comprehensive understanding of the way these various limits, whether on land or energy or water, are likely to operate in concert. Much of the scientific work tends to focus on single issues, treating scarcity of water, or energy, or a shift in climate as discrete, individual crises. It is becoming clear, however, that these factors are often intensely interrelated; a crisis involving one input often magnifies the negative effects of a shortage in another area, resulting in a complexity that will very likely exacerbate food insecurity in ways that we cannot yet fully imagine. Climate change, for example, will accelerate water shortages and insect problems in regions like Africa and Asia, and these events will likely hurt crop yields, forcing governments to import more grain. But because climate and water shortages are also likely to reduce the world's *total* grain output and thus push up global grain prices to nearly twice the current levels by 2025, according to the International Food Policy Research Institute, these same developing countries will be even less able to afford to imported grain.[100] Similar dynamics will likely exist around shortages of energy, fertilizers, and any other input: the more we need, the less we will be able to afford.

Clearly, what is needed is an aggressive and coordinated effort by *all* players in the food system itself — not just ecologists or sustainability hand-wringers, but food companies, food scientists, and food policymakers — to assess how and whether the current food economy can adapt to so many challenges on so many fronts, and if it can't, what alternatives might be feasible in a world that will be fundamentally more constrained than the one we acknowledge today. "We know that climate is changing, and we know that oil could very easily be at $250 a barrel tomorrow if the Middle East blows up," Kirschenmann told me. "So if we are really scientists, we should at least be asking ourselves what kind of agricultural system could produce the food and fiber we need in a world where oil is $250 and where we have twice the severe weather but only half the water that we have now. What kind of agriculture could we come up with? It's an entirely reasonable question to ask, and yet, no one wants to touch it, because when you get down to it, no one has a clue."

In some respects, such uncertainty is unavoidable, given the inherent challenges in feeding 10 billion people with a food model that is already unable to sustainably feed 6.5 billion. Over the next half a century, we're going to need to learn how to do more for less — that is, to feed more people with substantially fewer impacts per person than we're incurring now — a tall order in the best of circumstances, and one that will be incredibly challenging as inputs become steadily more problematic.

In fact, although scientists and companies are already scrambling to address this uncertainty with a new generation of food technologies and methods, as we'll see in coming chapters, even the most promising of these innovations faces substantial technological hurdles, as well as pervasive economic, political, and cultural resistance. Many of the biggest food companies have little interest in giving up lucrative practices. Consumers, too, aren't especially eager to "fix" the food system — not least because many of the worst problems are tied directly to the foods and practices consumers prefer most. Even without another day of research, we can be fairly confident that much of the "overdrafting" in the food system is being driven by a global diet that is increasingly carnivorous. And while there are ways to produce meat with fewer inputs or external costs, as we'll see in chapter 10, even with such methods we'll still need to slow, and perhaps even reverse, the trend toward greater per capita meat consumption. And unless we expect the developing world to bear the burden of this shift to a more sustainable food economy, it is consumers in the West, and especially North America and Europe, who will need to change their food practices — something that is hard to imagine occurring voluntarily.[101] Indeed, for all the bad press about obesity and food safety, per capita meat consumption in the United States continues to rise by about a pound a year, and Europe is not far behind.

What is especially worrying about such resistance is that every year we delay making changes, the problems we will eventually need to solve grow larger and more intractable; in fact, many of the current fixes are actually making things worse. Most mainstream producers and policymakers continue to address the symptoms of an unsustainable food system and not its underlying problems, and thus the solutions they offer are really just refinements of the same unsustainable practices that led to the symptoms in the first place.

Consider how seemingly advanced countries like the United States are responding to warning signs in the ever tighter grain markets. In the rush to cash in on high prices, grain farmers are not only stealing acres from other crops but are also plowing up many tens of millions of acres of marginal, easily damaged farmlands that the federal government, under its Conservation Reserve Program, had been paying farmers *not* to farm. And farmers are getting a lot of help. Ordinarily, a farmer who plants on his program acres risks losing federal farming aid. But in late 2007, meat-industry trade groups such as the American Meat Institute, the National Cattlemen's Beef Association, the National Pork Producers Council, and the National Chicken Council asked the USDA to permit farming on program land "without penalty"[102] — a proposal that soil experts fear will unleash a wave of erosion and degradation to rival the debacle that followed the "fence row to fence row" campaign in the 1970s.

Soil experts aren't alone in their fears about the coming decades. At the Des Moines Water Works, staff scientist Chris Jones says the facility is now extracting an unprecedented volume of nitrates from the two rivers that supply the city's drinking water. Although Jones is reluctant to blame those higher nitrate levels on the corn boom under way upstream, he did tell me that the river concentrations of nitrates in January 2007 "were the highest they've ever been for that month, and the same was true for February. One suspects this is going to be a bad year."

And for all the attention being paid to the impacts of rising grain demand on American consumers and American environments, most of the effects of the coming expansion will occur not in the United States or other developed countries, where land prices are prohibitively high and environmental protections relatively strong, but in the developing world, where both prices and laws are weak. Even before the ethanol boom, forecasters were anticipating that by 2030, developing countries would expand their farm base by about 468,000 square miles.[103] Today, such predictions seem conservative, especially in a place like South America, which possesses most of the remaining potentially arable land and whose governments are eagerly converting that land base into export-oriented farm systems. While this rapid expansion has generated billions of dollars in export revenues (much of it from meat-hungry China), it is exacting a steep price. Just as corn is displacing other crops in the U.S. Midwest, Brazil's soybean farms, which are expanding at a rate of five thousand square miles a year,[104] are doing so largely at the expense of existing small sustenance farms or pasturelands, which in turn must be moved deeper into the remaining forest. All told, nearly a fifth of the Brazilian Amazon rain forest — or around an eighth of the entire Amazon Basin — has already been deforested, and what remains is being chopped down or burned at the rate of about eight thousand square miles every twelve months.[105]

And South America is merely the leading edge of a much larger transformation of the developing world, as rising population and changing diets push the high-volume, low-cost model of food production into high gear. Since 1980, according to the World Wildlife Fund, more than 1.1 million square miles of forest — an area larger than India — has been cleared, much of it to make way for pasturelands and croplands, especially soybeans, corn, and palm oil plantations. Beyond the impact on such intangibles as wildlife (the rate of species extinctions is ten to a hundred times higher now than thirty years ago), it's not at all clear that such expansion will bring the kind of output increases that will be needed. Unlike the American Midwest or the Black Sea region, where farmers still have the benefits of deep topsoil, much of the forest soils in South America are thin and highly acidic with less organic matter, and once cleared and planted, they don't hold up well under conventional intensive agriculture. Loss of organic matter occurs rapidly, which causes gradually declining yields and leaves areas vulnerable to erosion. Farmers abandon the land and move to a new tract, which almost always requires the clearing of forests. In a sense, each bushel of soybeans or crate of frozen chicken that Brazil exports represents not only Brazil's cheap labor but a share of its water and soil — resources already under considerable pressure. "The idea Brazil can plant more crops for the next hundred years is crazy," says Charles Brummer, a University of Georgia plant breeder who has done research in Latin America. "They can drain swamps and cut down more trees, but they're already on borrowed time."

PART III

9

Magic Pills

I N   A   S L E E K   N E W brick-and-glass building on the edge of the Iowa State University campus in Ames, plant geneticist Patrick Schnable spends the better part of each very long day trying to disprove doomsayers like Charlie Brummer and Lester Brown. A compact, slender man with dark hair, wire-rims, and a friendly if somewhat rushed demeanor, Schnable runs the ISU Maize Genetic Mapping Project, part of an ambitious federal program to create a DNA blueprint for the maize plant. It is a monumental task: whereas the human genome has roughly 26,000 individual genes, the lowly maize possesses more than 50,000, the mapping of which has required a completely new set of analytical methods and tools and many tens of millions of dollars in funding. But the payoff, says Schnable, will be huge.

The reason maize needs so many more genes than humans do, Schnable tells me, is that unlike animals, which can *think* about how to adapt to environmental changes, "plants need a hard-wired response for every possible environmental eventuality" — that is, a specific preprogrammed response to cope with a shift in temperature, say, or moisture, or insect attack, or any of the thousands of conditions that determine a plant's success. And because these hard-wired responses, or traits, are controlled by a single gene or clusters of genes, which scientists such as Schnable can manipulate at the molecular level, the mapping of the connections between specific traits and specific genes is unleashing the capacity to control plant behavior, and to design plants (and animals and other organisms) for precise end uses and markets. "The corn genome is like a computer program," Schnable tells me. And once he and others completely map it out, "we really can do almost anything."

In fact, although genetically manipulated, or transgenic, foods aren't new — these high-tech crops account for more than a quarter of total corn acreage and more than half of soybean acreage[1] worldwide — enthusiasts like Schnable say innovations are coming that will completely transform food production. Plants will be designed to manufacture pharmaceuticals; animals will be engineered to grow leaner meat. More fundamentally, breeders promise varieties of corn, soybeans, and other staple crops tailored to the constraints of future food production: plants bred to tolerate heat and drought and salty soils; plants bred to use nitrogen more efficiently and thus more sparingly; plants bred to produce vastly more edible weight — and all delivered much more quickly and cheaply than would be possible with conventional breeding. To cite one example average corn yields in the United States are expected to reach barely two hundred bushels per acre by 2035, but researchers at Monsanto, one of the leading players in the booming transgenic market, believe that with genetic manipulation, that number could easily be pushed to three hundred bushels and beyond.

Yet not everyone is prepared to be saved by this brave new technology. Ever since the first commercial transgenic food product — the Flavr Savr tomato — was rolled out in 1994, these high-tech foods have provoked the kind of controversy normally reserved for debates over abortion or war. Critics warn that transgenic plants and animals (often referred to, incorrectly, as "genetically modified," or GM)* may contain novel substances that could harm humans or intermingle with and destroy native species, while the more action-oriented opponents have taken to sabotaging transgenic farm fields and[2] destroying research facilities — to the point where many European and African governments now ban the production or import of transgenic products.

Such opposition has provoked an equally vitriolic response from backers of transgenic food, who accuse opponents not only of exaggerating and even fabricating the technology's risks but also of obstructing the emergence of the only technologies that can feed the billions of newcomers

---

* All plants that undergo breeding of any kind are "genetically modified"; thus, "transgenic" distinguishes those varieties whose genetic material has been altered through molecular manipulation.

expected by 2050. This "blind opposition" to transgenic crops, wrote one indignant proponent, "is the triumph of dogma over reason."[3]

In the controversy over transgenic technology, one can see the outlines of a much deeper and more fundamental disagreement over the future of food. There may be a consensus of sorts that the current system of food production is ailing — that its economic, ecological, and even nutritional foundations are not as firm as they were once thought to be. There may be agreement that the coming challenge won't be simply making more calories for more people (many of them poor), but doing so with less water and energy, poorer soils, and a shifting climate. Where agreement breaks down, however, and where the debate becomes as fractious and polarized as any in modern politics is in how that new food system should be built.

The dominant view, offered by most economists and food policymakers since the emergence of the modern food system, is that the food economy is more or less self-correcting. That is, when the economy grows unbalanced — for example, when surging populations exceed existing production capacities, as during Europe's industrialization or in Asia a century later — the sector becomes unstable, prices rise, and consumers suffer. But eventually, these dislocations induce new technologies, such as mechanization, and especially new biological technologies, such as better seeds and synthetic fertilizers, that bring on the additional supply and push the Malthusian monster back into its box.

In this view, today's high prices and rising external costs are indeed signals not that we've reached our productive limits but that we've reached the limits of existing crop technologies and need a new generation of technologies, among them transgenic technology, to restore our historic upward curve in productivity. "When you look around and ask what kind of technology is going to get us there," says Mike Phillips, with the Biotechnology Industry Organization, "it's hard to see anything else with this potential."[4]

But this is not the only vision for the future of food. For almost as long as proponents of industrial agriculture have celebrated its self-correcting nature, critics of that model have presented a very different view: namely, that the very technologies by which the food economy supposedly corrects itself are actually destroying that economy. Although natural systems do indeed tend to correct imbalances, this capacity has been steadily eroded by a century of intrusive agricultural practices. At best, our recent food triumphs are temporary reprieves, because they rely on the unsustainable withdrawals of energy, water, soil, and other increments of the natural capital upon which food production depends.

In this view, known variously as organic, sustainable agriculture, agroecological farming, among others, the simultaneous emergence of so many food problems — safety, insecurity, declining productivity — is a signal that industrial agricultural has nearly exhausted the underlying system's restorative capacities. What is needed is not simply a new round of technology but an entirely new model of sustainable agriculture that is conscious of natural limits, is mindful of external costs, and, above all, seeks to moderate our obsession with technological fixes, or what Larry Yee, a small-farms proponent from Ventura County, California, calls the silver-bullet syndrome. "After World War Two, chemicals were the silver bullet that would solve all our food problems," Yee says. "Now that that hasn't worked, we're switching to biotech. But the paradigm hasn't changed; it's still the hope that there is a silver bullet out there that will solve all our problems."

These are more or less the same arguments that appear in the larger debate over the sustainability of all economic growth, so it is tempting to leave the combatants to what appears to be an intractable impasse — except that we don't have that luxury. With the urgency and complexity of the challenges facing the food system, we really *do* need something of a magic bullet, or at least a sense of whether any of the proposed solutions — whether high tech or low tech — are really going to work. Are there functional technologies out there that can supply the amount of food we'll need? Do they offer substantially different approaches or simply replace old problems with new ones? More fundamentally, can these alternatives, whether biotechnology, organic farming, or any of a coterie of other ideas, truly address the larger challenges of producing more food in a world of constrained resources, or do they merely put off the day of reckoning?

In a sense, the industrial food economy has always been about silver bullets. Since the nineteenth century, a steady parade of technological success stories, from high-yield hybrids to high-speed processing, has taught us to expect that even the hoariest of food crises will be solved in a manner that simultaneously improves lives, generates profits, and makes us even better at producing more calories for less money.

Nowhere have such expectations run higher than in the field of plant and animal breeding, which has long thrived at the intersection of science and commerce and which by the 1970s seemed the perfect candidate for the mother of all scientific and commercial breakthroughs: genetic engineering. With the 1974 discovery of artificial, or recombinant, DNA, geneticists had the ability to transplant a gene, and whatever traits that gene influenced, from one organism to another. They could even move a gene from one species into an entirely different one — bacteria, plant, or animal — thereby cracking a barrier that had always confounded conventional breeders. With recombinant DNA, or rDNA, science could theoretically create an infinite number of new organisms designed for practically any objective that could be thought of — not least being profit. When the U.S. Supreme Court ruled several years later that a living organism could be patented,[5] and therefore that these new organisms could be owned and sold, the gene revolution was born.

Although the first commercial transgenic product, synthetic insulin, was pharmaceutical, transgenic foods were close behind. In the 1980s, Monsanto developed a recombinant cow hormone, known as bovine somatotropin, or rBST, that boosted milk output by 25 percent.[6] Soon after, Monsanto took a gene from the bacterium *Bacillus thuringiensis,* which secretes a natural insecticide, and placed the gene in a corn plant, resulting in a plant known as Bt corn that kills pests. But Monsanto's real killer product was a plant engineered to tolerate exposure to glyphosate, Monsanto's widely used herbicide that sells under the name Roundup. Because farmers could now spray Roundup directly onto their RoundupReady corn and soybeans, instead of laboriously applying the herbicide between each plant, these new seeds vastly simplified weed management, freeing up labor hours that could be used to, among other things, farm additional acres. The ideal input for an agricultural model geared for high output and lower costs, herbicide-tolerant seeds now account for half of all corn[7] and more than 93 percent of all soybean[8] acreage in the United States. According to industry officials, these seeds are merely the advance guard for the new generation of supercrops that will transform agriculture as surely as hybrids did a century ago.

Although transgenic technology is extraordinarily complex, what it offers plant breeders is quite simple: greater control. In conventional breeding, researchers mate or cross existing plants or animals in the hopes of producing offspring with desirable traits; these offspring are then crossed and recrossed until the traits of interest — strong stalks, say, or large seeds, or rapid growth — are sufficiently pronounced. This method has produced outstanding results — modern corn yields are around six times what they were in 1930. But it also has major limitations. First, the process is largely random: because breeders can see only the resulting traits (strong stocks, for example) but not the underlying genes that influence those traits, breeders have little control over the experimental outcome; each cross is another random mixing of tens of thousands of genes, essentially a throw of the genetic dice, except that you can't actually see the dice, only whether you won or lost.

Second, because breeders must wait to see how offspring develop before selecting the best of the litter — which may take months, in the case of plants, or years, in the case of animals — breeding traditionally has been a drawn-out process. Third, because breeders have been working with the same genetic material for more than a century and have made millions of crosses with that material, the odds of finding some new genetic combination that produces a significant gain in yield or drought tolerance or any other trait diminishes by the year. Corn offers the perfect example: yields rose quickly early on, doubling in the first decade after hybrid techniques were perfected, but they have since settled to around 2 percent a year — despite the fact that corn, being the dominant commodity, gets more research money than any other crop.

In this context, one can grasp the excitement of geneticists like Schnable, who believe that these limitations can be overcome with transgenic technology. Instead of random crosses, geneticists will be able to select superior parents by looking directly at their genes and then crossing only those plants, dramatically boosting the odds of superior offspring. And whereas conventional breeders must cross and recross the same genetic material, transgene companies can now essentially upgrade that genetic material, by either supplementing one organism's DNA with genetic material from another organism or manipulating an organism's existing DNA. At a Seattle company called Targeted Growth, researchers say they've figured out how to modify the corn gene that instructs cells in corn grains to rapidly divide — a change the company says will raise yields by 25 to 30 percent, from the current 150 bushels per acre to perhaps as high as 195.[9]

Longer-term, transgenic proponents believe vastly larger yield increases are all but inevitable. As Theodore Crosbie, Monsanto's vice president of global breeding,[10] told me, "If you consider the increase in gain we think we've demonstrated from molecular technology, and then lay on top of that the effects we expect from the second- and third-generation biotech traits, our calculation shows a U.S. average yield of three hundred bushels an acre" — or nearly twice the current average. And this is only the *average*: just as today's top corn farmers now routinely get three hundred bushels per acre,* proponents say transgenic seeds will let American farmers regularly produce in the high four hundreds, with considerably higher yields a distinct possibility. As Crosbie reminds me, "the theoretical maximum yield on corn based on photosynthesis rates is around six hundred bushels."

Predictions like these, combined with the commercial success of existing transgenic traits such as herbicide tolerance, have given the technology a powerful forward momentum. The governments of the United States and the United Kingdom regard the technology as essential in meeting future food challenges and routinely chastise skeptics as putting future generations at risk. "Countries that choose to turn away from biotechnology should recognize the consequences of their actions to the world," warned Secretary of Agriculture Dan Glickman at the World Food Summit in 1997.[11] And despite steady protests by consumer safety groups, credible scientific organizations, among them the FAO, the World Health Organization, and the U.S. National Academy of Sciences, have found no evidence that transgenic foods ever caused adverse human health effects.

And, of course, the companies themselves — diversified chemical giants like Monsanto, DuPont, and Dow, along with farm input companies like Swiss-based Syngenta — are eager to move into a marketplace that is not only growing (seed sales are expected to hit $20 billion by 2010, up from $450 million in 1995)[12] but boasts products that can be touted as helping to save the environment while fighting poverty. This is seen as particularly helpful for companies like Monsanto, the St. Louis–based chemical giant formerly known for such public-relations nightmares as PCBs and Agent Orange. As *The Economist* put it, "Rather than having to discuss toxic spills, Monsanto now talks about feeding the world."[13]

Granted, this road to the future hasn't been without bumps. In 2001, for example, Monsanto shelved its transgenic potatoes after McDonald's,

---

* The top producer for 2006, a Missouri farmer named Kip Culler, grew 347 bushels per acre.

Burger King, and Pringles refused to buy them. Further, after years of pro-test by opponents and refusals by some governments to import transgenic products, some industry analysts worry that the still-new technology may be too financially risky,[14] and some chemical company shareholders have campaigned to force the companies to get out of transgenic products alto-gether. Despite such obstacles, most input companies regard transgenic ag-riculture as a wave they can't afford to miss. When activist shareholders challenged Dow Chemical's transgenic program, Peter Siggelko, who runs Dow's biotech division, was unmoved. "We don't intend to budge," he ex-plained in an interview. "This is better and safer for farmers."[15]

Five minutes from Patrick Schnable's laboratory, and light-years from the gene revolution, agronomist Kathleen Delate watches the transgenic train with a kind of weary defiance. Delate runs Iowa State University's organic agriculture lab, which means that for the last ten years she has devoted con-siderable effort to trying to convince farmers that transgenic seeds may in fact *not* be better or safer, for farmers or anyone. This is, she admits, a mi-nority opinion. Nearly every acre of soybeans and corn in Iowa is trans-genic, and programs like Schnable's suck up much of the funding for farm research, while alternative programs like hers have a hard time getting bud-get dollars, much less professional respect. When Delate, a forty-seven-year-old lifelong organic farmer, arrived on campus in 1996, just as the first transgenic soybean crops were being planted, her rejection of the new technology was seen by many colleagues as "not only Luddite, but as a wholesale repudiation of their life's work." Even today, Delate told me, "there are probably ten professors at ISU who will use the word 'organic' in public."

But neither Delate nor any of the hundreds of thousands of farmers who make up the organic movement show any signs of budging. They are not persuaded that transgenic foods are safe: the lack of reported health problems, these skeptics suggest, may have something to do with the fact that consumers wouldn't know to report an adverse reaction in the first place, since, courtesy of industry lobbying, transgenic foods aren't labeled as such. Nor are they convinced that researchers fully understand the cas-cade of molecular events that occur inside a cell when a gene is manipu-lated, thus making it impossible to predict all potential health or envi-ronmental effects. And on a philosophical level, few organic farmers are interested in a product that helps a small number of very large companies further propagate an unsustainable model of an industrial agricultural system. Transgenic seed, Delate tells me, "is absolutely an extension of the agribusiness status quo."

Such defiance is hardly limited to the organic community. Organic farming, or simply "organic," is only the most visible (and commercially successful) facet of a larger, more complex social movement that is centered on alternative food production. It emerged more than a century ago as an opposite, if not quite equal, reaction to the noxious consequences of industrial agriculture — from the devastation of the Dust Bowl to the additive excesses of the chemo-gastric revolution. Some of the earlier variants in this revolt, such as agrarianism, which envisioned a reprise of a nation of small farmers, have receded; others, especially organic farming and "sustainable agriculture," are still going strong, with vast numbers of practitioners, an evolving literature, conferences, academic departments, and growing commercial prospects. And despite occasional differences of philosophy — especially in terms of the "correct" use of technology — most of these communities share an abiding antipathy toward both the methods and ideals of agribusiness.

On the surface, this antipathy takes the form of a rejection of synthetic inputs, such as pesticides, chemical fertilizers, and transgenic seeds. But most advocates of alternative agriculture oppose industrial farming on a much deeper level. Many are affronted by agribusiness's implicit confidence that food can (and should) be produced like any other commodity, with efficiency and cost the guiding priorities. Many are dismayed by the way the rationalization of agriculture has divorced people from both the land and the process of making food, and has replaced local production with a global supply chain. More fundamentally, most adherents of alternative agriculture reject the underlying economic rationale of agribusiness, especially the reductionist idea that agriculture is most efficient and productive and simply better when it is broken into its components — livestock here, crops here, fertilizers here — and each one of these is ramped up to an industrial scale.

In strictly monetary terms, industrial farming *is* more efficient. Farmers who focus on a single species (corn, say, or hogs), and who simply buy their inputs from other low-cost producers and sell their raw output to low-cost processors, can indeed generate more calories at lower cost than can farmers who try to manage the entire food-production cycle themselves (growing corn to feed their hogs and spreading the manure in their cornfields). But as critics argue, these larger industrial operations are more efficient and their costs lower in only the narrowest sense and only if one's calculations exclude external costs, such as contaminated water or, say, erosion. In reality, because these external costs erode the natural capital on which all food production depends, industrial agriculture is efficient only in the very short term. Granted, this critique is hardly limited to agriculture. Alternative economists such as John Ikerd and Herman Daly have long chastised the entire industrial system as being unsustainable and "short run."[16] But the critique has been especially resonant among opponents of agribusiness, because it suggests, among other things, that the large, hyperefficient factory farms are actually *less* efficient than the millions of small, diversified family farms they have replaced.

Although some critics of agribusiness see this as an argument for moving back to an older form of farming, the more pragmatic recognize that preindustrial agriculture is nothing to be yearned for, and that the greater inherent efficiencies of small farms were largely forced: lacking synthetic fertilizers, tractors, or chemical pesticides, farmers had no choice but to operate their farms as closed systems — keeping animals for traction and manure, spreading that "fertility" manually, and weeding and harvesting by hand — all of which were tremendously labor intensive and not terribly productive. What many alternative advocates do argue, however, is that some of the ideas implicit in these past practices are highly relevant today and, with the aid of research and prudent use of technology, can and should be brought back into usage.

The ancient practice of cover crops, for example, though partly sidelined by synthetic fertilizers, is by no means obsolete: alternating a cash crop with a crop such as alfalfa, whose roots work symbiotically with soil bacteria to pull nitrogen from the air and fix it in the dirt, can be just as replenishing as an application of synthetic urea. (Even ultraindustrial corn farmers use alternating crops of soybeans, a nitrogen-fixing legume.) As a side benefit, by rotating a field through three or four different crops over successive years, farmers actually slow weeds and insects from establishing themselves, as they invariably do when farmers grow the same crop in the same field year after year, and rely instead on pesticides.

The point of this closed loop, or "natural systems agriculture," as Wes

Jackson at the Land Institute calls it, isn't simply to replace nasty synthetic inputs with kinder, gentler variants. Rather, the goal is to replace the underlying system that required the synthetic inputs with a system that does not — a system modeled on nature's own methods for circulating energy and nutrients, interrupting pest populations, and maintaining internal balance. Under such a model, livestock and crops are reintegrated: animals generate manure to fertilize crops to feed to livestock. Farmers choose crops not only for maximal yields but also for their capacity to encourage the complex but crucial nutrient cycle among plants, soil, and soil microorganisms. In short, where agribusiness sought to mimic the methods and structures of the factory, the next generation of farms will be what David Holmgren, codeveloper of an alternative system known as permaculture, calls "consciously designed landscapes which mimic the patterns and relationships found in nature."[17]

Importantly, this new farming is based on more than biological systems. Given the socially destructive side effects of industrial agriculture, proponents of alternative systems want farming to be intensely social as well, with new models emphasizing community-based production and consumption. Whatever inputs a farmer couldn't produce on-site would, as much as possible, be procured locally, and most of a farm's production would also be sold locally, in local or regional markets. Such a restoration of agriculture's traditional social dimension would not only reduce energy use but would also help reforge fading connections between consumers and producers and restore the idea of food as something made by people, for people. Farming "needs to be redefined in terms of biological, ecological and social principles," says Kirschenmann, "and not the purely chemical and physical principles that industrial agriculture is now based on."

Such a shift would hardly be painless. By moving out of a low-cost model — using more labor instead of pesticides, for example, or putting a priority on local production, even when foods might be grown cheaper elsewhere — the new agriculture gives up some of the short-term economic advantages of large-scale industrial farming. This would mean higher food prices for consumers. It would mean less year-round availability. And because it would be replacing some of the large industrial farms with a larger number of smaller operations, with greater labor requirements, it would require the rehabilitation of farming as a valuable and desirable occupation, not an occupation that parents now have trouble encouraging their children to take up. "If we really want to be productive, we need to re-develop some aspects of an agrarian society," argues Kent Mullinix, a specialist in sustainable horticulture with Kwantlen University in British Columbia. "I think a lot of people would enjoy being farmers, but somehow, as a society, we've decided that farming is inappropriate. You can be a fire-man or an engineer or a bond trader, but for some reason, you can't be a farmer."

With such preconditions as higher food costs and more farm labor, it's not surprising that alternative farming remained on the margins for so long; after decades of industrialization, consumers, too, had come to measure the success and efficiency of the food system by its ability to deliver rising convenience, relentless novelty, and declining prices. By the 1970s and 1980s, however, even mainstream consumers were uneasy with the industrial model. The media were filled with stories about the health risks of industrial food additives and the ecological impacts of farm chemicals. The abysmal conditions of farm laborers were gaining attention. And if few consumers had read the National Academy of Science's critique of agribusiness in 1989, many were riveted by news reports that same year about how apple farmers routinely used Alar, a suspected carcinogen, to "color up" Red Delicious apples.

Almost overnight, mainstream America — or at least a significant minority of that mainstream — became *very* interested in alternative agriculture. These new consumers might not have understood much about sustainability, natural cycles, or the deeper flaws of industrial agribusiness. But they were now acutely aware of farm chemicals and wanted food that did not have them — and, importantly, they were willing to pay for it. Thus, while the entire alternative agriculture movement was moving slowly but surely toward the food mainstream, it would be organic farming, which had a specific product to offer, that would get there first, and do so profitably.

This new fame came at a price. As demand for organic foods grew — by 20 percent a year, doubling every forty-eight months — the nascent organics industry struggled to supply the necessary volume. Booms and busts were frequent, and products varied enormously in quality and could rarely be found outside big coastal urban markets. (As one early advocate complained, organic foods meant "gnarly apples and wormy turnips attractive only to live-off-the-land hippies and urban extremists."[18]) There were also heated disputes over what qualified, and didn't, as organic — at one point, more than thirty different organizations were offering competing standards.[19]

In the early 1990s, the U.S. Department of Agriculture created an advisory panel to write a single set of organic-industry standards, which, when finally adopted in 1997, paved the way for larger players, more investment, and a much larger variety of products. Big food processors like Dole saw opportunities, as did retailers (even Wal-Mart now sells organic), and they created hundreds of new organic products in nearly every category, sending ripples up the supply chain. Where "organic" had once been confined to produce and fruit, the emergence of organic processed foods and meats spurred demand for organic grains and soybeans, which began to attract farmers even in conservative places like Iowa.

As important, this new demand prompted rapid technical innovation. Although early organic yields had been woefully short of conventional crops, methods improved quickly — to the point that in certain crops, organic farmers can now almost match their nonorganic neighbors. According to surveys by Delate, organic-corn yields in Iowa are now between 90 and 92 percent of conventional yields, while soybeans are at 94 percent. And if organic crops still cost more to grow due to the higher prices of nonsynthetic fertilizers, for example, the hefty premium that consumers are willing to pay (organic corn sells for roughly twice the price of conventional)[20] not only covers those higher costs but, on a per-acre basis, generates more profits than do conventional crops, even RoundupReady varieties.[21] As Lynn Clarkson,[22] an organic grain farmer from Illinois, told a congressional hearing on organic farming in 2007, "[S]ome of the happiest row crop farmers in the U.S. regularly produce over 200 bushels of organic corn per acre and gross well over $1,000 per acre. They do so without contaminating the environment."

As important as profits have been, Delate argues, there were other reasons farmers came into organic. Many of the farmers Delate works with have lost family or friends to some of the various cancers that afflict farming regions and are routinely associated with farm chemicals. Many also worried about the physical condition of their farms, especially the declining quality of the soils. And without question, many organic farmers shared a traditional populist anger at an agricultural model in which producers competed with one another for fractions of a cent per bushel while the big consolidated buyers enjoyed a virtual monopoly on prices. In conventional farming, "farmers are price-takers and buyers are price-makers," says Delate. "Organic producers saw an opportunity to have a bit more control over the pricing of their production."

Considering their general antipathy toward industrial methods, it is hardly surprising that organic farmers are so openly hostile to transgenic products. Beyond the purely pragmatic concerns (if an organic farmer's corn crop is contaminated by pollen from a nearby transgenic corn field, he loses his price premium and suffers a huge loss), many farmers regard transgenic as the latest attempt by the "agro-industrial complex" to take over the food chain. Many organic farmers also feel that the U.S. government is pushing transgenic crops, not because they are safer or better, but because transgenic foods are now central to the business strategies of politically powerful chemical and seed companies. In 1997, for example, as the USDA finalized its organic standards, it sought to include transgenic seeds on the list of approved organic inputs — in large part, it later turned out, to avoid hurting foreign sales of transgenic crops. According to a leaked USDA memo, U.S. officials, who were then lobbying European governments to accept imports of American transgenic crops, became "concerned that our trading partners will point to a USDA organic standard that excludes GMOs as evidence of the Department's concern about the safety of bioengineered commodities."[23]

Advocates of transgenic foods counter that the organic industry itself is rife with dishonesty and cynicism. Some organic trade associations fill their marketing materials and websites with health warnings about transgenic foods that more credible skeptics, such as the Union of Concerned Scientists, say have no scientific basis. And many organic producers have been quite content to exploit consumer fears of transgenic products for their own commercial gain. In the early 1990s, when a Massachusetts biotech company announced plans for a fast-growing transgenic salmon, many grocery chains and high-end restaurants not only refused to sell the "frankenfish" but did a brisk trade in "organic" salmon, a product whose lack of reality (there being no official standards for organic fish) didn't stop retailers or restaurants from charging consumers a hefty premium.

In truth, however, the transgenic industry has brought much of this enmity on itself. Big transgenic seed companies do regard organic farming as a walking reproach to transgenic methods and products, and some have worked to undermine the organic market by, among other tactics, planting antiorganic news stories. In 1999, an analyst at the industry-funded Hudson Institute published an article, "The Hidden Dangers in Organic Food," in which he claimed to have data from the U.S. Centers for Disease Control and Prevention proving that consumers of "organic and natural foods" were eight times more likely to contract E. coli 0157:H7 than were those who ate conventional foods. When contacted, CDC officials responded that the agency had no such data and that the article's claims were untrue.[24]

Input companies are in fact heavy-handed with *anyone* who criticizes transgenic technology. Monsanto has in the past sued a dairy that marketed its milk as "rBST free," calling such claims "deceptive" and "misleading."[25] The company has also been implicated in smear campaigns against uncooperative researchers. When the journal *Nature* published a controversial study by two Berkeley scientists claiming to have found transgenic contamination in native Mexican corn (and, implicitly, evidence that engineered DNA could "flow" from transgenic crops into other crops), several of the "experts" who wrote letters criticizing the *Nature* article were later revealed to be not from academia but associates of a public-relations firm hired by Monsanto.[26]

Further, if organic farmers and other transgenic skeptics have overstated fears of transgenic foods, they've also forced the airing of more legitimate questions about the technology that neither companies nor regulators have satisfactorily addressed. For example, most of the questions about human health risks center on the not entirely unreasonable concern that a gene transplanted into a new plant or other organism might produce an allergic or toxic reaction in someone who consumed that new organism. In the early 1980s, in fact, a soybean implanted with a gene from a Brazil nut acquired the nut's allergenic properties. The project was canceled, and transgenic companies now avoid transplanting obvious allergens; they also run numerous tests on all transgenic products to catch any possible allergens or toxins. Such safeguards, backers say, explain why no credible reports of adverse reactions to transgenic foods have ever been confirmed, and thus why such august entities as the World Health Organization, the Food and Agriculture Organization, the National Academies of Science, and the American Medical Association all offer some degree of support for transgenic foods.

There are, however, several important caveats. First, many of these expert organizations confine their approval to *existing* transgenic products — things such as Bt corn or herbicide-tolerant soybeans, which involve the manipulation of one or two traits — but do not necessarily vouch for the more ambitious multitrait products under discussion. Second, and more fundamentally, because genes influence cellular behavior in such biochemically complex ways, there are concerns that the effect of engineering a single gene could in fact spread beyond the targeted trait.

One such concern centers on the proteins that transplanted genes create in their new host cells. Proteins are the gene's physical expression; the gene's DNA code "tells" the cell to assemble the protein from various building blocks in the cell. This "expressed" protein then initiates a cascade of molecular events within the cell that result in some trait, such as growth, which engineers hope to influence. The complication here is that this expression-and-cascade process is governed by many factors — by the protein's DNA-determined structure, to be sure, but also by the sugars, fats, and other chemical compounds that are present in the cell when the protein arrives — a chemical soup that varies from cell to cell. Because different kinds of cells contain different chemical compounds, the same protein may produce substantially different results when expressed in different cells. Thus, by transferring a gene from one species to another and by causing proteins normally expressed in one cell to be expressed in a completely different cellular environment, transgenic modification may raise the risk of unanticipated, and potentially unwanted, effects that weren't observed in the donor organism. The potential for such surprises was underlined in 2005, when researcher Vanessa Prescott and her colleagues found that when a gene from a pinto bean was transplanted to a pea, the pea acquired an allergenic effect that hadn't been present, or even expected, in the bean.[27]

Proponents of transgenic foods acknowledge the implausibility of anticipating *all* possible outcomes from a single genetic manipulation, but they nonetheless insist that product testing will still catch any dangerous results. Yet such confidence isn't universal. As a report by the Institute of Medicine at the National Academies of Science noted in 2004, despite advances in testing and epidemiological techniques, "there remain sizable gaps in our ability to identify compositional changes that result from genetic modification of organisms intended for food; to determine biological relevance of such changes to human health; and to devise appropriate scientific methods to predict and assess unintended adverse effects on human health."[28]

Perhaps most important, the National Academies and other expert organizations found serious gaps in governmental systems for regulating transgenic foods, especially in the United States, where oversight is split among three agencies. The USDA, for example, oversees only the crop trials for proposed transgenic foods; the Environmental Protection Agency regulates the pesticides produced by transgenic crops (such as Bt in Bt corn) but looks only at the gene and gene products, not any potential health impacts; those health impacts are left to the U.S. Food and Drug Administration. But unlike European regulators, who treat transgenic foods as food additives and subject them to mandatory testing *before* products reach the market, the FDA does not test transgenic foods prior to their release. Instead, the agency has classified transgenic foods as being substantially equivalent to the nonmodified counterpart and, under a policy of voluntary consultation, lets transgenic companies run their own safety tests and then request the agency's opinion about the product's likely safety.[29] It's only if problems develop postmarket — that is, after the product is being sold — that the agency takes action, much as happens in the case of food-borne illnesses. And because the costs to a company of a post-market problem would be so high, says Eric Flamm, the FDA's top biotech expert, even without mandatory premarket testing, transgenic companies have every incentive to screen out problem products.

The problem, counters Jane Rissler, a plant pathologist with the Union of Concerned Scientists, is that with transgenic foods, such postmarket action is often too late: if some future transgenic food is found to cause health or other problems *after* it gets into the food supply, containment is next to impossible. This weakness was highlighted in 1998 when the U.S. government approved a Bt corn variety known as StarLink for sale as animal feed but not as human food. (The agency was waiting for human health studies.) Despite this stipulation, the StarLink Bt toxins turned up in more than three hundred consumer[30] products in grocery stores, including Kraft taco shells — in large part, experts believe, because in the modern commodity industry, product streams are almost impossible to segregate. The StarLink debacle led to a massive program of mill closures, export bans, and product recalls that cost StarLink's maker, Aventis, an estimated $1 billion,[31] demonstrating that even those producers or consumers who wanted to shun the gene revolution might not have a choice in the matter. Even today, consumers wishing to avoid transgenic foods cannot, because the industry has successfully blocked any requirement that transgenic foods be labeled — despite surveys showing that nine out of ten consumers want such labels.[32]

The environmental record of transgenic technology is similarly clouded. Transgenic crops have long been touted as promoting sustainability because they reduce the need for herbicides and insecticides, but the evidence is mixed. Although these crops do allow farmers to use substantially less insecticide,[33] some research suggests that the use of herbicides has actually increased. Why? Farmers who use a herbicide-tolerant crop, like RoundupReady soybeans, tend to use only the associated herbicide, like glyphosate-based Roundup, on that crop, which then causes weed populations to adapt to the herbicide and eventually become resistant to it. Once these superweeds emerge (and since RoundupReady crops were introduced, in 1996, thirteen weed species in fourteen states have become resistant to glyphosate),[34] farmers must find a new herbicide, and in some cases they've been switching to older, more potent products, such as paraquat and 2,4-D.[35]

A second environmental risk centers on something called gene flow. Early on, skeptics argued that the transgenic plants would mate, or outcross, with related wild species and pass their transgenic traits to the offspring, effectively allowing the engineered trait to escape any controlled usage and move into the wild. In fact, because outcrossing is likely only if the transgenic crop is near a population of wild relatives (which is rare with modern farm crops like soybeans or corn), gene flow into wild plant populations hasn't been a big problem.[36] Gene flow does occur, however, between transgenic varieties and *non*transgenic varieties of the same farm crops, as seeds and pollen are carried from field to field. This variety of gene flow is a great worry, and not just for organic farmers. Given consumer anxieties about transgenic foods, *all* farmers worry about the fallout of transgenic contamination — particularly because some transgenic plants are designed not for food, but to manufacture pharmaceutical compounds, raising the possibility, according to a National Academy of Sciences report, that "crops [genetically] transformed to produce pharmaceutical or other industrial compounds might mate with [crops] grown for human consumption, with the unanticipated result of novel chemicals in the human food supply." Or, as an executive with Pfizer Pharmaceuticals put it to an FDA panel, "[W]e've seen it on the vaccine side where modified seeds have wandered off and have appeared in other products."[37]

All food plants, even conventionally bred varieties, carry some degree of risk, and because of the challenges facing the future food system and the urgent need for new solutions, the potential for health and environmental harm of transgenic foods has to be weighed against the potential benefits, such as dramatically higher yields, or salt tolerance, or nitrogen efficiency, which proponents promise are imminent. But such promises, it turns out, may not be so easy to keep. On the question of yields, for example, transgenic crops have shown only fractional gains over their nontransgenic counterparts, and in some cases, transgenic yields are lower.[38] Proponents counter that transgenic breeders haven't been focused on yields but on "money" traits, such as herbicide-tolerance resistance, which farmers have been willing to pay for. With grain prices soaring and with the new prominence of food insecurity among policymakers, research will focus on yields, and results will follow.

But here, too, there is reason for skepticism. Kendall Lamkey, a breeding expert and chair of ISU's agronomy department, argues that the massive yield gains the industry promises will require a level of technical mastery that even transgenic technology may be hard-pressed to deliver. Whereas existing transgenic successes, such as herbicide tolerance, involve the manipulation of just one or two genes, a meta-trait like yield is vastly more complex. Yield is essentially an index of the plant's ability to reproduce itself, and reproduction requires a series of thousands of specific events, from germination to the setting of ears, during which the plant must employ every biological skill in its genetic arsenal, from nutrient uptake to adapting to temperature change. Yield, says Lamkey, is the result of "the entire plant functioning together, and using every resource it has to reproduce itself."

To get large, sustainable boosts in yield, Lamkey says, companies will need to manipulate not just a few maize genes, but a large proportion of maize's fifty thousand genes. What's more, companies will need to do this in such a way that the newly engineered trait is stable and the new plant can deliver its higher yields, not just in the laboratory, but in the field, across the same range of conditions that conventionally bred plants can tolerate. And to date, such stability has proven elusive — largely, says Lamkey, because as trait complexity increases, trait stability naturally falls. In a simple trait, such as herbicide tolerance, which involves just a few genes and just a few chemical reactions influenced by those genes, the gene-trait connection is much more like an on-off switch and, as such, will express itself under most conditions; a RoundupReady seed's herbicide tolerance will persist regardless of changes in temperature or rainfall. Yield, by contrast, involves so many different traits and underlying chemical processes that, so far, most of the big yield gains researchers have engineered transgenically hold up only in the laboratory or in a narrow set of field conditions. "If you move the new plant to a new environment," Lamkey told me, "the yield effect just goes away."

Findings such as these have led some industry observers to seriously question the conventional heroic narrative of the gene revolution as a savior of the food economy — and to propose an alterative narrative more in keeping with industry tendencies toward consolidation and risk reduction. Pat Mooney, director of ETC, a food advocacy group, and a veteran observer of the seed business, says input companies have long known that engineering supertraits such as yield would be enormously difficult. He argues that the real goal of transgenic technologies wasn't so much to develop supertraits, but to transform the seed business into the same value-added model that is so profitable in other food sectors.

Until late in the twentieth century, Mooney points out, the global seed business was the antithesis of rational industrial agriculture — more than seven thousand different companies, and none with more than 1 percent of the market. In the early 1980s, the big farm chemical companies began buying up seed companies by the hundreds, hoping to become fully integrated input suppliers that could sell farmers a complete package — seeds, fertilizers, and pesticides. Companies also saw seeds as a way to add value to their core farm products — pesticides — which, like everything else, had become such low-margin commodities that companies like Monsanto were actually losing money on former cash cows like Roundup.[39] And the gene revolution would make these strategic goals much easier to accomplish.

First, because transgenic technologies are so costly, Mooney says, the relentless hype about an imminent gene revolution persuaded small seed companies that they wouldn't be able to compete, which allowed the big input firms to more easily consolidate market share. Second, after decades in which seeds had changed little, transgenic technologies allowed input companies to create truly novel seed traits, such as herbicide tolerance, and thus justify charging premium prices. Third, transgenic let companies tailor those new seed traits to the companies' existing farm chemicals and create seed-chemical "platforms," like Roundup and RoundupReady soybeans. This was key, because developing a new farm chemical like Roundup costs a company hundreds of millions of dollars, but developing a single seed trait to match that chemical is much cheaper. Once these seed-chemical platforms had been accepted by the marketplace, input companies could significantly extend the shelf life and market share of their farm chemicals (even after the patents on the chemicals themselves expired, as is the case with Roundup) by continuously bringing out new transgenic crops, such as RoundupReady cotton, that were dependent on the core chemical.

But the biggest benefit of transgenic technologies is in many ways the least obvious: input companies can now own a piece of the food chain that had previously rebuffed all attempts at corporate possession — the seeds themselves. For centuries, seeds have lacked the kind of patent protection accorded other inventions, such as farm tractors or herbicide formulations; once a new seed trait was on the market, it was fair game for farmers and other breeders. Part of this vulnerability was intentional: governments had traditionally regarded seed traits as a public good, to be saved and traded by farmers as a means of assuring food security.* But part of it was technical: until the gene revolution, breeders had no means of proving that they'd bred a particular trait. Transgenic technology changed that.† Because companies can now alter a seed's DNA and then describe that alteration pre-

---

* The public-good concept had been a driving force behind the early breeding programs at universities, government research centers, and international consortiums like CGIAR, all of which tended to give their seeds away to farmers — much to the consternation of seed companies. In 1970, the seed industry persuaded Congress to grant seeds some patent protection, but even here, farmers were allowed to save seed, a practice that input companies argued cost them hundreds of millions of dollars in lost sales.

† In the 1990s, a U.S. seed company, Delta and Pine Land, developed a trait that instructed a transgenic plant to become sterile upon maturity; with this so-called terminator trait, farmers could no longer save seeds but would have to buy new ones each year — a potentially devastating change for developing countries. Monsanto, which bought Delta and Pine Land, came under heavy fire from NGOs and eventually promised not to use the terminator in food crops.

cisely, traits have become a biological form of intellectual property, just like books or software, and just as worthy of protection — especially given the massive financial risks input companies take to launch new traits. This argument has been persuasive with patent agencies in the United States and Europe, which have begun granting more protections for seeds. As important, during global trade talks, negotiators from the United States and Europe, partly at the behest of input companies, have insisted that trading partners honor patent protection to seeds.

For the input industry, such protection has been critical. In 2007, nearly three-quarters of Monsanto's profits came from transgenic seeds and traits[40] — in no small measure because Monsanto is the absolute owner of traits such as Roundup resistance. This means the company can insert those traits in its own seeds or out-license the traits to other companies, like Syngenta; it can also require farmers who buy the RoundupReady traits to promise contractually not to save the seeds, and it can (and does) sue those farmers who save seeds anyway. Between this proprietary control of seeds and the company's relentless efforts at consolidation (abetted significantly by seed ownership), Monsanto now controls a fifth of the $20 billion global proprietary seed market (and just three companies, Monsanto, DuPont, and Syngenta, account for 44 percent of market).[41] Further, with its various licensing agreements, Monsanto owns 90 percent of the transgenic traits sold worldwide.

Before the 1980s, such a concentration of economic power, especially in a sector so central to food security, would have set off "antitrust" alarms among the government regulators who prohibit monopolies. But since that time, when the Reagan administration initiated a trend away from antitrust enforcement (ostensibly because larger companies, being more efficient, are better for consumers), most of the attention that the consolidation of the seed industry has attracted has been positive. In 1998, after Monsanto bought DeKalb, the number two corn seed producer, and announced plans to acquire another seed firm, one analyst admiringly declared that "Monsanto is creating a giant tollbooth in front of the cotton market and the soybean market and canola market and the corn market."[42] But for industry critics, this concentration of power is far more unnerving than any health concerns about the transgenic foods themselves.[43] "Seeds are the first link of the food chain," argues Hope Shand, another analyst with the ETC group. "And whoever controls the seeds controls the food supply.

We're not talking about software or mousetraps. We're talking about the very basis of the food system, and the fact that it is in fewer and fewer hands is something that should concern everyone."

Proponents of transgenic technology, naturally, take a different view as to the meaning of their success. Officials within the input industry argue that the huge market share and economic power of a company like Monsanto, coupled with the assurances of proprietary power over its seeds, have created the conditions under which significant new traits can finally be developed — traits whose power will simply leave old-school skeptics like Mooney and Lamkey gasping. And such potential cannot be ruled out, given the kinds of innovations that companies like Monsanto and scientists like Patrick Schnable are working on.

What even optimists acknowledge, however, is that although considerable gains are almost inevitable, it is less certain *when* these improvements will take place, or under what financial conditions. Even single-trait seeds can take anywhere from six to thirteen years to bring to market, and more complex traits like yield or drought tolerance could take much longer. When I asked Monsanto's Crosbie when he expected to see the beginning of the surge that will take corn yields past three hundred bushels by 2030, he admitted that no one knows. "You can graph [yield growth] as a straight line, but the growth will actually come in jumps, in stair steps," he said. "And it's hard to predict exactly when the next stair-step will happen, until it happens."[44]

Significantly, development costs for these complex supertraits, such as yield, are widely predicted to be high — a fact that is hard to square with a current industry business model focused on developing least-cost traits built for existing farm chemicals like Roundup. The conventional strategy, says Mooney, is that "once you have a chemical accepted by the market, you spend most of your time working to expand the geography of the crops that can use it." What this means is that companies looking to maximize the return on the money they've already sunk into a core chemical, like Roundup, can be expected to choose seed traits that support that chemical and that can be inserted into the most widely planted commercial crops, such as corn and soybeans. Not only do these crops offer the largest potential sales volume, but they are grown by large-scale farmers who can afford to buy herbicides, transgenic seeds, and other high-tech inputs. Farmers like Mango and Janet Mutisya are not a big part of Monsanto's transgenic seed strategy.

This is the main reason nearly all transgenic crops are now planted in just six countries — the United States, Argentina, Brazil, Canada, China, and South Africa — and almost none in the least-developed countries, who cannot afford such inputs. It is also why, according to the FAO, just four crops (corn, canola, cotton, and soybeans) and just two traits (herbicide tolerance and pest resistance) now account for 99 percent of the transgenic crops planted worldwide.[45] By contrast, the FAO says, because most of the transgenic research and development is being carried out by private companies, there have been "no serious investments" in creating transgenic varieties of crops such as sorghum, millet, pigeon pea, chickpea, and groundnut, which are far less commercially viable than soybeans or corn but are the five most important crops for poorer farmers in semiarid tropical regions like sub-Saharan Africa and India.[46]

Even the seed industry itself has shifted in the way it talks about transgenic foods. Although company executives still come before congressional hearings with promises of world-saving technologies (and although Congress continues to respond with hundreds of millions of dollars in research funding), the companies have begun taking a considerably more cautious approach. Monsanto's recent comeback from financial troubles, notes *Forbes* writer Mark Tatge, was widely attributed to the jettisoning of both CEO Robert B. Shapiro and Shapiro's "save-the-world mission." Since then, Tatge notes, Monsanto "has turned inward." Or as Monsanto's new CEO, Hugh Grant, puts it: "We decided that, near term, our biggest block should be the Americas, where the majority of the crops are grown."[47] And in fact, much of the research funding, both private and public, being poured into transgenic development today isn't going into designing new crops at all, but into designing varieties of corn that can convert more efficiently into ethanol.

While many antitransgenic activists continue to get worked up over patent protection as the main threat to poor farmers, the larger risk is that the transgenic industry will simply bypass the developing world altogether. Genetically modified crops, complained Louise Fresco, the former head of FAO's agricultural division, were created "to reduce input and labor costs in large-scale production systems, not to feed the developing world or increase food quality."[48] Not only is transgenic research missing poor-country crops, Fresco continued, but it is largely avoiding the traits needed by the developing world, especially drought and salt tolerance and resistance to tropical diseases.

In recent years, philanthropic organizations, most notably the Rockefeller Foundation and the Gates Foundation, have made multimillion-dollar commitments to programs that include research in transgenic crops specifically tailored to the developing world. But here again, proponents must confront the gap between what they can pay for and what these new traits will cost. "It comes down to economics," says ISU's Schnable. Although genetic technologies allow the industry to do "almost anything" in terms of potential new traits, companies "are struggling with the question of where to put their limited R&D money." And at any for-profit company, Schnable says, investments will always be targeted toward products with the highest probability for repayment: "It would be hard for a seed company to recapture its investment in traits for people in the developing world, who don't have a lot of money to spend."

Even some in the seed industry itself concede that launching a new generation of world-saving traits will require the kind of massive public investment that launched the Green Revolution. Mike Phillips, with the industry group BIO, acknowledges that while seed companies might profitably develop and market drought tolerance to corn farmers in Nebraska or salt tolerance to farmers in Utah, "you're going to have to work with governments as far as getting those traits available in the developing world."

This reality runs counter to the oft-stated forecasts that a gene revolution will restore food security in places like Africa and southern Asia and has fostered the view among critics that the industry is simply using problems like hunger to gain political support for transgenic technologies it can't afford to offer to the hungry. But it's also true that transgenic technologies, whatever their real potential turns out to be, are themselves being crushed under magic-bullet expectations. That is, transgenic technology is being touted as a fix for complex problems, such as food insecurity or drought tolerance or chemical overdependence, whose root causes go far beyond a lack of quality seeds. Hunger is a complex social, political, economic, and ecological problem that requires social, political, economic, and ecological solutions, none of which can be genetically engineered. Drought tolerance is relevant only when crops can expect at least *some* rain.

Similarly, building a truly sustainable agriculture will take far more than a corn plant bred to lower pesticide use or to use less nitrogen. Companies may well genetically engineer corn to fix some of the nitrogen it needs, says ISU's Lamkey, but in a place like Iowa, the real nitrogen problem isn't caused by plants but by, among other things, a corn-soybean regimen that leaves soils exposed during the wet winter months and a drainage system that sucks water and nitrogen from those fields and deposits it in the groundwater. "You can modify all the genes you want," Lamkey says, "and it won't solve that problem."

As the heroic narrative of transgenic technology begins to unravel, one natural beneficiary would seem to be the organic movement, whose products are not only unengineered but also address many of the issues, such as sustainability and the social importance of farming, that the transgene industry largely misses. But in a strange way, the organic movement now finds itself in a position not altogether different from that of its transgenic rivals, as many of its core "alternative" values are forced to conform to an economic and political model that sees only what it can attach a monetary value to.

Most visibly, the arrival of big retailers like Wal-Mart is pushing this once fractious and individualistic industry into the familiar supply-chain format. As these big retailers exert their famous price pressure, the price premium that now allows smaller, less cost-efficient players to participate will narrow and the market will swing toward larger, low-cost producers. In some cases, these larger players will be veteran organic producers who have scaled up their operations. But in most cases, says David Swenson, an agricultural economist who studies the organic sector, more and more of the organic production will be generated by large conventional producers, who, by simply adjusting their existing large-scale methods to meet federal organic standards, have been able to capture both the higher price premium that organic provides and the cost-cutting economies of scale implicit in the industrial model. And while the current organic price premium will allow smaller players to stay in business for a while, that premium will eventually shrink as organic becomes "normalized" and as larger producers, in the familiar drive to maximize sales volume and market share, cut prices to the bone. "The point where these large producers stop making a profit is considerably lower than it is for small producers," says Swenson, who believes that the organic price premium will become too small to keep smaller players in business.

And consolidation isn't the only familiar trend creeping into organic. Already, organic production, once a prime locus of the locally grown movement, is following the conventional pattern of regional specialization dominated by agro-zones like the Central Valley in California. In fact, big buyers like Wal-Mart, General Mills, and Dean Foods have been so successful at spurring demand for low-cost organic ingredients that domestic suppliers cannot meet it all, and big buyers are turning to the big, low-cost producers in Brazil, Argentina, and China. As these imports increase, they will drive down domestic organic prices farther, which, while benefiting consumers, will also accelerate the organic industry's evolution toward a large-scale, low-cost model that would seem to leave little room for smaller farmers, local production, and many of the other values once associated with the organic movement.

On one level, these are predictable and perhaps inevitable developments. As organic moves further into the mainstream market, organic producers must adopt more of the rules of that market. And that market doesn't know how to attach a price to "small scale" or "local" because whereas a growing number of consumers will pay more for food that is pesticide-free, most aren't yet ready to pay more for food that is produced locally or on smaller farms. During the early 1990s, as the USDA developed its organic standards, some organic advocates lobbied to formalize values such as "local" and "small scale" by including them in the federal standards. But as the Leopold Center's Fred Kirschenmann, who sat on the USDA's standards advisory board, recalls, these softer values were too nebulous for a federal agency accustomed to specific numerical standards. As Kirschenmann puts it, "[T]he lawyers at the USDA said the regulation needs to be based on a 'yes' or a 'no.'"

This, too, is hardly surprising: the USDA works primarily with large, industrial food companies, which demand specific, measurable targets — minimum pesticide residue levels, for example. Once such numerical standards are set, companies are then free to meet them as cheaply as they can by whatever means are at their disposal — moving to large-scale production, for example, or buying cheaply overseas. Under such a system, "local" or "family farmed" or other softer, less easily quantified ideas have as little meaning as "homemade."

Yet if the mainstream market hasn't been able to fully engage with the deeper ideas of sustainability, organic farming has also become somewhat disengaged. For example, because organic farmers can't use chemical herbicides, they often control weeds mechanically — that is, by repeated tillings of the soil before planting and after harvest — a practice that keeps weeds in check but in some areas so disrupts the soil structure that it actually accelerates erosion and nitrogen loss (and thus climate impact), while adding considerably to fuel use.[49] But because the federal organic standards do not specifically address erosion, climate, or energy use, organic farmers may not be as financially inclined to worry about these externalities.

Similarly, where federal organic standards forbid the use of synthetic fertilizers on organic fields, these standards don't speak to the deeper problems that synthetic fertilizers represent, such as the disappearance of farming systems that once replenished nutrients on-site. Traditionally, farmers replaced soil nitrogen by planting cover crops or spreading manure. But in today's ultraefficient market, even many organic farmers are unwilling to run diversified operations. Many commercial organic vegetable or row-crop farmers can't afford the time or land to raise livestock to make manure; they are, in effect, under the same pressure to specialize as are their conventional competitors. By the same token, few commercial organic farmers want to alternate their cash crops with nitrogen-fixing cover crops, since doing so would mean fewer "turns" of their higher-value cash crops. Thus, when ISU's Delate encourages her organic farmers to plant cover crops, many just shrug their shoulders. "They tell me they can't afford to plant cover crops, because they are selling every vegetable crop they can plant," Delate told me. "It's cheaper for them to plant another crop and simply buy the manure" from some off-site producer.

Here is the heart of the problem. Although using off-site manure doesn't technically violate "organic" standards, it does contravene an important concept of alternative agriculture: by relying on off-site manure, these farms are committing what many sustainability advocates decry as "input substitution" — that is, replacing one problematic input with another that is only slightly less problematic. Manure may not be synthetic, but it often must be trucked in from a distant livestock operation, which requires considerable fuel (especially if there are no livestock operations nearby, which, with the growing regionalization of livestock, is increasingly likely). In this sense, manure is only a temporary solution because the underlying problem — a reliance on off-site inputs — remains unchanged.

Stuart Hill, a sustainable agriculture expert at the University of Western Sydney, sees input substitution as one of the key flaws in modern organic farming and much of what passes for sustainable agriculture, not least because substitution gives the appearance of having solved the problem while possibly making it worse. The better our substitution strategies become, writes Hill, "the more we are unintentionally protecting and perpetuating the design and management features of the agro-ecosystem that are the underlying causes of the problem."[50]

The more cynical of organic's many critics contend that the movement has sold out: organic, the saying goes, is now little better than a product line of the food industry it once seriously challenged. "The reason organic is so successful today is because the food industry is totally comfortable with it," says one Washington-based sustainability lobbyist. "The USDA has these easy standards, all the big food companies have their lines of organic foods, and it's just not radical anymore. If you're an environmentalist, you can get the most hard-core rightwing Congressman to talk to you just by uttering the word 'organic.'" A less cynical view is that the organic movement, under the pressures of the market, has become too rigidly centered on well-intended standards that are now blinding some of its members to the need for a fresher, more flexible approach. "Organic," says Yee, the small-farms advocate, "has fallen into the prescription trap, where practitioners say, 'OK, if you do this, this, and this, all will be right in the world.' Well, it's not true."

Ultimately, the most pressing challenge for organic farming, or for any of the alternatives, isn't whether they can remain true to their conceptual roots but whether they can, in any form, meet the needs of a planet soon to have substantially more people and substantially fewer resources. Despite booming growth, the organic farming sector is still a tiny share of the food market — less than 2 percent in the United States. And while current growth rates have that share doubling every five years or so, there are legitimate questions about how large organic can become before it, too, runs into limits.

Beyond the extra fuel and farm labor that many modern organic methods require, there are concerns about other inputs, notably fertilizers. Manure may seem to be in oversupply today, given the sewage problems at industrial CAFOs, but if organic farming was somehow boosted to the scale of conventional farming, those poop surpluses would quickly turn into deficits. Vaclav Smil, the University of Manitoba resource economist, calculates that feeding ten billion people using natural fertilizing methods — mainly manure and cover crops — would require anywhere from a doubling to a tripling of the current area of global farmland, which, given the current scarcity of arable acres, could occur only with massive additional losses of forests and other ecologically critical lands.[51] Other researchers have challenged Smil's findings, arguing that he underestimates several key potential sources of nonsynthetic nitrogen: more encouraging studies suggest the possibility of crops and cropping systems that "fix" nitrogen far faster than current cover crops do — which we'll explore more fully in the next chapter.

Yet even these optimistic forecasters admit that future fertilizer supplies will be tighter than they are today and will necessitate substantial changes, both in terms of how we produce food and how (and in what form) we consume it. For example, nearly all scenarios for sustainable future food production assume large reductions in meat consumption — assumptions that may sit well with diehard sustainability advocates but will likely fare poorly with mainstream consumers and producers; indeed, even the organic community itself may have trouble swallowing a less carnivorous future, considering organic meats are now one of the hottest sellers.

In that sense, the movement toward organic and other alternative forms of food production is colliding with many of the same limits that their high-tech rival, the transgenic industry, has already encountered. Like transgenics, alternative agriculture suffers from an ideological purism: a kind of my-way-or-the-highway attitude that brooks little dissent or compromise. The organic movement has also been co-opted by the market it set out to change. More to the point, neither organic farming nor any of the other alternatives vying for attention and acreage can by themselves hope to address the coming "food problem." What is becoming clearer by the week is that the food challenges of the future — a rising population, degrading soils, declining quantities of energy and water, climatic instability, and a host of food-related heath problems — now exceed the capacity of any single technology or school of thought. Instead, such challenges will require not only new technologies and methods but an openness to new ideas about what constitutes success and failure — ideas that could be as foreign to many consumers and policymakers as either transgenic foods or organic farming was when they emerged decades ago. In building the next food economy, making room for those ideas may prove to be the most challenging step.

10

Food Fight

O N   T H E   M O R N I N G of Wednesday, September 13, 2006, employees at the Kames Marine Fish Farm, near Kilmelford, Scotland, arrived at work to find that the huge halibut farm had been attacked the night before. Offices had been ransacked, a boat had been wrecked, and a huge crane disabled. More seriously, the massive underwater pens that had held some fifteen thousand farm-raised halibut had been torn open, and the fish — some of them forty pounds in size and all promised to a large British grocery retailer — had vanished.[1] Police had no individual suspects; the autumn night had been quite dark, and the only strange activity any witnesses could recall was a blue or possibly burgundy minibus seen driving in the vicinity the evening before. But investigators had few doubts as to the attackers' motives. A wall had been spray-painted with the letters "ALF," the acronym for the Animal Liberation Front; some of the damage, such as the discharging of fire extinguishers to cover fingerprints, was trademark ALF, whose large but shadowy membership holds virtually all human uses of animals, whether for food or research, as a form of cruelty and a justification for sabotage. And indeed, later that day, a Canadian animal-rights website posted an anonymous communiqué from an ALF cell in Scotland that claimed responsibility for the incursion. "All the pens were destroyed & sunk," the communiqué stated, "& we saw hundreds, if not thousands of fish swimming free towards the sea."[2]

What ALF failed to mention (perhaps because cell members hadn't waited around long enough to notice) is that many of the liberated halibut did not have long to enjoy their freedom. Thousands of the pen-raised fish, apparently disoriented by their first experience in open water, became stranded on nearby beaches and died. Others were tangled in seaweed or were eaten by herring gulls and otters; industry experts speculated that many of the rest were either consumed by other predators or had simply starved to death. "They claim they liberated them into the sea," Stuart Cannon, Kames Farms' managing director, told the *London Times.* "But sadly, as we all know, farmed animals, whether they are fish or any animals, don't survive unless they are looked after." Then, perhaps aware of the incident's symbolic importance, he added, "We farm them in a sustainable way. The welfare of the fish is at the forefront of our minds. Isn't it better to have farmed fish than to be pillaging the seas where stocks are declining dramatically?"[3]

Cannon may or may not care one whit for the welfare of his fish, but the attack on his fish farm does reflect the complex and often bizarre challenges of building a new food system. Cannon is quite correct that the world's natural fisheries have been overfished to the point of exhaustion (the global catch hasn't risen above ninety million tons a year since the mid-1990s, despite more aggressive commercial fishing operations). He is also right to suggest that farm-raised fish offer an appealing means to supply a meat-hungry world with cheap protein. Cannon's halibut, for example, need just 1.25 pounds of feed for each pound of weight gain, and farm-raised salmon need even less, which make fish a far better protein proposition than even chicken and which helps explain aquaculture's extraordinary growth: more than a third of the world's fish supplies are raised on farms, and by 2025, that fraction could be more than half, depending on how rapidly the wild fishery collapses and how well people like Stuart Cannon continue to position farm-raised fish as the food of the future.

What Cannon doesn't say — and what groups like ALF offer as a justification for what they do[4] — is that aquaculture's massive forward momentum also masks some extraordinary externalities. Fish are indeed efficient feed converters, but carnivorous species like salmon and halibut are fed on fishmeal made from herring and other smaller species harvested by the same industrial methods that depleted other fisheries; nearly a sixth of the entire commercial fish catch is fed to farmed fish.[5] And inputs aren't the only troublesome issue. Fish farms are essentially floating CAFOs: A large-scale operation generates as much nitrogen-rich fecal matter as a human town of sixty-five thousand, creating enormous water-quality problems in the bays and inlets where these facilities are built.[6] And, as animal-rights groups argue, the fish themselves may not find aquaculture a vast improvement over life on the open sea, due to the crowded conditions of the pens, the high incidence of disease, and the often rough handling and processing methods.

Some of aquaculture's external costs can be addressed by better practices — for example, moving pens into deeper water, where steady currents can dilute sewage to the point of innocuousness. But because modern aquaculture is driven by the same least-cost model that governs all meat production, fish farmers tend to favor the cheaper inshore facilities and (as with terrestrial livestock) often "solve" their water-quality problems simply by moving from countries that have relatively strict regulations, like the United States, to countries that do not, like Chile. And because of the world's eagerness for cheap, efficient protein, most forecasters expect high-impact aquaculture operations to become even more prevalent over the next three decades — a prospect that, while it hardly justifies the actions of groups like ALF, explains something of the frustration industry critics feel in their inability to effect change through normal channels. As Keith Mann, a British ALF activist, explained in a 2006 documentary, "We all desperately want to fight animal abuse legally. It just doesn't work like that. The people who run this country, they have shares, they have investments. . . . [S]o to think that you can write to these people, and say 'we don't like what you're doing, we want you to change,' and expect them to do so, it's not going to happen."[7]

You needn't be an ALF sympathizer (and given how their tactical extremism alienates all but a small fringe of consumers, it's unlikely that you are) to understand the frustration such groups feel at the apparently unstoppable momentum of the modern food economy. Despite a growing number of highly visible weaknesses and vulnerabilities, our system of food production continues not only to expand at an accelerating clip but also to thwart the emergence of any real alternative systems.

Much of this momentum is brute economic force: food demand is rising rapidly and it is simply far easier to meet that need with the production system already in place. And yet, as the debate between organics and transgenic foods makes clear, there are other sources of resistance to new ideas and alternative models. Stakeholders in the current system, such as large industrial producers and input suppliers who have billions of dollars invested in product lines, processing facilities, and other pieces of physical and virtual infrastructure, routinely use their substantial economic power and political influence either to block alternatives or co-opt them. Lawmakers are regularly corrupted by the system (witness Congress's repeated inability to reform the U.S. farm program), and even policymakers who haven't been co-opted are nonetheless reluctant to force an industry so large and economically important to change too quickly.

But there are other, less easily vilified contributors to the momentum of the status quo. Mainstream consumers, for all their celebrated insistence on novelty and change, show little interest in a new food economy that might require them to pay substantially more for their food (to cover its external costs) or to eat substantially less of something they enjoy (such as meat).

More subtly, there is a strong component of institutional momentum within the entire food-making community — among the food producers and processors, of course, but also the food scientists and technicians whose job it is to develop and implement new ideas — that makes it very difficult to even consider alternatives. Cynics point to the massive amounts of money and expertise that the food industry routinely lavishes upon universities, government research centers, and other ostensibly public institutions — funding that, in an era of dwindling public budgets, often leaves researchers reluctant to critique mainstream ideas and technologies. But a deeper intellectual inertia is operating here. In a world of hyper-cost-consciousness, food companies naturally tend to spend their research dollars on familiar technologies with assured returns, thus avoiding high-risk ideas that might be truly revolutionary. And, as we've seen, many players in the modern food business don't believe that the system really is broken: after a century of dramatic successes at nearly every level of the modern food system, there is a profound unwillingness to leave those methods behind and gamble on alternatives that appear to violate every sound principle of food production. In this context, to suggest that large-scale monocultures aren't always superior; or that efficiency needs to be recast as a means, not an end; or that external costs can't be deferred indefinitely; or that a sustainable food system may need to be driven by something other than the low-cost imperative is akin to Galileo's suggesting to church officials that the Earth may not be at the center of the universe.

The point here is that transforming food production into something more sustainable isn't simply a matter of exchanging one set of inputs for another or finding some new technology, but of developing a new way to *think* about food and food production. And given the political and intellectual inertia behind the existing system, what is becoming clearer all the time is that the battle over the next food economy will be as much about ideas as economics, and that the route to a truly sustainable food system isn't likely to be the path of least resistance.

On the southern Japanese island of Kyushu, one of the key battles in the shaping of the next food economy is taking place on a seven-acre rice farm. Each June, Takao Furuno[8] releases hundreds of ducklings into his newly planted rice paddies; the ducklings ignore the seedlings (which contain too much of the mineral silica for their liking) but gobble up insects and weeds. Their waste fertilizes the rice, while their constant churning of the paddy bed stimulates the roots of rice plants to grow faster. As the season progresses, Furuno fills the paddies with a freshwater fish species known as loach, which he protects from the voracious ducks with a layer of duckweed, an aquatic fern. The duckweed uses solar energy to fix nitrogen into the paddy soils to fertilize the rice; it also harbors a blue-green algae that feeds a species of worm that is eaten by the fish, whose wastes also fertilize the rice. In the fall, Furuno removes the ducks (who would otherwise eat the mature rice) to a barn, where they produce eggs and grow to market weight. He harvests the rice, plants the paddies in a cover crop of wheat, and rotates his entire property in several dozen varieties of intensively cultivated vegetables,[9] which he sells, along with his rice, ducks, eggs, and fish, to his neighbors.

Furuno's system, dubbed the Aigamo method, is everything an alternative model of food production should be: solar-powered and closed-looped (all nutrients are produced on-site); free of synthetic herbicides and insecticides; and, with the exception of some grain to fatten his ducks and some student laborers, requiring no outside inputs. Most fundamentally, it is astonishingly productive. Although systems like these — known variously as integrated farming, adaptive management, and polyculture — are praised for their environmental friendliness, Furuno gets 3.5 tons of rice per acre (about as much as his conventional farm neighbors earn), along with enough vegetables, duck meat, duck eggs, and fish to supply one hundred local families year-round. Such bounty generates an income of $136,000[10] and justifies the inscription on his business card: "The world where one duck creates boundless treasures."

Furuno is, in other words, soundly refuting the conventional wisdom that high agricultural output can't occur without strict adherence to a model of monoculture. In the process, he is highlighting one of the key debates about the future of food: namely, whether food systems of the future will be characterized by the uniformity of the present or by a variation on the more diverse systems that dominated the past. For Furuno and other adherents of polyculture, the answer is clear: only by moving to a more diverse mode of agriculture can a resource-constrained world hope to generate the food it will need without incurring overwhelming external costs.

For all its potential, however, the concept of a reintegrated agriculture has yet to be embraced by more than a tiny minority of mainstream producers — for reasons that demonstrate some of the fundamental obstacles to changing the food system. Intensive agriculture of this kind favors smaller farm sizes, a requirement that may conflict with Japan's current push to consolidate its millions of tiny rice farms (the current average size is three acres) into larger, more conventionally managed units. Polyculture is also quite labor intensive, which makes it ideal for places like China and Africa, where farm labor is plentiful, but challenging in an advanced economy like Japan's, where industrialism has drained the countryside of its workers. Rural labor is so scarce that many Japanese farmers must use student interns, volunteers, and even retirees — a reality that helps explain why in Japan, a country that has practiced intensive polyculture for centuries (and which, being mountainous and densely populated, would be an ideal candidate for *more* polyculture), just one in twenty farmers actually practices it.[11] "When you go to these greenhouses," says John Dyck, an Asian specialist at the U.S. Economic Research Service,[12] "many of the hired workers are women in their 70s and 80s." Such workers have a wealth of knowledge about farming practices, says Dyck, but "that's hardly a sustainable labor pool."

And labor is but one potential problem. In Boone County, Iowa, the very heart of industrial monoculture, agronomist Matt Liebman and his colleagues at Iowa State University have crafted an American variant of Furuno's duck-rice system that dramatically cuts the need for farm chemistry while minimizing erosion. By using a four-year crop rotation, in which the usual corn-soybean regime is followed by nitrogen-fixing cover crops like alfalfa, and by encouraging populations of deer mice, crickets, and other seed predators that eat weed seeds, Liebman has been able to cut herbicide use by 85 percent and nitrogen inputs by 75 percent — a significant success during a time of rising input costs — without hurting yields. And because the soils are protected year-round by cover crops, this model dramatically reduces nitrogen leaching and volatilization; according to one study, an intensive legume rotation can cut a farm operation's contribution to global warming by more than 60 percent.[13]

But here, too, Liebman is encountering resistance to his ideas. Because his model assumes that farmers will always have the option to use small amounts of synthetic pesticides and fertilizers (an option some sustainability advocates believe will probably always be necessary), the crops cannot earn the organic price premium, reducing incentives for conventional farmers to try it. Similarly, because these diversified farms are selling a variety of products, not all of which have developed markets or government support, these farmers will be at an economic disadvantage at harvest time. In the United States, cover crops like alfalfa don't earn federal subsidies; if a farmer grows more alfalfa than he can feed to his own livestock (assuming he has any) or can sell to his neighbors, he's not likely to get the same return per acre as would a farmer who simply grows the amply subsidized and vastly more marketable soybeans and corn.

Perhaps more fundamentally, integrated farming systems like Liebman's or Furuno's largely contradict one of the central principles of modern industrial agriculture — the trend toward simplicity. Most models of polyculture are not only labor intensive but what Liebman calls "thought-intensive" as well. Because polyculturalists are attempting to re-integrate components, such as fertility, livestock, and cropping, that have been kept separate for nearly a century, these new systems are by necessity much more complex and experimental than the monocultures they hope to replace. Furuno, for example, must carefully monitor the performance of each crop and apply any new insights the following season. Similarly, farmers using natural weed predators like mice must ensure that the little animals have ample plant cover under which to hide from hawks.

More problematically, because these models invariably require more labor hours, they don't appeal to today's high-volume, low-margin farmers. With commodity profits so narrow, the average U.S. farm family now earns most of its income from nonfarming activities, especially off-farm jobs — extracurricular activities that are possible largely because labor-saving synthetic pesticides, fertilizers, and machines have dramatically cut the hours farmers must spend in their own fields.[14] With all the technological advances built around the single-crop model, "it really has become possible to be a three-month farmer today," says Ferd Hoefner, of the Washington-based Sustainable Agriculture Coalition. "But the more diverse your farm system, the more that management goes back to being year-round." When farmers are considering whether to adopt an integrated model or stick with a simpler monoculture, Hoefner says, the potential loss of off-farm income is "a huge motivator."

Last, polyculture is far riskier than its conventional rival. One of the advantages of conventional farming and its reductionist principles is that they break agriculture into series of relatively predictable problems — declining fertility, say, or bugs, or weeds — and then provide farmers with readily available solutions, like synthetic nitrogen and Malathion. By contrast, polyculture, with its interlocking components and its various loops and cycles, is not only more complicated but offers few of the assurances that conventional farmers now obtain from standardized commodity contracts or factory-specified doses of fertilizers and pesticides. Agriculture has always been a high-risk sector, says Liebman, and the evolution of modern farming has been driven in many respects by "farmers looking for ways to get comfortable with risk. And one way you do that is to use the set of practices that most of the people in your community use, and that your parents used, and that you've modified to make work for you. So when you ask people to move away from those practices and to step into this big information void, they get pretty uncomfortable."

Some advocates of polyculture hope that as the science of integrated farming improves, researchers will be able to offer farmers a broad suite of sustainable technologies and methods that might make it easier to leave the current chemically intensive model. Liebman, however, isn't holding his breath. Whereas industrial farming is premised on the somewhat illusory belief that we really can design a prepackaged solution for food production, polyculture assumes that such a package is by definition unobtainable. Farms, like all living systems, are unique collections of human, biological, and ecological elements and are constantly evolving; thus, sustainable farming is farming that adapts itself to specific landscapes and seasons and conditions, and has the capacity to change as these elements change. "Adaptive management is never prepackaged," Liebman told me. "So unless farmers want to engage in that process and have the information and technical support necessary to do so, wide-scale adoption of integrated systems won't happen in the Corn Belt."

As significantly, says Liebman, even if researchers can reduce the risk and complexity of alternative methods sufficiently to attract mainstream farmers in places such as Iowa, the true challenge will be to make these more sustainable methods feasible for places such as Africa, where farmers are understandably loath to take any risk whatsoever. As Liebman notes, agriculture's problems and its potential solutions are "distributed in a very heterogeneous manner" globally, and the chances for "finding solutions in Iowa are probably a lot higher than in the Sahel."

A few hours northeast of Rod Van Graff's feedlots, near the town of Reardon, Washington, wheat farmer Fred Fleming swings his pickup truck off the highway and onto a farm field to show me his plan to save the world. It's early March, and the field is already showing the green blades of a crop of winter wheat, a hard white variety that bakers prefer and that, with the ripple effects of the ethanol boom, looks to be a money maker for the first time in decades. What makes this field special, however, isn't what Fleming planted but *how*. Where most of his neighbors till their fields after harvest, making the dirt clean and neat for the next planting, Fleming leaves the stubble and other plant matter from the previous crop and, with a machine known as a direct seeder, pokes his seeds through the sod and into the soil beneath. The results of this "no-till" farming aren't pretty: the fields look so bedraggled and unkempt that when Fleming, a tall, gregarious fifty-nine-year-old whose family has grown wheat here since the 1920s, switched to no-till in 2000, his neighbors "would drive by and assume I was having marital problems."

But looks can be deceiving. Because no-till farming leaves the dirt largely intact, the topsoil in Fleming's fields has been able to knit together into a thick mat of roots, rhizomes, bugs, worms, and decaying organic matter — a living layer that retains nutrients and moisture, helps keep yields high, and protects the soil from erosion, much as the sod did when these fields were all in native prairie grass. This seemingly minor difference from conventional farming becomes glaringly apparent a few minutes later when we drive to another farmer's fields some miles away. Because this farmer uses conventional methods, the field has been tilled so regularly — as many as a dozen different times a year — that the soil is almost inert; when I reach down and dig out a handful, it crumbles on my palm like a moist brown talc with a cakelike texture, no bugs, and not much of a smell. Of course, with enough of what Fleming calls "farm chemistry" — mainly, synthetic fertilizers — this sorry dirt will still produce big yields, but not without a substantial cost. Because these overworked soils have lost their structure, they're highly vulnerable to erosion; dust storms here still shut down airports, and a heavy winter rain can be a disaster: three weeks before I showed up, a downpour cut huge gullies across the field, taking tons of soil with it. "And all that soil, and all that farm chemistry, ends up in the Spokane River," says Fleming, who often tells visitors that erosion has cost this region thousands of tons of topsoil over the last century. The claim is hard to believe until Fleming points to the ridge top where this farmer's land butts up against land that has been idle for decades in a federal conservation program: even from a distance, I can see that the ground level in the conservation area is more than two feet above the surface of the farm field.

Fleming isn't the only one touting the benefits of no-till wheat farming. Soil scientists from nearby Washington State University say the layer of sod that builds up after several years of no-till not only minimizes erosion but may reduce the amount of nitrogen leaching from the soil. As important, the method is also economically sustainable: even before the current grain boom, Fleming discovered that socially concerned commercial buyers, especially urban bakeries catering to upscale "correct" consumers, would pay a premium for "sustainably grown" wheat flour. In 2002, Fleming and a few like-minded farmers formed Shepherd's Grain, a marketing co-op that bypasses big commodity traders and sells directly to bakeries and other commercial customers. Their pitch was simple: not only is Shepherd's Grain flour cheaper than organic, but because it is grown with no-till methods, it actually disrupts the soil less than organic does. It's an argument that is gaining traction: in the four years since its launch, Shepherd's Grain's sales have jumped from ten thousand to nearly half a million bushels a year — and all, crows Fleming, "while finding a way to protect an ecosystem."

Advocates of sustainable food production point to Shepherd's Grain as proof that the trend toward commodity food production can be reversed or at least redirected: by taking their grain directly to consumers rather than selling to a middleman, Fleming and his partners have transformed wheat from a low-cost commodity into a upscale consumer product whose premium price allows it to be produced in a much more ecologically sustainable fashion. As important, however, is the way this venture has allowed midsize producers to compete in a food market increasingly ruled by very large players. By working in a cooperative organization, Fleming and his fifteen partners have sufficient scale and market power to strike favorable deals with buyers and other players in a supply chain that has historically squeezed such midsize producers to the point of bankruptcy or consolidation. "Individually, we're tiny," says Fleming. "But collectively, no one else even comes close."

Shepherd's Grain's success highlights a second stress point in the evolving food economy: the battle over scale. Just as greater crop diversity can reduce externalities at the farm level, greater size diversity could allow producers to balance the economic demands of their particular market with the agronomic and ecological capacities of their landscape.

In practice, of course, the trends in production scales are moving the other way. In the industrialized world and, increasingly, in many emerging economies as well, the farm sector is morphing into a two-tier system, with a small number of very large producers and a very large number of very small producers — neither of which has much capacity for sustainable food production.

The downsides of large producers are quite familiar: although they are exceptional at generating huge volumes of cheap food (in the United States, for example, just 163,000 mega-farms, representing a third of the agricultural land base, generate 60 percent of our food),[15] the need for continuous cost reductions leaves these large-scale, low-cost operations structurally disinclined to incorporate, or even acknowledge, external costs — a disinclination that will grow stronger as they face competition from foreign low-cost producers.*

What is often surprising, however, is that *small* farmers and producers are also challenged as long-term contributors of sustainable food. To be sure, small farmers can (and often do) produce food in a sustainable fashion. Small farmers have also been critical in bringing the ideas of sustainability into the mainstream and in linking rural and urban communities, thus reminding consumers that food is, or can be, something made by a person. For all that, the small-farm sector lacks the structural capacity to generate the volumes of food that will be necessary in coming decades.

---

* There are important exceptions; companies like AgriNorthwest, a massive grain and produce grower in Washington State, recently shifted to direct seeding as a means of cutting operating costs.

Few small farmers have time or the skills to develop a Furuno kind of output; in the United States, for example, most of the nation's 1.3 million small farms are part-timers or even hobbyists, relying on off-farm income to support what is less a profession than an avocation. And even if small farmers were somehow able to raise their output substantially, the nation no longer has a distribution system to transfer the yields of hundreds of thousands of small producers to retailers and consumers. All told, small farms, though they are the fastest-growing segment in the United States, generate less than 10 percent of the food supply. "We tend to lionize the small producers," says Scott Exo, executive director of the Food Alliance, a Portland, Oregon–based nonprofit that acts as a third-party certifier for sustainable producers. "There are undeniably opportunities for small farmers to deal directly with consumers in farmers' markets, restaurants, and other new markets, but in terms of actual food production and consumption, it's a real drop in the bucket."

It's for these reasons that many sustainable food advocates are looking to the area *between* large and small producers — and specifically, to midsize farmers, whose scale (between fifty and five hundred acres) and numbers (half a million in the United State alone) theoretically make them the ideal basis of sustainable food production. Many of these farms and ranches are large enough to generate sizable volumes at a reasonable cost. But they are not so large that farmers can't work each acre with some degree of knowledge and deliberateness — a critical distinction, according to sustainability advocates like Wendell Berry, who famously argues that sustainable production can't happen unless farmers "know and love" their land. And, as agricultural economist John Ikerd puts it, "Each farmer can only know and love so much land."[16] Most midsize family farmers and ranchers might not put it in exactly those terms, but many would likely agree that any kind of sustainable production is a lot more plausible when farmers and ranchers are actually familiar with soil qualities, hydrology patterns, and other strengths and weaknesses of their land. Thus we can understand why sustainability advocates are so excited by the success of Shepherd's Grain and other ventures that have been structured to give midsize producers the necessary marketing clout while preserving their capacity for stewardship.

Despite the apparently huge potential for this model of midsize food production, defending the middle ground is no easy feat. Half a century of falling commodity prices has left many midsize farmers unable to compete with their large-scale rivals; so many have sold out (often to those same low-cost rivals) that, since the 1970s, even as large and small farms have grown, midsize operations have declined markedly[17] — a trend agricultural experts bemoan as the "hollowing of the middle." Even with today's higher commodity prices, large producers continue to reap advantages from their scale economies and market power; and, of course, large commodity operations also benefit enormously from government programs that compensate growers by the acre or the bushel but pay little or nothing to ranchers and farmers who produce in ways that actually conserve their land and other natural resources.

Further, although some successful midsize producers have managed to convert the easier manageability of their lands into a sustainable practice that can justify a price premium, the process has added significantly to the complexity of their work. When Fleming switched from commodity wheat to a premium-priced sustainable flour, he expected the hardest part would be mastering the new production methods. In fact, the more challenging task was to learn how to sell to consumers, a task Fleming, like most commodity producers, had been content to leave to the big grain buyers and food processors. Fleming now had to understand the entire supply chain, study grocery retailing, develop a brand and a product story, and familiarize himself with something he'd never bothered with before: consumers. As a result, this fourth-generation rural conservative found himself spending long days in upscale urban supermarkets, trying to discern how left-of-center, socially concerned shoppers made their flour-buying decisions. "We had to learn a new language," says Fleming. "It was a hard mind shift to make, and I still have flashbacks to my commodity side of life."

Being a pioneer has other costs, as well. Some of Fleming's neighbors regard his alternative methods as a thousand-acre rebuke to their own faith in industrial farming, and some alternative farming advocates aren't much happier. Because tilling helps keep weeds under control, no-till farmers such as Fleming must compensate with "judicious" applications of synthetic herbicide (usually Roundup); they must also apply some synthetic nitrogen. As a result, not only is Fleming's wheat excluded from the organic market, but his methods earn the scorn of some in the alternative farming movement. Invited recently to address a statewide meeting of alternative producers, Fleming, a Republican from a very conservative town, found himself in a room filled with flute music, odd clothing, and organic "purists." "They were polite," he recalls. "But some of them definitely gave me the cold shoulder."

There are other penalties for those seeking payment for producing food sustainability. Because value-adding producers like Fleming are essentially maintaining ownership of their product as it moves down the supply chain — not just handing it off to an intermediary as soon as it leaves the "farm gate," as conventional commodity producers do — they get to keep a larger share of the value being added. For example, members of an Oregon-based cooperative ranching venture, Country Natural Beef, whose cattle are raised on chemically untreated pasture that is allowed to fully recover naturally between each season and who carefully trace each animal as it moves from pasture to slaughterhouse to grocery display case, receive 96 percent of the wholesale price — substantially more than conventional producers get. But in so doing, these new producers expose themselves to a new set of risks. Where ranchers and farmers once dealt only with rapacious middlemen, now they must work in a retail market that is often far more competitive. Big grocery chains will gladly appropriate new value-adding ideas, like "organic," which they then will seek to produce more cheaply using conventional, large-scale producers. Further, the rapid consolidation of the natural-food retail market (in 2007, Whole Foods, the largest natural-food chain, announced a half-billion-dollar buyout of its nearest competitor, Wild Oats)[18] will give these socially correct retailers even more price leverage over their suppliers.

Finally, if traditional producers thought that commodity markets could be unpredictable, with wild price swings, that is nothing compared to the fickleness of consumers. "Today's value added is tomorrow's commodity," says Doc Hatfield, who, as a cofounder of Country Natural Beef, knows how quickly a consumer trend can change. Country Natural Beef spent years finding a way to compete with traditional organic beef by offering consumers a different value proposition — beef that was ecologically sustainable — only to watch consumer preferences shift from environmentally correct food to food produced locally. And while Hatfield is all for local food, it's a challenging criterion for grass-fed beef. In order to sell fresh beef year-round to progressive consumers in a key urban market like Portland, Oregon, Country Natural Beef must source its cattle not only from in-state ranches but occasionally from northern California, because that's the nearest place where pasture grass grows all winter and, thus, the nearest place where cattle can be grazed sustainably. Yet conveying this geographically complex reality to the "well-intentioned, moderate-income consumer" hasn't always been easy. "Local is the new organic," says a mildly exasperated Hatfield, echoing what has become the lament of the alternative food universe. "But you have to be at least a hundred miles from Portland before you can find an environmentally friendly place to raise cattle."

In the tiny mountain village of Sambucano, Italy, just a few miles south of the Swiss border, I'm sitting in a restaurant above an enormous alpine meadow, chewing on a piece of locally raised lamb and learning, between sips of red wine, how gourmet food can reconnect consumers to the countryside. My hosts are all associates of Slow Food, an international organization that began as a jocose response to McDonaldization but has since become the elite guard of the eat-local movement. Just now, agronomist Antonio Brignone, a diminutive, elfish man with bright green eyes and silver hair, is explaining how the flavor of the Sambucano lamb, a breed that was nearly extinct ten years ago, cannot be found anywhere else on earth because it feeds on nothing but local grass, which is the product of local soils, local weather conditions, and even the local angle of the sun. "You can't find Sambucano lamb in New York," Brignone tells me. "If you want it, you have to come here." Brignone's point is that when it comes to real food a particular taste is inseparable from its place of origin, and if consumers wish to preserve that taste, they must preserve the countryside from which it springs. "Once you start to understand how gastronomic food is produced," says Renato Sardo, a former Slow Food official, "you realize you have to do something to protect the ecology of the place where it is produced."

Granted, the endangered Sambucano lamb may not be the linchpin of the next food economy. But the move toward local food, for all its trendiness (the more adamant adherents, known as "localvores," strive to buy products that have traveled the least "food miles"),[19] highlights one of the problematic pieces of the modern food economy: the increasing reliance on foods shipped halfway round the world. Because long-distance food shipments promote profligate fuel use and the exploitation of cheap labor (which compensates for the profligate fuel use), shifting back to a more locally sourced food economy is often touted as a fairly straightforward way to cut externalities, restore some measure of equity between producers and consumers, and put the food economy on a more sustainable footing. "Such a shift would bring back diversity to land that has been all but destroyed by chemical-intensive mono-cropping, provide much-needed jobs at a local level, and help to rebuild community," argues the UK-based International Society for Ecology and Culture, one of the leading lights in the localvore movement. "Moreover, it would allow farmers to make a decent living while giving consumers access to healthy, fresh food at affordable prices."[20]

While localvorism sounds superb in theory, it is proving quite difficult in practice. To begin with, there are dozens of different definitions as to what local is, with some advocates arguing for political boundaries (as in Texas-grown, for example), others using quasi-geographic terms like food sheds, and still others laying out somewhat arbitrarily drawn food circles with radii of 100 or 150 or 500 miles. Further, whereas some areas might find it fairly easy to eat locally (in Washington State, for example, I'm less than fifty miles from industrial quantities of fresh produce, corn, wheat, beef, and milk), people in other parts of the country and the world would have to look farther afield. And what counts as local? Does food need to be purchased directly from the producer? Does it still count when it's distributed through a mass marketer, as with Wal-Mart's Salute to America's Farmer program," which is now periodically showcasing local growers?[21]

The larger problem is that although decentralized food systems function well in decentralized societies — like the United States was a century ago, or like many developing nations still are — they're a poor fit in modern urbanized societies. The same economic forces that helped food production become centralized and regionalized did the same thing to our population: in the United States, 80 percent of us live in large, densely populated urban areas,[22] usually on the coast, and typically hundreds of miles, often thousands of miles, from the major centers of food production.

Some agricultural economists have advocated the wholesale shifting of our food centers closer to urban centers, as is happening by default in many fast-growing Asian countries. Such an arrangement would impose far fewer costs in terms of transportation energy but poses extraordinary technical and economic challenges. In developing countries, the intermingling of food production and dense urban populations is now seen as a driver for flu outbreaks. As well, even in mature economies such as the

United States, where biosecurity is much higher, the high cost of urban land precludes all but the most high-value food production. This is one reason so much of the farmland in the rapidly urbanizing Salinas Valley has shifted from low-value pursuits, such as ranching, to high-value crops, such as spinach and tomatoes; it's also why most low-value crops, like corn and soybeans, can be grown economically only on relatively cheap rural acres. Even in regions where farmlands still exist near cities, the roads, rails, and short-haul supply chains that once linked farms to nearby urban markets has been largely usurped by national and global supply chains, managed by retailers and distributors who prefer sourcing from large suppliers, not from hundreds or even thousands of local producers. These are powerful economic forces that would work against any large-scale shift to local food production.

The more practical advocates of local food insist that they aren't asking for wholesale localization. "I know perfectly well that every product cannot be made on a small farm," says Sardo, formerly with Slow Food. "But we have to start somewhere." Indeed, the more nuanced view of localism doesn't seek to completely do away with nonlocal foods — at least, not right away — but to recover some degree of geographic diversity in food sources, based on region, food type, and season, while striving to moderate the more egregious impacts of globalization — soy protein from China, for example, or planeloads of year-round fresh produce from South America.

But here, too, the benefits of going local aren't always clear. Although reducing the distance traveled by a food product would seem to be an automatic gain for sustainability, this isn't always the case. A semi driving several tons of produce 312 miles from a mega-farm in the Salinas Valley to a Wal-Mart in Reno may seem an egregious waste of energy, but it actually burns less fuel than would the dozens of pickup trucks needed to haul the same quantity of produce to a farmers' market in Reno from local farms just twenty miles away. One advantage of the centralized industrialized food system, with its carefully scheduled deliveries and obsessive focus on efficiency, is that it can keep food-transportation energy costs down.

The more fundamental problem with the food-mile concept is the same one that plagues organic: it's a simplistic solution to an extraordinarily complex problem. In the same way a pesticide-free head of lettuce may still not be environmentally friendly, distance isn't always the most important determinant in a particular food product's sustainability. Organic food produced in Chile and flown to the United States may represent massive food miles, but it also represents a shift in farming practices in Chile — fewer pesticides and synthetic fertilizers — which might be beneficial to the Chilean environment and people.

Even if we omit such fuzzy, hard-to-quantify social benefits and focus strictly on tangible costs, such as energy savings, climate impact, or water use, "local" doesn't always represent a win. According to researchers at the University of Wales Institute, the shipping of food from the farm to the grocery accounts for, on average, just 2 percent of that product's total environmental impact. Far more significant contributors are the way the food is processed, packaged, and especially the way it is farmed, because modern agriculture and livestock methods rely so heavily on energy-intensive and ecologically dubious fertilizers, irrigated water, and imported grain.[23]

To capture this complexity, many sustainability advocates want to replace food miles with the more detailed ecological footprint concept, which tries to account for *all* of the costs of a particular food product — usually expressed in the number of theoretical acres needed to generate all the materials, energy, plant matter, and other inputs necessary for making and moving a particular food product. "I'm a bit worried about the food miles [debate] because it is educating the consumer in the wrong way," Ruth Fairchild, one of the authors of the University of Wales study, told *The Guardian* newspaper. "It is such an insignificant point. . . . If you just take food miles, it is the tiny bit on the end."[24]

This point was made succinctly when local-food advocates in the United Kingdom complained about the practice of importing meat and dairy products from New Zealand. In response, researchers at Lincoln University in New Zealand showed that because New Zealand farmers use far less fertilizer than their counterparts in the United Kingdom do and because New Zealand sheep feed almost entirely on grass whereas UK livestock are mainly grain-fed, consumers in the UK importing New Zealand mutton and dairy products actually cut energy use and climate impacts by 75 and 50 percent, respectively, over locally produced items.[25]

This isn't to dismiss distance as an important factor in gauging the sustainability of a particular product or practice or to reject the idea that local food might be worth promoting for other, less quantifiable values. Reconnecting consumers with producers, and helping those consumers become more aware of the specific place where their food is grown, or with specific traditions, would seem to offer an important counterweight to a food system characterized by increasing uniformity and separation. As well, a robust local-food movement might help revitalize an environmental movement that has become almost bloodless. As Sardo puts it, "The gourmet needs to be an ecologist, because without the right ecology, you will lose the flavor. But we also know that the ecologist needs to be a gourmet, in order to be less sad, less apocalyptic."

Nor is it to suggest that just because global sourcing of some foods is cost-effective and perhaps even sustainable today, this will always be the case. A sharp rise in energy prices or tough new regulations on carbon emissions, to say nothing of new revelations about food safety in places like China, could rapidly shift that balance, rendering some or most of our system of global sourcing obsolete and making it far more economically viable to grow apples and potatoes in places like Iowa and Nebraska again.

What is true, however, is that for all the benefits of local food, the realities of the local-food debate are far more complex than they often appear — and certainly will not be resolved by a single number or definition. "The emphasis on speed and convenience has seduced a lot of us into thinking that we can boil down the implications of our decisions into a single word, and that if we embrace the word, then we can go merrily on our way and feel good about it," says Food Alliance's Exo. In the next food economy, Exo says, consumers will "still have to think."

Not surprisingly, most strategies for reforming the food system rely heavily on "thinking" consumers. Many of the groups advocating a more sustainable food system, improved nutrition, or related issues have created elaborate education campaigns to tell consumers about their food, the idea being that the more one knows about what is wrong with food, the more likely one is to take steps to make it right. This is the rationale behind the numerous efforts to force companies to label their products with information not just about the ingredients used but also where the ingredients originated and whether the conditions of production were sustainable or equitable. "Food consumption is an area where individual decisions can make a difference — supply will follow demand," declares the Sierra Club on its website "The True Cost of Food," which invites readers to take a pledge: "We, the consumers, through our food choices, can stop the practices that harm our health, our planet, and our quality of life."[26]

The question is just how much action consumers can take. Although demand does indeed follow supply, and although consumers are indeed the ultimate arbiters, it's not clear how much choice consumers really have. Whereas surveys indicate a large percentage of consumers have some degree of awareness of sustainability — studies by the Seattle-based Hartmann Group show that 96 percent of consumers operate with what it calls "sustainability consciousness" — a much smaller fraction is sufficiently motivated to change the kinds of foods they buy. For all the attention paid to organic foods, the category still makes up less than 2 percent of food purchases in the United States, and not all of this food qualifies as "sustainable." According to the Kellogg Foundation, which devotes much of its budget to promoting sustainable food systems, the share of U.S. food production that could be classified as "sustainable," which the foundation defines as safe, nutritious, produced in an environmentally sound manner, and affordable, is just over 1 percent. Further, reaching the foundation's goal of a 10 percent share within a decade will require a growth rate of 30 percent a year,[27] which is about half as fast as organic is growing.

Why has it been so challenging to motivate consumers? One obstacle, obviously, is the extra costs that sustainably produced foods entail. It may be true, as farm-labor advocates claim, that a tomato picker could earn a living wage if consumers paid just a penny more a pound for tomatoes — something few consumers would begrudge. But how much more would consumers need to pay to ensure a living wage for *all* pickers, packers, chicken deboners, and other food workers? Would consumers ever be willing to fork over enough to make these very hard jobs respectable or desirable or safe, or, more generally, to relieve the food industry of its long-standing dependence on immigrant labor? (A dependence, it should be noted, that is far more central to the current immigration debate than are the stated concerns over security or population.) And cheap labor is actually one of the modern food system's lesser external costs; others will be far more expensive to ameliorate. If Corn Belt farmers began growing more cover crops of nitrogen-fixing alfalfa, they would by necessity be growing less corn, which would diminish overall corn supplies and drive up prices for anything made from corn (as the ethanol boom is now aptly demonstrating). True, some consumers seem perfectly willing to pay a premium for sustainably produced food, and for many food activists, the key to success centers in large part on persuading more people that slightly higher foods costs are a reasonable tradeoff for *real* food. What the campaign comes down to, says Sardo, is convincing consumers "that they should be paying *more* for their food. They need to understand that if a chicken costs only a dollar, it's probably not a chicken."

But again, how receptive mainstream consumers will be to that logic isn't clear. Outside of upper-income groups, price remains a huge determinant in product choices. That's one reason fresh produce consumption is so low among lower-income groups, while consumption of vastly cheaper junk food is quite high. Industry observers say that is also why Wal-Mart, the bellwether of mainstream consumer consciousness, scaled back its ambitious organic venture less than six months after it was launched. Unlike their upper-income counterparts at Whole Foods, Wal-Mart shoppers were simply unwilling to pay the 10 percent premium the company was charging for organic. As Peter Ricker, a Maine orchardist who supplied organic apples to Wal-Mart, groused to *BusinessWeek*, "The Whole Foods customer is walking in there to buy organic and is more concerned about how the fruit was farmed. But the Wal-Mart customer is used to shopping with a calculator."[28] And Wal-mart's 10 percent premium is relatively tame. Cardiff University's Ruth Fairchild found that if consumers shifted to an organic diet (with the idea of avoiding energy-intensive farm chemicals), they would raise household food expenditures by 31 percent.

Or consider the other costs of moving to a more sustainable food system. Year-round fresh produce, for example, requires such a massive energy-intensive supply chain that it is hard to imagine maintaining such a luxury in a world of higher energy prices or rising concerns about climate-changing emissions. And yet, even upscale, environmentally concerned shoppers who frequent farmers' markets show little interest in giving up access to year-round produce, which means retailers don't, either. "When locally produced items are available, we *do* carry them," says Amy Schaefer, a spokeswoman for Whole Foods. "But our produce department would be mostly bare for many months of the year if we only bought locally."[29] Even in Italy, a country obsessed with its homegrown foods, domestic retailers are winning over consumers with low-cost year-round produce from Spain and Morocco.

And then, of course, there is the question of meat. The good news is that meat can indeed be produced sustainably — for example, with diversified operations that integrate livestock and crops. Such operations generate fewer externalities, such as poop lagoons or massive nitrate leaching; they consume less grain, which reduces demand for so much chemically intensive grain production; and they can turn out a much more nutritious product: Country Natural Beef has better fatty acid ratios than most conventionally raised beef. That said, it is hard to imagine producing meat sustainably *and* in the same volumes that we do currently — even in a food economy that is increasingly geared toward the relatively efficient white meats.

In fact, most of the proposed models for sustainable meat production have an output that is nowhere near current levels. For example, when Tim Crews, at Prescott College in Arizona, and Mark Peoples, at CSIRO Plant Industry in Australia, created a model for a U.S. meat industry that doesn't require synthetic nitrogen, the changes in output were stark.[30] By their estimates, to replace synthetic nitrogen with naturally fixed nitrogen, the United States would need to convert at least half its current grain acreage into nitrogen-fixing legume crops. This shift, in turn, would reduce total grain output so drastically that the country would need to either eliminate grain exports or lower its per capita food supply by 25 percent. Leaving aside the no-export option (which, in a grain-short global market, would entail considerable geopolitical impacts), the question then becomes, could Americans live with a 25 percent reduction in food supply, and the answer is: not likely. To be sure, Americans could obviously *survive* on fewer calories than we're producing — given that our current per capita food supply is about 44 percent more than we need nutritionally[31] and much of that "extra" is actually the food wasted during harvest, processing, in food service, and at home. But Crews and Peoples' calculations suggest that even if Americans did eliminate most food waste, the lower grain production in their scenario would still require consumers to reduce their meat consumption to roughly an *eighth* of the current level[32] and to get most of the daily protein requirement from plant sources. Quite aside from the massive resistance such an idea would provoke from grain farmers, meat companies, retailers, and food-service chains, it's hard to imagine that U.S. consumers who aren't already vegetarians would voluntarily cut their meat consumption so dramatically — no matter how persuasively change advocates make the case that eating meat is killing us.

This reluctance is going to be a key factor in the rebuilding of the food economy. The idea that consumers, once they've been made aware of the true costs of their diet, will act spontaneously to reduce those costs seems farfetched in the extreme: after all, just because smoking, heavy drinking, and indolence are widely understood to be harmful, they haven't stopped billions of consumers from engaging in those same behaviors frequently. And, in fact, many of the researchers who are busily defining what is and is not a sustainable food seem quite cognizant of the possible futility of their calculations. Under Fairchild's ecological footprint model, for example, practitioners of a truly sustainable diet would need to reduce or eliminate not only meat but wine, spirits, chocolate, cheese, and ice cream[33] — deletions that, Fairchild acknowledges in a serious case of understatement, "may preclude many from adopting this type of diet."

Most sustainability advocates, in the end, believe consumers won't change without sufficient encouragement. Cornell University ecologist David Pimentel, for example, calls for a special sustainability tax that would be applied to foods on the basis of their external costs: foods like meat, dairy products, and eggs, which have the highest externals (and, Pimentel argues, the highest health costs), would bear the highest tax rate, while foods at the bottom of the food chain, including grains, legumes, produce, and nuts, wouldn't be taxed at all. Such an idea has little support among change advocates — largely because anything remotely resembling a tax would be a political near impossibility in the United States. But most advocates do agree that some form of government intervention — new regulations on food makers and farmers, for example, or the removal of surplus-encouraging farm subsidies, coupled with an intensive education program for consumers — will be necessary to "force" the market to address issues of sustainability and health.

And the market will indeed need to be forced. Whether the goal is more diversified farming, more local production, or more nutritionally sound snacks, the free market can carry these ventures only so far: only a certain percentage of consumers will pay extra for food that meets the criteria for a sustainable, healthy food economy. Once this level of demand is met, creating more demand will require additional and substantial incentives that originate from *outside* the market — which in most cases means some form of government policy. Just as Roosevelt-era lawmakers sought to dampen natural (and destabilizing) price swings by changing the way the market operated (in that case, by trying to limit supply), governments will now have to deliberately manipulate food markets in order to encourage the food system to move toward a more sustainable, but possibly less profitable, form of production. In other words, to change the food system, we will need to transform food policy, which, given the complexity of the issues and the investment that many stakeholders have in current policies, will be extraordinarily difficult.

Few policy matters are as complex or difficult as food, in part because food involves everything from agricultural practices and trade to consumer health and nutrition. Policymakers brave enough to take on food must grapple with dozens of different sectors; they must react to the constantly changing state of knowledge about everything from breeding genetics and hormonal activity to the intricacies of trade, the risks of new technologies, and, with the boom in biofuels, the links between the food business and the energy sector. Further, policymakers must do all this in an atmosphere heavily charged by the political and economic power of various players in the food economy — otherwise known as the food lobby. And if this lobby is more diverse than one might expect, with a great many advocacy groups pushing issues like sustainability or food safety, it is dominated by companies and trade groups that have enormous reserves of financial and political capital and whose agendas, though quite logical in a purely business context, often run counter to prospects for long-term sustainability.

The most familiar form of influence is, of course, the campaign contribution, and although the food industry isn't the most generous of all industries, it is by no means miserly. Between 1990 and 2006, companies directly involved in the production, processing, and marketing of food and drink — which is to say, everyone from seed sellers to restaurant chains and grocery store chains — contributed $459 million, according to the Center for Responsive Politics, which tracks campaign contribution filings. Within this group, the most generous sectors included commodities producers, who gave $91 million, livestock ($31 million), and food processors ($27 million). The most munificent single entities were Philip Morris/Altria, which sells both food and tobacco ($20 million), ADM ($7.6 million), and the American Crystal sugar company ($5.6 million). Grocery retailers, meanwhile, under the trade group Food Marketing Institute, gave $4.85 million.

What this largesse actually buys donors is a matter of considerable dispute. But as contributions have risen, there has been no shortage of developments in the regulatory and policy arenas that have been favorable to some of the more heavily contributing industries.

In 2000, for example, a bill by Congressman Richard Pombo, a California Republican, that would have eased farm pesticide regulations was found to be almost identical to a proposal written by lobbyists for pesticide manufacturers.[34] Food companies and trade groups also have achieved a remarkable degree of influence in the ways that governors and presidents appoint the regulatory officials who oversee the food industry, with the result that top-echelon officials at agencies like the USDA and the FDA often come straight from the ranks of the same food industries those agencies are supposed to regulate. John Block, Ronald Reagan's agriculture secretary, owned an industrial-scale hog operation in Illinois.[35] Block was followed in the position by Richard Lyng, a seed-industry executive. Ann Veneman, agriculture secretary under George W. Bush, was on the board of directors at Calgene, creator of the Flavr Savr tomato, and later a part of Monsanto. Carlos Gutierrez, a former president of Kellogg, turned down an offer to run Coca-Cola before becoming secretary of commerce under George W. Bush.[36]

Certain industries and companies are particularly well represented in the halls of government. Former Montano executives have served as the U.S. deputy to the World Trade Organization,[37] deputy director of the U.S. Environmental Protection Agency,[38] and head of the FDA's New Animal Drug Evaluation Office,[39] among other offices. In many cases, influential individuals move back and forth between government and industry. Clayton Yuetter, Bush senior's agriculture secretary, later joined the ConAgra board of directors.[40] Michael Taylor, who ran the USDA's Food and Safety and Inspection Service under President Clinton, then worked as an attorney for Monsanto, then took a job at the FDA, then rejoined Monsanto as a lobbyist. Charles Conner, a one-time congressional agriculture committee staffer, went to work for the Corn Refiners Association, a commodity trade group, for four years before becoming in 2005 deputy secretary of agriculture, where he has helped guide biotechnology policy.[41]

The full effect of so many financial and professional connections between the food industry and Congress, the White House, and the various regulating agencies is impossible to gauge, but we can get a sense from the few cases that have been made public. Meat-industry lobbyists were able to block proposals to reclassify *E. coli* as an adulterant until well after the public furor over the Jack in the Box outbreak (and even then they fought reclassification), and the industry is still pushing to prevent the USDA from reclassifying salmonella. The input and pharmaceutical industries have thwarted efforts to label foods that contain transgenic ingredients — despite surveys showing that most American consumers favor such a requirement.[42] Similarly, proposals to force food companies to disclose their products' countries of origin have been defeated repeatedly. In 2007, a law already on the books requiring country-of-origin labels for meats and perishable agricultural commodities was delayed until 2010 — largely because food companies that rely heavily on imported foods fear such labels would give an economic advantage to competitors who source domestically.[43] (USDA officials bravely if somewhat lamely explained the delays as being necessary to spare the meat industry the "potentially burdensome requirements" of printing new labels and to avoid "unduly exposing consumers to outdated labels."[44])

Retailers, too, have benefited from political connections. Despite long-standing calls by industry critics for a return to tougher federal antitrust actions as the only sure way to slow rampant consolidation, such action has been slow and inconsistently applied. Even as some of Wal-Mart's competitors have been sued by the government for abusing pricing power, notes *Harper's* writer Barry Lynn, federal regulators have left the politically well-connected Wal-Mart "free to extend its domain in whatever direction and to whatever extent it wished."[45] In fact, Wal-Mart CEO Lee Scott feels so safe from regulators that he could recently admonish British regulators to go after British grocery rival Tesco for being too large. In Scott's words, once a firm's market share approached 30 percent, "there is a point where government is compelled to intervene" — and this from a man whose own company has at least 30 percent of the U.S. grocery market. And for those curious about how Wal-Mart might fare under a Democratic White House, Lynn notes that Hillary Clinton was a Wal-Mart board member.[46]

Another arena in which industry influence has had an enormous effect on food policy is that of food trade. The fact that many federal trade officials are former executives in the pharmaceutical and input industries (the current top agricultural negotiator, Richard Crowder, ran the American Seed Trade Association and has worked for ConAgra, Pillsbury, and Monsanto's DeKalb Genetics Corporation) is probably not unrelated to

Washington's aggressive promotion of trade in transgenic foods and its pugnacious tactics against countries that seek to block imports of transgenic foods. In 2003, when the United States prepared to sue the European Union over its ban on transgenic imports, the government of Egypt, under pressure from Washington, joined the lawsuit but later withdrew so as not to offend its trading partners in Europe. In retaliation, Robert Zoellick, then the U.S. trade representative (and now head of the World Bank), punished Cairo by withdrawing a long-promised offer of a U.S.-Egypt trade deal.[47]

Most recently, Washington's close relationship with big U.S. food companies has been implicated in its reluctance to crack down on contaminated Chinese imports and its willingness to consider further imports of products, like chicken, that many experts fear China can't produce safely. Lester Crawford, the former FDA head and veteran of the food-safety debate, points out that in the past, Washington has been swift to punish countries who fail to ensure food safety; when Mexican shipments of meat and produce to the United States were found to have serious safety problems in 1980s and again in the early 2000s, Crawford says, "the United States didn't hesitate to cut them off, and Mexico is one of our largest trading partners. We should be doing the same thing with China — that would get their attention." But instead, argues Robert Cassidy, a former U.S. trade official, in an interview with the *Washington Post*, "so many U.S. companies are directly or indirectly involved in China now, the commercial interest of the United States these days has become to allow imports to come in as quickly and smoothly as possible." Washington, Cassidy said, is "kowtowing to China."[48]

Ultimately, this growing resistance to change affects not only the food economy we have but also the way that food economy will develop. It is widely understood, for example, that the U.S. farm program, and its subsidies in particular, helps perpetuate a host of unsustainable practices, from overproduction and artificially cheap food to the promotion of monoculture cropping, that only make changes harder to bring about. And yet, despite decades of promises by lawmakers to reform U.S. farm policy, the task is once again proving impossible — even with the recent alignment of key political and economic forces. In 2007, for example, there was broad bipartisan support in Congress for changing farm subsidies. Commodity prices were high, which tends to make farmers worry less about losing federal price supports. There was heavy pressure from public-heath groups, who argued that subsidized grain is a factor in cheap junk food and, hence, obesity.[49] There was even heavy support from the Bush White House, which regards farm subsidies as a hindrance in trade talks but also, correctly, sees subsidies as being grossly unfair to developing countries. Reform momentum was riding so high in the summer of 2007 that lawmakers were even considering a measure to ban farms with incomes of $200,000 or more from receiving any federal payments, down considerably from the current maximum of $1 million.

But despite such a massive alignment of interests, Congress was unable to pass a single bill that would have phased out farm subsidies. One factor was the intensive counterlobbying by large agribusiness operations, commodity traders, livestock producers, and state and local governments that now benefit from farm subsidies. (More than half the total federal farm payout went to farms in just twenty congressional districts.[50]) But another factor was the Democratic leadership, which feared losing freshman Democratic congressional members from farming districts in the next election, without whom Democrats would lose control of the House. By the end of 2007, Congress had voted not only to extend the current farm program another five years but actually *added* money to some existing crop programs; they even threw in subsidies for a whole new category of crops — fresh fruit and vegetables — that had somehow managed until now to flourish without government support.[51]

And subsidies are only the most famous of the ways government food policy encourages bad practices while discouraging potential alternatives. Under the federal Environmental Quality Incentives Program (EQIP), for example, taxpayers help the owners of large livestock operations pay to upgrade their massive manure lagoons. Such funds help prevent the kind of disastrous poop spills of the kind that soiled North Carolina in 1995 and have strong support from environmental groups. But by helping cover the external costs of CAFO, complains Sustainable Agriculture Coalition's Ferd Hoefner, EQIP, whose budget has climbed from $200 million to $1 billion in the last decade, also helps perpetuate a model of food production that is in nearly every respect unsustainable.

Similarly, federal dollars for agriculture research continue to be shunted away from the development of alternative farming practices and toward conventional large-scale farming — a dangerous trend, since research funding has an enormous influence on the character and success of future food systems. For more than a century, advances in food-production technologies — plant and animal breeding, soil science, preservation, and transportation — have been driven in no small part by public research programs and public dollars. And while many critics of the modern food system regard that funding as responsible for the current problems, they also understand that without more public dollars, there simply won't be an alternative model.

What is desperately needed, say advocates, is funding for research into crops and methods that might make alternative models feasible and profitable — plant varieties that more efficiently fix nitrogen into the soil, for example, or more efficient irrigation technologies — innovations that require substantial and continued investment before they can become viable enough for the market to be interested. "In terms of long-term change, nothing is more important than the research being done at the public expense," says Hoefner. "The public research agenda is the key determinate as to whether a technology is a decade away from being ready or is a generation or more removed."

Given the depth of the food system's resistance to change and the many forces — political, economic, and cultural — aligned in favor of the status quo — it's hardly surprising to find that many reformers believe real change can come only from *outside* the system — in the form of some crisis or shock that forces the system to evolve. For militant critics of the food industry, that crisis must be orchestrated intentionally, through acts of "resistance" that publicize problems in ways the public and industry cannot ignore. But other, more pragmatic critics believe the crisis is far more likely to show up in the form of some unplanned and potentially catastrophic failure — a massive outbreak of avian flu, for example, or the huge crop failure in India and China, or an acceleration of the collapse of North Africa's water systems, or an oil-price spike that effectively shuts down the movement of food. All of these are plausible scenarios; any of them would likely force rapid and aggressive transformations in the way we make and use food — although not necessarily in ways that many of us would like. Yet as the inertia of the current system grows, it is more and more likely that whatever changes we make to our systems of food production won't be the result of a carefully considered vision for how the future of food ought to be but a series of emergency reactions to a system in overload.

Epilogue: Nouvelle Cuisine

THE FACT THAT consumers and lawmakers in an advanced country like the United States have yet to decide that the food economy needs fixing is hardly surprising, given how diligently the food industry has worked to keep the visible surfaces of that economy looking fit and healthy. Despite everything I know about the food system now, I can still drive down to my favorite mega-grocery outlet, walk through the sliding glass doors, and experience, at least momentarily, the palpable sensation of unquestioned, unending abundance and security. The produce section is still overflowing; the meat and fish cases still brimming; the shelves in the store's center still fully stocked. Aside from the fairly large number of overweight customers moving up and down the extrawide aisles, there are no obvious portents of the system's pending breakdown — no empty gaps on the shelves, no "temporarily out of stock" labels in the produce department — nothing to remind me of the recent scandals over melamine or the ongoing concerns about *E. coli* and salmonella, and no suggestion that this extraordinary bounty won't automatically replenish itself next week, or next year, or a century from now.

But that feeling of safety and certainty quickly fades. I know that if it were possible to look behind the displays and shelves, to trace the chain of transactions and reactions represented by each ripe melon or freshly baked bagel, by each box of cereal or tray of boneless, skinless chicken breasts, the confident picture would change dramatically. I know I would see a vast and overworked system that is straining to satisfy a market that wants its food fresher, more varied, and cheaper every week. I would see feedlots with their thousands of identical animals, sprawling factory farms with their acres of identical plants. I would see the enormous inflows of feed and fertilizers, of Atrazine and Roundup, and the massive outflows of farm chemistry. I would see soils that are eroding, insects that are adapting, forests that are being converted to farms, and farms being turned into shopping malls. I would see irrigation wells reaching deeper for falling water tables, and air-cargo routes reaching farther for falling wages. I would see the declining margins and the lean inventories, the supply chains that grow simultaneously longer and thinner, with steadily rising throughputs and steadily declining margins for error. In fact, more and more these days, I find myself picturing how rapidly this system could shut down and how quickly these shelves and displays would be picked clean were the food economy to stumble across an "event" that exceeds its narrowing tolerance for disruption.

Suppose, for example, that we did finally get the outbreak of avian flu that, given our overburdened meat industry, most experts say is inevitable. Not a reprise of the Spanish flu, with its tens of millions dead, but something more virologically plausible, like the Asian flu that swept the globe in 1957. Credible scenarios, such as the one prepared by the Lowy Institute for International Policy in Sydney,[1] Australia, suggest that a similarly moderate outbreak today, likely originating on some Asian duck farm, would kill perhaps fourteen million people worldwide and create significant economic damage, much of it in the food sector itself.

Havoc would be most acute in sub-Saharan Africa and Asia, where populations are concentrated and governments and medical systems grossly ill equipped. Some three million people would die in the least-developed nations, mainly in sub-Saharan Africa. China and India would together lose more than five million and see their high-growth economies slow considerably as workers stayed home, factories closed, and skittish investors took their capital and fled for the relatively safe havens of Europe and North America.

But those safe havens would by no means escape unscathed. Although the United States would suffer far fewer deaths — the Lowy study envisions U.S. mortality at two hundred thousand — the nation would nonetheless pay a severe economic price due to our service economy's inherent vulnerability to any decline in consumer confidence or spending. And while all service sectors would be hard hit, for the food sector, with its reliance on huge throughputs of perishable goods, its overstretched global systems of supply, and its massive sensitivity to consumer safety fears, the impacts would be catastrophic. Restaurants would essentially shut down — one Congressional Budget Office study[2] sees food-service sales dropping to a fifth of current levels — as consumers avoided public eating places. Grocery stores, by contrast, would be unable to keep shelves stocked as thinly staffed just-in-time supply chains bogged down, and truck drivers, warehouse workers, and other key employees refused to come to work. In weeks or even days, stores in many areas would be empty, and food suppliers would be forced to beef up security to prevent warehouses from being looted and trucks hijacked.[3] The Lowy study estimates that the outbreak would cost the U.S. economy alone nearly 2 percent of its gross domestic product, with the CBO estimating a per-household loss of around $2,200.

Of course, cautious optimists have argued that the odds of such an outbreak are relatively small, both because the virus itself is more stable and because conditions that made past outbreaks so severe, such as poor medical knowledge and the chaos of a world war, are different today. Further, Asian governments, eager to maintain their access to Western markets, are making some small progress toward improving bio-safety. By this line of reasoning, the longer we go without an outbreak, and the more time we have to modernize our production systems and improve our bio-security capabilities, the greater our chances of dodging the bird flu bullet altogether.

What this argument ignores, however, is that bird flu is only one of a number of bullets that could plausibly strike the modern food system. A sharp spike in the price of oil, a series of extreme weather events, an outbreak of some new plant disease, the depletion of some critical aquifer — all would send massive and potentially disrupting shock waves through a system that, despite advances in such areas as bio-security, is losing more and more of its overall flexibility and resilience by the week. On this, time is most assuredly not on our side. Each year that we avoid a pandemic may indeed allow us to stockpile vaccines. But it also means that the number of other threats we face will rise — as, for example, when warming temperatures from climate change raise the risks of massive crop failure by insect attack or flood or drought. And as the number of risks increase, so too does the probability that at least one of these "bullets," and probably more than one, will strike home.

What should also be clear is that our evolving system of food production doesn't even need a bullet to break down. For all that a pandemic or some other specific catalyst could indeed push the food system toward collapse, we now understand that the system is *already* moving toward collapse: without any help from H5N1, we are already growing fatter (and hungrier), depleting more soil organic matter, drawing down more water tables, using more fertilizers and pesticides, losing more acres of forests and farmland. We are, in other words, already on a trajectory that if simply left to itself — analysts call this the business-as-usual scenario — will sooner or later push our food system, or a critical piece of that food system, across some crucial and irrevocable threshold. It may be only one sector or one country that fails — at least at first. But because our global industrialized food system is now so tightly integrated and interdependent, so reliant on the constant flows of material between regions and the ceaseless transactions among input industries, producers, processors, and distributors, there is no longer the possibility of discrete failure: a collapse in one part of the system will have extraordinary ramifications for everyone else.

Put another way, we no longer live in a world of single threats to the food economy. Given that such a disruption grows more and more likely by the year, while our capacity to respond to the inevitable cascade of consequences declines, we may well be on course for a perfect storm of sequential or even simultaneous food-related calamities that fundamentally change our ability to maintain food security.

The epicenter of this perfect storm will very likely be Asia. Although the catalyst could occur anywhere on the planet, Asia's massive population, the rapid growth in its food sector, and the yawning gap between that sector and the capabilities of its medical and political systems suggest that Asia's odds of being the lead domino are quite high. Most of the attention has been focused on the threat of avian flu. But in fact, in countries like China, India, Vietnam, and Indonesia, with their tight food markets and poor preparedness, a chain reaction of food-related catastrophes could be sparked by any number of events.

Suppose that while most of the world is waiting for the feared mutation of H5N1, the Ugandan wheat rust now decimating East Africa yields spreads, as most experts expect it to, through the Arabian peninsula and on to the wheat regions in Pakistan, India, Bangladesh, and, eventually, China. Based on past rust migrations, the fungus could be in China by as early as 2013, which could easily be before breeders have developed and distributed new rust-resistant varieties. (Starting today, breeders would have five years to develop a new variety for China, and even less time for Pakistan or India.[4]) The Ugandan rust, which can wipe out up to three-quarters of a crop in any given field, would devastate the wheat-intensive food economies of India and China in particular.

Under such a scenario, Asian grain markets, already tight from years of rapid livestock expansion, would turn explosive, pushing up prices to record levels and sending a shock wave through global markets. Just as they have today, high prices would ignite a planting spree by farmers worldwide — except that in this future scenario, such efforts are undermined by worsening climatic conditions. In Australia, the American Midwest, and southern Europe, the continued drought produces a succession of crop failures. In China, the shifting climate has exacerbated a pattern of extreme weather events, including drought and heavy flooding so severe that the corn crop also suffers a poor harvest. With grain prices higher still, livestock operators across Asia come under growing financial pressure to slash costs, slowing the already glacial pace of safety reforms. Such concerns are sidelined, however, as Beijing and New Delhi struggle to rebalance grain markets so volatile they are dragging down other sectors. China's central government ends all remaining grain exports and tries to reassert control over national grain production, touching off sharp political tensions between Beijing and provincial governments. By 2015, China, India, and neighboring nations have rapidly stepped up grain purchases on a world market that is already falling behind demand and pushing up consumer prices for meat, milk, and other staples.

In the United States, Congress enacts emergency measures to boost grain supplies. Subsidies for ethanol production are suspended. All idled acres held in various conservation programs are released, and farmers respond with the largest grain planting in U.S. history. But as in China, a gradually warming climate takes its toll on U.S. yields. Although rising temperatures are generally predicted to help midwestern crops like corn, climate change is also expected to push the El Niño southern oscillation into greater activity, resulting in a succession of droughts, rainstorms, and widespread floods that hammer the Corn Belt. Yields fall by between 20 and 30 percent. Farmers respond with higher applications of fertilizers, but the effects are blunted: warmer temperatures have accelerated the loss of soil organic material and reduced the land's capacity to use the extra fertilizers. In western states, meanwhile, declining snowpack and soaring urban water needs have dramatically reduced farm acreage and pushed up demand for imports. The higher food prices lead the big grocery chains to abandon their lines of pricier organic and sustainable foods.

With grain markets remaining tight, the World Food Program declares itself unable to feed any of the 120 million people, mainly in sub-Saharan Africa, who require assistance. The United States, Canada, and France push to create an international body to help coordinate global grain markets and relief efforts, but such efforts are undercut by China, India, and other big Asian importers, who are pressuring Brazil and Argentina for preferential access to their grain and soy production. Brazilian and Argentinean farmers, meanwhile, goaded by the highest sustained grain prices in history — and with financial backing from China and India — embark on a massive expansion of acreage. All earlier efforts to regulate farm expansion in the Amazon forest and other sensitive lands are abandoned, despite protests that these losses will exacerbate climate change. With similar expansions under way in Malaysia, Indonesia, and parts of Africa, global forest cover is falling faster than even pessimistic forecasters had predicted: over the ensuing five years, nearly half of the almost ten million square miles of remaining forest are burned or logged to make way for farms, blanketing much of South America and Asia in smoke.

In the United States, meanwhile, the rapid expansion of farming into erosion-prone lands, coupled with the increased rainfall, has touched off huge erosion problems. Chemical runoff is endemic. Warming temperatures have worsened insect outbreaks, leading farmers to step up applications of pesticide, but wetter conditions carry much of the chemicals, along with nitrogen fertilizers, into groundwater and surface water; coupled with increased runoff from livestock operations, there is massive contamination of drinking water systems across the Midwest and a major increase in eutrophication in all major waterways and estuaries.[5]

Although these environmental impacts will eventually harm food production, they are increasingly ignored as Washington and other governments struggle to manage the humanitarian crisis sparked by high grain prices in the developing world. In Central America, millions of urban dwellers begin moving northward, toward the U.S. border. In South Asia and Africa, the already fragile food economies are dealt a series of blows by climate-induced catastrophes. Between incessant drought and a monstrous acceleration of pest outbreaks, sub-Saharan Africa sees its grain production plummet. Demand for imports skyrockets, yet African consumers and governments are unable to pay for them. As farm workers become so malnourished they can no longer work their fields, the vicious cycle of hunger and poverty begins its slow spiral. Relief efforts mushroom, but as hunger spreads, teeming slums become too violent to enter, and in the rural areas, conflicts over water and rangeland mushroom into border wars between tribes and eventually between states, forcing the last relief agencies to withdraw. By 2020, Africa is in the throes of the largest famine in history.

Asia is not too far behind. In China and India in particular, high food prices have so eroded economic growth that unemployment is soaring. Millions leave the urban areas, returning to the country to look for work on farms. The livestock industry and dairy industry, caught between soaring input costs and declining demand, implode, forcing the central government to take over operation of the largest livestock facilities. But meat production is now rapidly decentralizing, with tens of millions of farmers now producing pigs and poultry on tiny, low-tech operations where biosecurity is not feasible.

In late September 2018, public health officials in Vietnam report the deaths of several dozen people suffering severe flulike symptoms. A week later, lab technicians at the National Pediatric Hospital in Hanoi confirm that the deaths were caused by a highly pathogenic, highly contagious virus believed to be avian in origin.

There are, of course, any number of possible variants to such a scenario, with different opening moves and greater or lesser degrees of consequential misery. Certain forecasts for future oil supplies, for example, suggest that food production is now so inextricably linked to fossil fuels that a peak in oil output, and the subsequent decline in food supplies, would shrink the global population by several billion over the next two decades.

A joint analysis by the Stockholm Environmental Institute, the Santa Fe Institute, and the Brookings Institution sees the collapse of our natural production systems creating such conflict and chaos that our capacity to recover will be contingent on "the degree to which the prevailing power structure governments, transnational corporations, international organizations, and the armed forces manages to maintain some semblance of order."[6]

Such dire scenarios may indeed be helpful in persuading us that the risks to our food systems are real. But what we really want, after we've convinced ourselves that a collapse *is* possible, are scenarios in which things *don't* fall apart — scenarios in which we manage to overcome the current political, economic, and cultural inertia and prevent that collapse, or at the very least, minimize its impacts. And here, too, there are many possibilities to consider.

To date, only one country — Cuba — has made a serious and comprehensive effort to refashion its food economy around a more sustainable model. The catalyst wasn't epidemiological or environmental, but geopolitical — the fall of the Soviet Union — yet the effect on the system was similar. In the early 1990s, the oil, fertilizers, pesticides, and other elements of large-scale agribusiness that Moscow had supplied to Cuba in exchange for sugar and citrus suddenly stopped, as did the Soviet-bloc grain and other "people" food that Cubans no longer grew themselves. The Caribbean nation found itself trying to feed ten million inhabitants with a farm system and a cluster of comparative advantages that had been geared for participation in a global market; when deprived of that market,[7] the system collapsed. Tractors lay idle. Farm fields became choked with weeds, while livestock, accustomed to gorging on imported grain, slowly starved on pasture grass. Food prices soared (black-market prices for meat, cooking oil, and eggs jumped 1,000 percent a year), and consumers reacted as they always do in such circumstances: between 1989 and 1993, Cubans' daily per capita intake fell from three thousand calories to fewer than two thousand, which put Cuba below the Caribbean's perennial basket case, Haiti.[8]

Desperate times called for desperate measures. Deprived of industrial inputs, Cubans had little choice but to deindustrialize their food model, making it into something less mechanized and less chemically dependent, and far more focused on food for local consumption. The massive state-run farms were broken into cooperatives, and hundreds of thousands of laborers were "reallocated" from urban and factory jobs to work on farms; by some current estimates, as many as one in four Cubans is involved in food production. In Havana and other cities, thousands of produce gardens, ranging from large collective operations to small patio plots, were planted, and, importantly, farmers were allowed to sell their surpluses at hundreds of new agricultural markets. At universities and research centers, meanwhile, scientists scrambled to find ways to replace the heavy industrial farm inputs. Oxen-breeding programs were expanded to replace tractors. In place of synthetic fertilizers and pesticides, Cubans adapted numerous methods of integrated agro-ecological farming, including mixed livestock-crop operations, crop rotations, interplanting, and integrated pest management.

The results have been almost Chinalike. Although Cubans are still short of meat and dairy products, per capita intake has recovered so completely that the country now leads most developing nations in nearly all nutrition and food-security categories; indeed, the larger concern is that Cuba has resumed the march toward obesity that had been interrupted by the Soviet collapse. But outside Cuba, many advocates of alternative agriculture are pointing to the country's successful deindustrialization of its farm sector as an illustration of what conventional food systems everywhere might do. "The experiment in alternative agriculture currently under way in Cuba is unprecedented," wrote food activist Peter Rosset, "with potentially enormous implications for other countries suffering from the declining sustainability of conventional agricultural production."[9]

Rosset's prognostication may be a tad premature, given Cuba's rather unique comparative advantages: a warm, wet climate that is ideal for year-round growing, surplus labor that could be reallocated to agriculture, and an authoritarian political system that made such reallocations possible. As Bill McKibben dryly observed in a 2005 *Harper's* article, Cuba "is a one-party police state filled with political prisons, which may have some slight effect on its ability to mobilize its people — in any case, hardly an 'advantage' one would want to emulate elsewhere."[10]

But Cuba's story does offer some critical lessons. While Americans and Europeans today would no more voluntarily adopt a Cuban model of food production than they would an Indian level of meat consumption, the real question is no longer what a rich country would do *voluntarily* but what it might do if its other options were worse — if, say, its industrial food system was running low on some key input, or had exhausted its soils, or had suffered some catastrophic failure in production or safety. Because given the speed with which our food system *is* using up inputs, *is* degrading its natural assets, and *is* creating new opportunities for pathogens and contaminants — and given the inability of policymakers, industry leaders, and consumers to make any substantive changes proactively — the chances that we will be forced to react to some major food disruption grow by the month. In such circumstances, notes McKibben, "it's somehow useful to know that someone has already run the experiment."

So what might our own version of Cuba's experiment in sustainable food production look like? Suppose that instead of waiting for a domino effect to start in Asia, we were moved to take proactive steps by some more proximal shock: a large number of poisonings from a Chinese food shipment that had evaded inspectors. Or a U.S. outbreak of H5N1 that claimed only a small number of human lives but created substantial economic turmoil in the food system (not least being the depopulation of the nation's nine-billion-plus bird flock). Suppose further that even as this shock is striking the food system, popular support for serious food-policy reform — driven by rising energy costs, concerns about climate change and food safety, and a growing skepticism about traditional agribusiness — has become so strong that policymakers can no longer defend the status quo but have been forced to consider deeper and more fundamental changes. Suppose finally that the food industry and the farm lobby, after decades of increasingly bad public relations, has finally lost the political capacity to kill reform in the cradle. Under such circumstances, what kinds of action would we take?

One obvious direction would be toward a food system that was less dependent on global or even national suppliers and more reliant on regional food sources. Such a system, if carefully developed, could be more secure, more climate-friendly, and certainly more energy-efficient than current models. (Even now, high fuel prices are motivating food companies to reengineer their processes to cut energy use, especially in packaging and processing.) Granted, *local* production doesn't always translate into better energy savings or reductions in climate footprint. But for certain foods, and in certain geographic areas, companies might find it more cost-effective to source products or raw materials if not locally, then at least *regionally* — particularly if anxieties about food safety and nutrition were encouraging more consumers to actively explore alternatives to mainstream food supplies.

If such a regionalized food economy were to emerge, however, it would bear little resemblance to the existing local model of small farmers feeding urban farmers' markets and community-supported agriculture associations. Instead, it would consist of fully developed regional food systems in which urban consumers would be linked, by way of supply chains, to a variety of urban, suburban, and regional food producers. Such short-distance supply chains are already well established in other parts of the world. European consumers still rely heavily on regional production, especially fresh produce, while many of Asia's regional food systems haven't yet been supplanted by supermarket supply chains. The city of Hanoi, for example, gets four-fifths of its produce, half its meat and fish, and 40 percent of its eggs from producers that are inside the city or on the fringes. East Calcutta boasts more than thirteen square miles of fishponds that also serve as water-treatment facilities, and massive Shanghai draws more than half its produce and meat from farms either in or surrounding the city.[11]

In the more advanced food economy of the United States, by contrast, most of the old regional food systems have been superseded by national and global supply chains; by one estimate, the average American community produces just 5 percent of the food its citizens consume. In some cases, the reasons are agro-ecological: regional climate or soils are unfavorable, or water resources are limited. In others, nearby farmlands have been decimated by urban and suburban growth; or the necessary infrastructure — farm-to-market roads, rail spurs, warehouses, and other marketing and distribution systems — has been abandoned or removed.

Even in the United States, however, pieces of the old regional systems remain intact and could under the right economic and political circumstances be significantly expanded. In many large American cities, a nascent but ambitious movement is already under way to cultivate a new urban agriculture, with operations ranging from backyard and rooftop gardens to restaurant salad gardens to large community farms located in greenbelts and in reclaimed industrial areas that produce orchard fruit, vegetables, honey, even livestock and farmed fish. As significant, regional food advocates are now pushing to rebuild or expand agricultural zones just outside of urban areas, much as they exist in most Asian countries, and to link these regional producers to local urban markets, schools, hospitals, and other buyers.

Some of these efforts are purely commercial; many, however, operate with a declared social agenda, such as fighting urban blight, supplying fresh food for school lunch programs, providing inner-city residents access to fresh, affordable food, or preserving rural livelihoods. Until recently, such noncommercial objectives had been relegated to the margins of mainstream food politics. Now, however, with rising concerns about food safety, advocates of food regionalism have been gaining serious support from some state and even federal lawmakers who see regional farming as a means of protecting food security from outside supply disruptions — hurricanes, import restrictions, even terrorist attacks. "Ten years ago, food security was mainly a question of adequate calories, irrespective of where they came from, but that's completely changed," says Tom Forster, policy director with the Community Food Security Coalition. The coalition contends that communities should be able to produce at least a third of their own food supplies locally and is among a large contingent of advocacy groups hoping to leverage current food anxieties to significantly change the meaning and practice of local food. "This isn't just high-end farmers' markets," Forster told me. "We're talking about the ability to feed cities as the next major food policy objective."

Barriers to regional food systems are still extensive. Smaller farmers near urban areas are under intensive economic pressure from development; distribution systems are inadequate or nonexistent; demand for regionally produced food, though growing, has yet to spread meaningfully beyond high-end and activist consumers and into the mainstream. But in a not-too-distant future in which energy prices are high and consumers' anxieties about food safety haven't abated, regional food advocates say these barriers could fall. Already, concerns over obesity and the sorry state of children's nutrition have led many public schools to begin procuring locally produced food for their lunch and breakfast programs — a trend that regional food advocates are working hard to expand. In recent years, these efforts have been boosted by new laws in at least 17 states, including New York and Michigan, encouraging public schools to source foods locally. Public Schools in New York alone serve 885,000 meals a day — second only to the Pentagon.

Equally encouraging, congressional lawmakers, looking for ways to make up for an egregious farm bill, are considering a law that would allow federal agencies to give preferential treatment to regional producers when procuring food — treatment that is currently prohibited. (Trade-policy experts have suggested that such governmental preferences for regional food, while technically a violation of free-trade treaties, wouldn't be likely to upset trading partners, who will very likely be working to shore up their own regional food security.) Over the next decade, developments like these could create a massive new market for regional food. Already, many commercial food-service suppliers, like Sodexho and Bon Appetit, are beefing up their portfolios of local producers in anticipation of new demand.

There is, in fact, a real risk that demand could grow too quickly. If policymakers do make the promised changes in federal procurement regulations, for example, the resulting market for regionally produced food would be so great that even large commercial providers would be overwhelmed. To avoid such a catastrophe, advocates like Community Food Security Coalition are lobbying for federal funds to scale up the regional and local distribution systems that would move food from farm to market — essentially resuscitating the short-distance supply chains that were once common in and around nearly all American cities.

Initiatives like these would be hugely expensive. Existing national and global supply chains represent billions of dollars of investment, and creating, or re-creating, a regional system to work alongside these larger systems will require substantial capital, much of it public money. But such an outlay would hardly be unprecedented. Every previous step in the evolution of the modern U.S. food system — from early seed research, extension services, and land grant universities to the building of rail lines and port facilities, to the more recent push into transgenic foods — have depended on huge infusions of public money and have been justified by advocates as being critical to national food and economic security. The current need is no less urgent — a message that, advocates say, is finally sinking in. Despite tight budgets, Forster says, food security has a new resonance among lawmakers. With growing concerns about energy, climate, the decline in nutritious food, and even international politics, he says, "it's much easier to make the argument that every region needs a certain resilience in its food production capacity."

If regionalism can help bring more security and sustainability to the food system, the next and larger step will be to somehow maintain that security and sustainability while substantially boosting the system's output. For as we've seen, the challenge for the future economy isn't simply lowering the external costs of the current operation, but lowering those costs while simultaneously feeding another three to four billion people over the next half a century. In particular, we will need a means of providing substantially more protein than we now produce, yet with substantially lower external costs than we now incur. And given that our current livestock model is unlikely to ever reduce its external costs sufficiently to meet future demand (not even if we shifted entirely from high-impact cattle and pigs to the relatively low-impact chicken), we are faced with the prospect of a meat economy that cannot move forward without significant changes in both supply and demand.

On the supply side, the issue is fairly straightforward: because conventional, terrestrial livestock are constrained by biological and ecological limits, more of our protein will need to come from a more recently opened frontier — the sea — in what food-security optimists are calling a blue revolution. Fish are inherently efficient feed converters: cold-blooded and hydrodynamic, their weight supported by water, fish burn far fewer calories for body maintenance than terrestrial species do, and thus they can devote more calories to putting on weight. Fish are also more amenable to industrialization than any of their terrestrial counterparts; they can be raised en masse and are highly responsive to breeding. (In barely three decades, breeders have nearly tripled farmed Atlantic salmon feed efficiency.) Fish and other marine species are also far more diversified than terrestrial livestock; whereas the land-based meat industry is built around a handful of species, commercial aquaculture includes some 440 fish, shellfish, crustaceans, and others, in part because the creatures are relatively easy to domesticate. Most commercial species were domesticated in the last century, and a quarter of them in the last decade alone. This is one reason that global aquaculture production, which was near zero in 1950, now provides more than a third of the total commercial fish catch (most of it in Asia); it is also why many food-security experts believe that under a high-demand scenario, aquaculture could become a major player not only in seafood but in the larger meat economy.

Aquaculture as it is practiced today has serious drawbacks, not the least being the issue of sewage, the heavy reliance on antibiotics, and the unsustainability of feed supplies for carnivorous fish such as salmon or halibut. But recent years have seen the development of alternative methods of aquaculture, known as deep-water or open-water aquaculture, and new plant-based feed supplies — alternatives that, under the right circumstances, could become a massive new source of high-quality, relatively low-cost protein with far fewer externalities than today's dominant meats.

What would ignite this blue revolution? On a certain level, the revolution is already under way. Even without an explicit food crisis, demand for cheap protein, coupled with declining wild fish stocks, had already touched off a boom in the aquaculture industry. Rising feed costs will only continue to increase the competitiveness of the industry and thus draw more investment. Lawmakers could accelerate that expansion with any number of sub-sidies, such as tax breaks for fish farms, low-interest financing, and research funding. And of course any additional catalysts — for example, our H5N1 outbreak, and the culling of billions of chickens, which would drive up all meat costs substantially — would simply accelerate demand for fish, since beef and pork cannot compete in feed efficiency.

In the face of rising demand, the larger challenge would be to ensure that this new protein revolution actually is better than the old one, given that the aquaculture industry's current course is hardly sustainable. Most of the boom in fish farming is occurring in near-shore facilities, which are profoundly vulnerable to environmental problems; open-water farming, though it is practiced commercially, makes up only a minuscule part of the market and is still struggling to develop suitable technologies and methods. Feed remains a challenge; whereas the freshwater species, like carp and tilapia, which are dominant in Asia, eat grain or soybeans, the carnivorous species raised elsewhere are rapidly depleting wild fish stocks. However, if researchers can develop plant-based feeds for all farmed fish, aquaculture could, in theory, generate three times as much protein as terrestrial live-stock can from the same inputs of feed.

To push the aquaculture industry onto a sound trajectory, existing laws and policies need to be reshaped around a new objective of long-term sustainable production. Nations with large existing aquaculture sectors need to phase out any incentives, such as tax breaks, that now promote in-shore operations, while also tightening environmental and safety regula-tions governing, for example, the limits on sewage outflow. Operators also need to invest in better containment facilities to prevent escapes of farmed fish into the wild, where they can intermingle with, and weaken, wild fish stocks. Importantly, these regulatory changes need to occur within an in-ternational framework in order to prevent opportunistic farm operators from simply leaving nations with tough laws and setting up shop in more relaxed jurisdictions.

But the blue revolution's success depends even more on the acceler-ated development of alternative production methods. Current research programs into open-water aquaculture need substantial infusions of fund-ing, as do efforts to develop new plant-based feeds, replace fishmeal, and expand the number of herbivorous and omnivorous fish species that can be commercially raised. Likewise, a serious research effort is needed to shift aquaculture away from the current single-species monoculture model and toward a more integrated, closed-cycle polyculture system, similar to the traditional Asian model. Encouragingly, researchers have developed practical working models of intensive, multispecies aquaculture, in which the waste from the primary animal — salmon, say, or shrimp — is recycled as feed for secondary commercial crops, such as shellfish and seaweed. Studies suggest that such operations, many of which can work on both larger and smaller scales, can substantially lower the volume of nitrogen pollution while producing significant volumes of commercial protein; in one case, a 2.5-acre operation generated 35 tons of fish, 100 tons of oysters, and 125 tons of seaweed in a year.[12] With the right encouragement, argues Carlos Duarte, who works with the Instituto Mediterráneo de Estudios Avanzados in Mallorca, Spain, this new style of aquaculture would not only reduce pressure on land-based resources but also would produce "a fundamental change in the way humans relate to the oceans."[13]

A blue revolution, like its green predecessor, would not be without controversy. The need to rapidly domesticate new marine species and rapidly improve feed performance will renew the industry's efforts to develop better breeding technologies, not least transgenic technology. Elliott Entis, the entrepreneur behind the "frankenfish" salmon, says his transgenic breeding methods not only cut production times in half, from three years to eighteen months — essentially allowing farmers to double their output — but let breeders develop these new high-performing fish in as little as a fifth of the time that it would take using conventional breeding methods. Critics of transgenic technology continue to argue that these genetically manipulated traits may be unsafe for humans and that transgenic fish themselves could pose a serious gene flow risk to any native populations of the same species. But as demand for inexpensive protein skyrockets, the debate over the safety or necessity of transgenic technologies — a debate that has more or less stalled out on the margins of food policy — will move rapidly to center stage, where policymakers will be forced to weigh the potential rewards of transgenic technologies with the potential risks.

Given the intense political pressures on either side of the transgenic argument, policymakers need to adopt a credible, science-based, depoliticized approach to the question, with input from neutral scientific bodies and a thoroughly public debate — something that in the world of American food politics is almost impossible to imagine. And yet we will have to do more than imagine, because the debate over transgenic food engineering is just one of many questions that need to be settled, or at least carefully explored, over the next decade if we're to have any hope of meeting the looming challenges to the food system. The debate over organic versus synthetic, for example, or agribusiness versus family farm, or local versus global; the debate over diversity versus monoculture; the debate over nutrition and obesity, over food safety, and over food security — all need to be pulled from the shadows of advocacy politics and industry lobbies and exposed to the full light of a truly public process. These questions need to be discussed not simply in terms of their costs and benefits, but in terms of the ways they fit into a larger strategic vision. Without such an open discourse, we will never build a public consensus for action or develop any kind of coherent strategy that takes on the entirety of the challenge we face, but instead we will merely continue to nibble away at the problems piecemeal. Just as we long ago broke farming into its constituent pieces and are now suffering the consequences, our solutions have tended to follow a pattern that is no less reductionist, in which each problem (for example, synthetic farm chemicals) is met with its own discrete solution (organics). Yet just as most of our food challenges are now understood to be interrelated and evolving, our solutions, too, must be both comprehensive and capable of constantly adapting.

Consider the pivotal issue of trade. In recent years, trade reformers have focused their efforts on lowering import barriers in countries like the United States so as to give farmers in poor countries the chance to export more products. But under a broader strategy that seeks to balance fairness with expedience, trade reformers might work to help emphasize regional markets as a more practical outlet for poor farmers. Even in the rosiest of development scenarios, it will be years before developing-country farmers have the capacity to compete with their counterparts in the United States or Europe on anything but cash crops, like nuts.

Reformers have also pushed traditionally for greater emphasis on food sovereignty in the hopes of encouraging LDCs to bolster their own productive capacities and minimize their reliance on imported grains, and thus their exposure to grain price volatility. But these important objectives need to be weighed against broader goals, such as the global balance of water and other scarce resources. As we've seen, when water efficiency is considered, it can actually be more sustainable to ship grain from a water-efficient producer like the United States to a water-inefficient producer like Kenya or northern China. Such a comparative advantage may not always hold: future energy prices may be so prohibitively high as to outweigh the water savings. And today's big grain importers may eventually improve their water efficiency by upgrading wasteful irrigation systems or developing rainfed crops better suited to their drier conditions. But, argues Alexander Zehnder, the water trade expert at the Swiss Federal Institutes of Technology, getting "significant increases in water efficiency will take more time and more investment and more education — it cannot be done in a few years' time." Such improvements might take two to three decades; in the meantime, the grain trade may remain the most efficient way to distribute the world's finite water. What is not settled, Zehnder warns, is the emerging global politics of water. Of the five big virtual water exporters — the United States, Canada, Argentina, Australia, and France — four are in the industrialized north and are well known for attaching political conditions to their exports. In the coming decades, says Zehnder, the challenge will be to develop a global water market "without political attachments."

Ultimately, the more fundamental obstacle between where our food system is and where it needs to go isn't the challenge of boosting food supplies but of reducing food demand, especially for meat. Even with enormously productive polyculture farms, massive expansions of aquaculture, and Nobel Prize–wining breakthroughs in transgenic feed grains, whatever increases we might possibly achieve in sustainable protein production won't be large enough to satisfy any future demand for meat unless current trends in meat consumption are reversed and the global average for per capita meat-eating begins to fall.

Such a proposition, by now well accepted within advocacy and many scientific circles, has traditionally been untouchable in the political and cultural mainstreams, especially in regions like North America and Europe, where high rates of meat consumption are built in to both personal expectations and corporate strategies. And yet, as pressures mount on the food system, that proposition will be pushed inexorably from the fringes of the debate toward center stage.

The impetus will come from many directions. Rising grain prices may help make meat less attractive, as would any major outbreak of a food-borne illness, whether from pathogens like *E. coli* and salmonella, contaminated imports, or even a bird flu outbreak. A full-blown pandemic on American soil "will turn a lot of people into vegetarians," suggests the blog Avian Flu Investor.[14] But that would be temporary: Americans have always returned to meat — witness the recovery of beef, and ground beef in particular, following the Jack in the Box scandal.

What is needed is a strong public position by credible authorities — a position that lays out the huge external costs of a heavy meat diet and articulates the way in which the current system of subsidies and other government support keeps meat artificially cheap. Advocates for a vastly smaller meat economy believe this reformist message could be couched in terms of its benefits for health or its reductions in greenhouse gases. With lawmakers in the United States and elsewhere more and more amenable to climate-related legislation, and with the links between meat and greenhouse gas emissions now gaining attention, some advocacy groups sense a window of opportunity in which to press for changes in laws that subsidize livestock and cheap feed, and — perhaps — even to raise the issue of meat itself. "We need to get organizations who are working on food-policy issues, like the UN and the USDA, to at least come out and say, 'Here are the health benefits, here are the environmental benefits'" to lowering per capita and total consumption of meat, Dawn Moncrief, director of the Farm Animal Reform Movement, told an interviewer in 2007, soon after the FAO blamed livestock for nearly a fifth of total global human-caused greenhouse gas emissions. "If we could get these governmental and quasi-governmental agencies to come out and say it, that would be a good first step."[15]

Of course, it is also a first step that is hard to imagine in today's political climate. Although some of the changes we've explored, such as the move to a more regional food system or a rapid expansion of aquaculture, are conceivable in the current atmosphere, the meat economy remains far too central and immovable an institution to be transformed — and, specifically, to be shrunk — without offering substantial resistance. It's not simply that large, politically influential livestock operators and meat companies would fight any such plan tooth and nail. It's also that hundreds of millions of wealthy consumers in the United States, Europe, and parts of emerging Asia have come to see a heavy meat diet as a birthright, while a billion or so small Third World farmers, many of them desperately poor, regard meat as an important source of income and food security.

In ten to twenty years, as conditions have deteriorated further, policymakers may be willing to take this issue on. But now, such action is difficult to envision. Despite the recognition of many mainstream institutions, such as the FAO, that the livestock industry's external costs will pose severe problems in the future, until those problems actually reach crisis proportions, those institutions aren't likely to propose any significant change to the status quo. The FAO, after excoriating the livestock industry as a "major player" in climate change, recommended a series of proposals that would minimize the industry's emissions but that made no mention of reducing meat consumption itself.

In a perverse way, the story of the modern food economy now turns back to where it began — with the challenge of meat. In the 1940s, scientists like Thomas Jukes made it seem possible that despite a soaring population and a declining land base, the world could still have a food economy centered almost exclusively on meat. A half-century later, we understand that such a vision was unsustainable, in large part because it failed to account for the true expenses of such an economy. But even as these expenses become more and more apparent, all aspects of the system, economic, political, and cultural, have become so entrenched that the prospects for proactive and timely change seem less and less likely.

The inertia of the modern meat economy is really just a variant of the enormous momentum that is now driving the larger food system on its perilous trajectory. As with meat, we can trace this momentum back to the political influence of the major players, or the apathy of disaffected consumers. But we also need to see this momentum as structural — that is, as a consequence of the food economy's modern shift from a diverse and decentralized system to one that is increasingly centralized, uniform, and concentrated. We've seen how the loss of diversity, whether at the level of the field, the factory, or the industrial sector, has created a system that is less economically viable and more vulnerable to disruption — concerns that have led to a new push to bring diversity back into the making of food. But paradoxically, this lack of diversity is itself also a powerful obstruction to change. A farm economy composed of millions of individual farmers, producing hundreds of different crops and animals by way of countless different strategies and ideas, was, however inefficient in a strictly commodity sense, nonetheless far more resilient and flexible and adaptive than is a system made up of large farms locked into a handful of deeply entrenched technologies and models. Similarly, an input sector that is largely in the hands of three companies, or a grain industry controlled by five companies, or a grocery sector divided among five chains (and dominated increasingly by one) will all be not just more reluctant to change (owing to the sheer scale of their investments in status quo technologies and models) but much more effective at resisting that change, due to their concentration of economic and political power and to their virtual monopoly over the very discourse of change: Wal-Mart, Tyson, Monsanto, and other mega-players are now not simply the providers of food but, increasingly, the arbiters of ideas and attitudes about food, for whom any discussion about the future of food begins and ends with the status quo.

Some social observers contend that even as this massive institutional resistance thwarts government efforts to effect change, it has also fostered a vast and volatile reservoir of popular resentment — among activists and advocates, but also among ordinary citizens who are dismayed by the state of the food system, who no longer trust government to solve these problems, and who have begun to take matters into their own hands. The movement is reportedly quite large (the writer Paul Hawken estimates that it contains as many as two million organizations and calls it "the largest social movement in human history, albeit unknown to itself and unknown to the media"), and it is now busily working on issues of environmentalism and social justice, including those centered on food production. One can imagine this massive movement eventually achieving a kind of critical mass, spilling into the mainstream, and creating an impetus for change that even recalcitrant lawmakers and industrial lobbyists will not be able to resist.

But here again the question of timing becomes critical, particularly for a system as vast and complex as that of food production. The U.S. farm program doesn't come up for reauthorization until 2012, and even assuming that lawmakers might by then have summoned the political will to truly reform the subsidy system, it would be years before the changes worked their way through the food economy.

The same could be said for nearly every area where the sustainability of the food system is in question. Even under the best circumstances, the FDA needs years to build up its capacity to test and monitor food imports. Until then, the agency will continue to rely primarily on the industry to police itself — a policy that, with countries like China, is patently unworkable. The Chinese government insists it is rapidly upgrading its food-safety regulations, and industry optimists continue to argue that China's desperate need to maintain export revenues assures that such promises will be kept. But this is fantasy. Even if Washington were to adopt a far tougher policy with China — for example, by threatening to cut off food shipments until Beijing can police its food industry, rather than by offering to buy *more* food from China, and especially more meat — China simply lacks the ability to meet such demands quickly.

Given this degree of systemic inertia, it may seem that we have little choice but to resume our position of waiting for the crisis. This, certainly, seems to be the attitude in Washington, where on issues such as food safety, government officials now speak about "consumer responsibility." Indeed, as Michael Leavitt, secretary of health and human services, recently declared in a speech on avian flu, "any community that fails to prepare, with the expectation that the federal government or, for that matter, even the state government will come to their rescue at the final moment, will be tragically wrong."[16]

And yet there is something perversely liberating and even energizing about such candor. On the surface, we can read a warning like Leavitt's as a specific suggestion to prepare for a flu pandemic, which is probably not a bad idea. But the warning can also be taken as a prescription for a new approach to the entire food system. From all that we have learned about the food economy and the economics of food, it is clear not only that the modern system of food production will be more and more prone to disruptions but also that there are no entities, public or private, with the capacity to prevent those disruptions. Government may indeed succeed in blocking certain imports of contaminated food, or in tracking down the cause of an *E. coli* outbreak in spinach; companies may offer to disclose the sources of their ingredients, or they may voluntarily stop selling candy and soda in elementary schools. But no government, no company, no army of activists can "make" the food system sustainable, or ensure that the grocery store shelves will always be full.

This message seems to be sinking in, at least on the margins. Anxieties about food supplies and pandemics are clearly behind the surge of books, magazines, websites, and even reality-TV shows devoted to personal food production. Some of this growing feed-yourself genre seems driven less by practical considerations than by the distinctly survivalist fantasies of a world "after the fall," when only those with solar-powered farms deep in the Appalachians survive. But some of it is more practical, or at least provocative; as more than one website has suggested, the next time you come home from the grocery store with bags of food, do a mental inventory of the items that, if the modern food system were temporarily turned off, you could replace locally, or even produce in your little backyard garden; my own list is always discouragingly small.

Yet between the dire visions of survivalists and equally dire warnings of officials like Leavitt, there is a broader survivalist message that needs to be understood and encouraged and brought into the mainstream. Even if the modern food system manages to go years or decades without a problem, even if it suffers no more than a minor disruption and continues to turn inputs into outputs with increasing efficiency, the costs — to our bodies, our minds, and our world — will nonetheless be extraordinary. A system that perforce gives priority to only the most profitable aspects of food will continue to underemphasize or ignore or actively destroy the unmonetizable but critical elements of food as surely as any pathogen or famine or terrorist attack. And no one — not the FDA, not Nestlé, not even the Animal Liberation Front — will protect us from that threat: ultimately, that battle is our own.

This is not to suggest that we begin staging raids on fish farms. It is, however, to argue for a kind of direct action, although perhaps a more constructive one. That means lobbying Congress to reform the farm program, even if it will take years to have any effect. It means demanding that Congress increase funding for research into alternative farming methods and enact commonsense measures like country-of-origin labeling. It means lobbying school boards to improve lunch programs and dump junk food. It means working with the plethora of local and regional groups who are already building regional food systems.

Ultimately, it means taking back control of your own food. I'm not advocating that we all move to the woods and live on nuts and berries, or that we pretend that the preindustrial food economy, with its low yields, rampant diseases, routine adulterations, and endless hours of backbreaking labor, is something to be yearned for. But I am suggesting that in turning over the making of our food to others, in allowing the parameters and priorities of what we eat and how we think about it to be determined increasingly by a quite distant economic model, we have both encouraged the decline of food and lost something profound from our own lives.

Food has, for better and for worse, served for millennia as a sort of umbilical link between us and the physical, natural realm. By diminishing this link between consumption and production, we have allowed ourselves to drift away from the real world, and to understand less, and to care less, about its functions and condition. The fact that so many of us are surprised to learn about the damage from soil erosion and nitrate runoff, or about the accelerating loss of Brazilian forests to cattle ranching, or about the breathtaking volume of tainted food that is *still* being shipped to us from China shows just how unconnected we have become from what is arguably the most essential human function.

The costs of this slow divorce go beyond the physical and economic damage that we are doing, or allowing to be done, when we eat. So many of the social and cultural and psychological challenges we now face — obesity, deteriorating family relationships, lack of connection to something larger — things we now attempt to manage separately and often pharmacologically, intersect at the table. By turning our food over to someone else's care, we have handed over much of the control of the rest of our lives as well. Conversely, by reclaiming even a piece of the task of food production, by changing food from the passive object it has become to the vital and active enterprise it once represented, we could restore a great measure of balance to our lives. It could reconnect us to what is ultimately real and useful, and help us make sense of, and regain some degree of control over, a world that now seems increasingly random, and where food in particular, once the supreme object of ambition and intentionality, seems more and more something that simply "happens" to us.

In his critique of another modern blight — urban sprawl — designer John Thackara offers a powerfully apt analogy for the food crisis. "Nobody seems to have designed urban sprawl, it just happens — or so it appears," writes Thackara in his book *In the Bubble.* "On closer inspection, however, urban sprawl is not mindless at all. There is nothing inevitable about its development. Sprawl is the result of zoning laws designed by legislators, low-density buildings designed by developers, marketing strategies designed by ad agencies, tax breaks designed by economists, credit lines designed by banks . . . data-mining software designed by hamburger chains, and automobiles designed by car designers. The interactions between all these systems and human behavior are complicated and hard to understand — but the policies are not the result of chance. 'Out of control,'" Thackara concludes, "is an ideology, not a fact."[17]

We need to take a similar approach to food — to recognize that what has happened to our food system, and, as a consequence, to us, has not been some random and inevitable process. The transformation of the food system has indeed been driven and shaped by one of the most powerful and brutally efficient of all human forces — the market. But that system is still very much a work in progress, a product of billions upon billions of human decisions. And if many of those decisions are made in places and contexts far beyond our control, many more are made much closer — in our regions, in our communities, and in our kitchens. For thousands of years, food has mirrored society. It provided the substance and ideas that brought forth civilization, as well as the mechanisms by which civilization now seems to be taking itself apart. At the start of the twenty-first century, we are closer to that precipice than we have ever been, yet perhaps more capable, ultimately, of stepping away. Hunger has always been an invitation to make a better world, and it remains so.

Afterword

ALTHOUGH LESS THAN a year has passed since *The End of Food* was published, the concerns it documented about the global food system have, if anything, only become more visible. Food safety remains front-page news — from the ongoing dairy scandal in China to the salmonella outbreak in peanut butter in the United States that sickened more than 450 people and killed 6. Grain prices, driven to record highs in 2008 by crop failures, biofuel production, and speculation, have since crashed and farmers are failing by the tens of thousands. Factor in the Bush administration's last-minute deregulation of factory farms, new evidence of climate's impact on food output, and UN forecasts of new food shortages by 2010 (and perhaps another round of food riots like those that shook Haiti and Bangladesh), and the picture of global food security seems to have darkened since last year.

Yet recent months have seen encouraging developments as well. In Barack Obama we have a president who "gets" the link between climate change and food security. Some food industry executives are acknowledging that certain production practices, such as feeding antibiotics to livestock, aren't sustainable. The press is scrutinizing the external costs of meat, vegetable seed companies can't keep up with demand by a new generation of home gardeners, and in Berkeley, California, two business school graduates are building hydroponic greenhouses on the rooftops of grocery stores.

No, we're nowhere close to giving up the status quo. Tom Vilsack, America's new agriculture secretary, represents a cautious compromise between food reformers and the commodities industry, whose political power is virtually undiminished. Mainstream consumers haven't ditched fast food (indeed, economic hard times have boosted sales of cheap junk food). They haven't stopped eating grain-fed beef and refined flours or, more generally, abandoned the free-market notion that if you can afford something, you damn well ought to be able to take it home and eat it.

Still, it's in the context of the food crisis that many consumers have begun to question whether free enterprise is the best arbiter of food security and whether the same market forces that spawned so many of today's problems offer the best means to fix them. And in this respect, nothing in recent months has restored our confidence that the market alone can solve such challenges as global hunger, food-borne illnesses, and the climate-altering effects of industrial farming. Despite the food industry's massive capacity for innovation and change, beneath the surface, the main drivers of the food economy are still by and large pointing in the wrong direction. Without a fundamentally new relationship among consumers, producers, and, yes, the state (expressed, ideally, in a carefully crafted set of carrots and sticks), it's even clearer now that a food system driven by obsessive attention to cost reduction will keep pushing food production in directions it should not go and places it should not be.

Thus we still rely on low-cost producers and processors in China, where, despite strenuous government efforts, a safe food supply is more theory than practice. We still push for more centralized food production systems in North America and Europe, even though it's precisely this centralization that makes problems like the recent peanut butter salmonella outbreak so hard to contain. We're also seeing the problems once confined to low-cost food migrate to upper-end, "alternative" food systems: last August, Whole Foods, the nation's largest chain of health food stores, was forced to recall ground beef because of an outbreak of *E. coli.*

If there's a bright side to this succession of gloomy headlines, it's this: consumers who never paid much attention to the food system are now doing so. They not only know about specific problems in the food chain, but have a growing awareness of the complexity and interconnectedness of modern food production. As recently as a year ago, it was possible for consumers to think of problems like food safety and obesity as being separate; today we're far more aware of the way that changes in one part of the system (year-round produce, for example) are bound up with changes elsewhere (spinach farming in cattle country).

We're also more aware of the downsides of a business model obsessed with efficiency and cost. Yes, it's cheaper and more efficient to make peanut butter in a single huge facility in Georgia and ship the product in bulk to more than eighty-five different end users — but only until that facility becomes contaminated and hundreds fall ill. Indeed, in such a complex and interlocking system, just identifying the cause of contamination is difficult. It took federal regulators months to determine that chilies imported from Mexico, and not American-grown tomatoes, were the source of a massive salmonella outbreak in 2008 — a delay that put consumers at risk and cost tomato growers and packers great financial losses.

More dramatically, the past year has shown just how inextricably bound the food sector is with events in nonfood sectors. The price of oil, always a critical factor in production costs for fuels and fertilizers, is even more significant today because so much grain now goes into biofuels, which compete directly with oil. (By the end of 2008, U.S. farmers were shipping more than a third of the U.S. corn crop [and an eighth of the global corn supply] to ethanol plants.) So when oil prices eventually recover, we'll see an increase in farm production costs as well as an increase in demand for biofuels, and thus for grain — all of which could add up to higher grain prices.

As significant, however, are the newly visible links between food and finance. The drying up of credit has been devastating to the entire food industry, but especially to farmers, many of whom cannot plant the next year's crop without a loan. And consider the growing impact of speculation. During the early 2000s, as grain prices began rising because of higher food demand in Asia, many large hedge funds invested in grain as a way to balance out, or hedge, their other, underperforming investments — a strategy that accelerated as speculators fled the collapsing derivatives market. Unfortunately, the more grain that hedge funds bought, the tighter grain markets became, to the point where markets were caught in a vicious inflationary cycle that helped triple and even quadruple prices. From mid-2007 to mid-2008, the price of spring wheat jumped from $5 a bushel to nearly $24, and corn became so costly that food riots broke out in Mexico.

Not all the blame can be laid at the feet of speculators. Bad weather and rising biofuel production pinched supplies in 2007 and early 2008, as did panicked moves by big grain exporters, like Thailand and Vietnam, to protect their own supplies by cutting off exports. But the growing presence of hedge funds in the grain sector, and the resulting increase in speculative cash flows into that sector, unquestionably amplified the market's normal movements — both on the way up and on the way down. Last autumn, when a better than expected harvest added to global grain reserves and grain prices began to soften, the massive selloff of grain contracts by hedge funds clearly accelerated the price collapse. By December 2008, spring wheat had fallen back to $5 a bushel.

Today we find ourselves in a kind of food price purgatory. Though grain prices are well below their peak in 2008, they are still 30 percent higher than in 2006, which helps explain why the number of hungry people worldwide is again increasing. At the same time, prices have fallen below the cost of actually growing the grain, which all but assures another price bubble. Already farmers in the United States, Europe, and especially the developing world are being crushed between the low selling price of their grain and the boom-inflated prices they must pay for seed, fertilizer, and land, which are more than twice as expensive as they were in 2006. (In fact, the only real estate in the United States that has increased in value since the housing crisis has been farmland.) Under such pressure, sharpened further by the credit crisis, many farmers are being forced to reduce the acreage they plant or are leaving the sector altogether — both of which may lead to "significant shortfalls" in the 2009–10 harvest, according to the UN Food and Agriculture Organization.

The problem will be particularly acute in developing countries, where farmers have little in cash reserves to tide them over a period of low prices. Worse, as global grain output slows, the FAO warns, today's low prices will reignite demand for grain, especially among livestock producers and biofuel refiners, potentially "unleashing even more severe food crises than those experienced recently."

Experts will debate the causes of the 2008 grain bubble for years to come. In the meantime, the severe price swings, coupled with continued concerns over food safety and other problems, have focused political attention on the weaknesses of the food system. Consumer fears have crossed a threshold, and many policymakers feel that government can no longer put off serious action. In China, the government has given jail time and even death sentences to those behind the food safety scandals. And though we're unlikely to see such draconian measures here, we can expect tougher regulations in the food sector. Just as the financial collapse is ushering in a return to stricter banking regulation, the recent succession of food crises may be the coup de grâce for the deregulatory spree that began in the 1970s. Having seen the market's repeated failure to control price volatility and ensure safety, consumers and even some food companies seem ready for greater government intervention. In his inaugural address, President Obama declared that the financial crisis "has reminded us that without a watchful eye, the market can spin out of control" — and he might well have said the same about the crises in the food sector.

But what exactly should we be watching for, and how will we intervene? Though we have yet to see a detailed plan to fix the food system from the Obama administration, any such reform must attack the problem on several fronts. Obviously we'll need tougher, more comprehensive regulations for food safety and more funding for inspections and monitoring. But it's also clear that reform must go beyond merely stopping bad practices and craven actors. To build a new food economy, we must actively promote a broad range of new products, practices, and ideas. Just as Obama is proposing huge new investments in a green energy infrastructure, fixing our food system will require a substantial buildup.

We need to commit real dollars to food research and development — to get not only a new generation of food safety technologies but new production methods and crops that require less water, fertilizers, energy, and other constrained resources. (And note: by "new crops" I do not mean transgenically engineered foods. There is plenty of potential to improve crop performance using conventional breeding methods — accelerated, admittedly, with sophisticated gene markers.) We also need to support production practices and crops that are truly sustainable, and discourage those that aren't. That means reducing or eliminating government subsidies for crops that don't require taxpayers' help — corn and soybeans, for example — and shifting those resources to support less profitable but potentially more sustainable crops (for example, clover and other cover crops, which restore fertility naturally) or more progressive production models (such as local and regional food distribution systems).

We need to revolutionize the way we *think* about food; we need nutrition education for preschoolers and elementary students. We need innovative outreach programs to help underserved communities, such as poor inner-city neighborhoods, to take advantage of better food resources and practices. And though we don't *all* have to go back to the land, we do need programs that help restore the kind of food awareness we had when most of us were involved in producing what we ate.

Last, we need to ensure that our solutions reflect the complexity of the system we're trying to heal. We must blunt speculation's pernicious impact on food security (perhaps by restoring state-supported grain reserves, which were cut during the restructuring craze of the 1980s). But such reforms mustn't eliminate the benefits that speculators provide, such as access to capital and the collective market intelligence: by betting on future grain prices, speculators send a key signal to farmers, consumers, and other players — or as one analyst put it, "One thing the food futures market is telling us is 'plant more food.'" Similarly, while we must make the global trade in grain and other foods fairer and more sustainable, we dare not slow the global trading system's capacity to deliver grain to countries, like China and India, that are growing too fast to feed themselves. Reforming the food system will require not only a watchful eye but a temperate hand.

These are hardly small steps. For all our surging interest in transforming the food system, we should be realistic about the obstacles any such transformation will face. Subsidized soy and corn farmers, for example, will fight to keep that support, particularly now that grain prices have tumbled and many farmers are again in the red. Lawmakers will be hard-pressed to enact tough new safety regulations, especially when the food companies to be regulated are in such financial straits. Indeed, in times like these, the political temptation is toward less regulation, not more.

And consumers themselves, despite their new concerns about food, are not eager to pay more for better or safer food — a reality that poses a major challenge, given that sustainably produced food will likely be more expensive, at least initially, than conventional food. During last summer's food price escalation, sales of fast (and cheap) food skyrocketed. Likewise, while marketing experts argued that the food scandals in China had turned off American consumers to edible Chinese imports, it's hardly clear that American consumers or companies are ready to give up the cost savings that Chinese raw materials provide. Americans may slowly be accepting the fact that cheap food isn't really cheap — that it often comes with hidden costs to their (or their planet's) health — but price continues to be a dominant factor in the foods we buy, and any policies that don't take this into account will fail.

Given such complex challenges, lawmakers, consumers, and food companies will be more tempted than ever to seek out magic bullets — simple, easy (and cheap) solutions that may cure the symptoms but don't address the deeper problems in the system. Irradiation, for example, can blast the pathogens off salad greens; hiring more border inspectors will help stem the flow of poor-quality food imports from China. But neither of these solutions gets at the systemic flaws in the cost-obsessed business model that created these problems in the first place. Until we substantially change that model, the problems will recur endlessly — a point made dramatically in the summer of 2008, after the Whole Foods ground beef recall. According to press accounts, Whole Foods' supplier, Coleman Natural Meats, had purchased *its* meat from Nebraska Beef, a large, low-cost processor known for numerous health violations and product recalls, including a recall of five million pounds of hamburger meat in July for *E. coli* contamination. At the end of the day, Whole Foods got busted for trying to sell a new value proposition — sustainable food — using a conventional supply chain.

More and more, when we talk about improving and extending our food system, it's with the understanding that the easy victories have already been won — the "low-hanging fruit" have been picked. The next steps will be complex, high risk, and quite expensive. Yes, the planet still has considerable potential to expand its farm sector — in countries such as Brazil and Mozambique — but tapping that potential in a sustainable fashion won't be cheap. Likewise, while we can make our industrial food practices less damaging — by shifting away from feedlots, for example, and from massive applications of pesticides and synthetic fertilizers — these were the very "miracles" that helped drive food costs down over the last half century.

Thus the fundamental challenge we face is to make the food system more sustainable without reversing a century of progress, especially in food prices. For if paying more for what we eat may induce beneficial changes in wealthy cultures, it will be disastrous in those parts of the world where food costs are already too high, as the events of the past twelve months make chillingly clear.

In short, this is hardly the time to lose our focus on the pressing needs of the food system. Without question, with the financial crisis preoccupying policymakers and consumers — and with falling food prices diluting the urgency of even a year ago — it will be hard to maintain this necessary attention on food. But if we've learned anything in the past year, it's that any reprieve is likely to be temporary. We know that until we address the systemic problems of our food supply, the symptoms will recur, and we can expect more shortages, more outbreaks of food-borne illness, more tainted products. We know, in other words, that our food system is still deeply flawed. The only question, in my view, is whether we've finally reached a threshold where this awareness can be translated into action.

Acknowledgments

THIS BOOK HAS BEEN an exploration and an education, and I owe a considerable debt to a small army of experts who gave copiously and, for the most part, graciously of their time and insights. They include Lester Crawford, John M. Connor, Neil Mann, Robert Fogel, Kenneth Kiple, Arnel Hallauer, Gary Reineccius, John McMillin, Paul Aho, John Nalivka, Ronald Curhan, Peter Leathwood, Peter van Bladeren, Chris Brimlow, Hans-Jorg Renk, Ralph Gifford, David Swenson, Scott Exo, Randy Seeley, Mark Pereira, Lin Lugoung, Fred Gale, Jun Jing, Paulette Sandene, Chad Hart, Tian Weiming, Tom Remington, Paul Omanga, Hilary Roxe, Hanna Dagnachew, and the other excellent people at Catholic Relief Services; Mango and Janet Mutisya; Chris Barrett, Melinda Smale, Shaun Brunner, Mark Simpkin, Stewart Ritchie, Aleina Tweed, Victoria Bowes, William Marler, Trevor Suslow, Rob Atwill, Robert Mandrell, Jim Prevor, Jan Slingenbergh, Rod Van de Graff, Lester Brown, Frederick Kirschenmann, Kent Mullinix, Alexander Zehnder, Chris Jones, Charles Brummer, Patrick Schnable, Larry Yee, Theodore Crosbie, Kathleen Delate, Elliot Entis, Eric Flamm, Jane Rissler, Kendall Lamkey, Pat Mooney, Mike Philips, Matt Liebman, Julia Olmstead, Doc Hatfield, Fred Fleming, Chad Kruger, Diana Roberts, Antonio Brignone, Renato Sardo, Alessandra Abbona, Claude Fischler, Egizio Valceschini, Yvon Salaun, Marie Tanguy-Moysan, Amy Bentley, and Chen Chunming.

Special thanks to my editor at Houghton Mifflin, Anton Mueller, who worked tirelessly both to get this project in the air and to bring it in for a smooth landing, and to Amanda Cook, Alia Hanna Habib, Michael Webb, Nicole Angeloro, Gretchen Needham, and Laura Noorda for much heavy lifting. Thanks as well to my agent, Heather Schroder at ICM, for guidance, perspective, and the occasional lashing.

Lastly, I want to acknowledge the support, enthusiasm, and occasional expertise of friends and family, among them Damian and Susie at Schocolat, Brick and Angie at Buck Bay Farms, Charley and Brian at Tri House, and the crew at Sage Mountain, as well as Molly Roberts, Matt Roberts, Lynn Roberts, Ann Lister, Karen Dickinson, and especially my children, Hannah and Isaac, who endured their father's "food thing" with the patience of a ward staff and for whom this book was written.

Notes

PROLOGUE

1. Associated Press, "Wild Pigs Eyed in *E. coli* Outbreak," October 26, 2006, http://www.cnn.com/2006/HEALTH/10/26/tainted.spinach.ap/.
2. Jim Prevor, "Buyer-Led Food Safety Effort Leaves Open Question of Buyer Commitment," *Jim Prevor's Perishable Pundit,* October 30, 2006, http://www.perishablepundit.com/DailyPundit/2006/October/pundit061030-1.htm.
3. California Food Emergency Response Team, "Investigation of an *E. coli* O157:H7 Outbreak Associated with Dole Pre-Packaged Spinach," final, March 21, 2007, California Department of Food and Agriculture, http://www.dhs.ca.gov/ps/fdb/local/PDF/2006%20Spinach%20Report%20Final%20redacted.PDF.
4. Andrew von Eschenbach, "Opposing View: We're on the Case," *USA Today,* May 4, 2007, http://blogs.usatoday.com/oped/2007/05/post_7.html.
5. A. Bridges, "Pet Food Recall Raises Questions about Safety of Imported Food," Associated Press, April 16, 2007, in the *Wenatchee World.*
6. Kofi Annan, "The Right to Food: Note by the Secretary-General," February 2002; http://www.unhchr.ch/Huridocda/Huridoca.nsf/0/990d43116ffe0be1c1256c5d00368067/$FILE/N0254654.doc.
7. *Future Harvest,* "Preventing Childhood Blindness in Africa with Sweet Potatoes: Root Crop on a Mission to Remote Areas," June 18, 2004, http://www.futureharvest.org/news/sweetpotato.bckgrnd.shtml.
8. Sustainable Food Laboratory, "Overview," http://www.sustainablefood.org/overview/; Sysco Food Systems, "Social Responsibility and Sustainability," company website, http://www.sysco.com/aboutus/aboutus_sresponsi bility.html.

1. STARVING FOR PROGRESS

1. This account is based on the following articles: R. H. Gustafson and R. E. Bowen, "Antibiotic Use in Animal Agriculture," Agricultural Research Division, American Cyanamid Co., Princeton, N.J., at www.blackwell-synergy

.com/doi/abs/10.1046/j.1365-2672.1997.00280.x; Shannon Brownlee, "Agri-business Threatens Public Health with Antibiotics in Animal Feed," *Washington Post*, May 21, 2000; William Laurence, "'Wonder Drug' Aureomycin Found to Spur Growth 50%," *New York Times*, April 9, 1950; T. Jukes, "Antibiotics in Animal Feeds and Animal Production," *Journal of BioScience* 22: 526–34; as well as conversations with Lester Crawford (on April 5, 2006), former director of the U.S. Food and Drug Administration, who worked with Jukes's associates a few years after the incident and heard it discussed.

2. Laurence, "'Wonder Drug' Aureomycin."

3. Pat Shipman in Kenneth Kiple and Kriemhild Ornelas, eds., *The Cambridge World History of Food*, vol. 2 (Cambridge: Cambridge University Press, 2000), 14.

4. Neil Mann (professor at the School of Applied Sciences, RMIT University, Melbourne, Australia), phone and e-mail communication with the author, September 5, 2005.

5. Chris Scarre, ed., *Smithsonian Timelines of the Ancient World: A Visual Chronology from the Origins of Life to A.D. 1500* (New York: Dorling Kindersley).

6. Loren Cordain, *The Paleo Diet* (Hoboken, NJ: John Wiley and Sons, 2002), 39.

7. Cordain, 39.

8. Personal communication, August 22, 2007.

9. Kiple and Ornelas, eds., *History of Food*, 1126.

10. Reay Tannahill, *Food in History* (New York: Three Rivers Press, 1995), 72.

11. United Nations, *The World at Six Billion*, table 1, http://www.un.org/esa/population/publications/sixbillion/sixbilpart1.pdf.

12. Fernand Braudel, *Civilization and Capitalism, 15th–18th Century*, vol. 1, *The Structure of Everyday Life* (Berkeley: University of California Press, 1992), 193.

13. The Applied History Research Group, "The European Voyages of Exploration," http://www.ucalgary.ca/applied_history/tutor/eurvoya/Trade.html.

14. Mark Overton, *Agricultural Revolution in England* (Cambridge: University of Cambridge Press, 1996), 78, 86.

15. Braudel, *Civilization and Capitalism*, 61.

16. United Nations, *World at Six Billion*.

17. Braudel, *Civilization and Capitalism*, 61.

18. Ibid., 74.

19. Gaston Roupnel, *La Ville et la campagne au XVII siecle*, cited in Braudel, *Civilization and Capitalism*.

20. Robert Fogel, "New Findings about Trends in Life Expectation and Chronic Disease," Graduate School of Business Selected Paper Series, no. 76 (Chicago: University of Chicago, 1996), 18.

21. Robert Fogel, Nobel Prize lecture, 1993, http://nobelprize.org/nobel_prizes/economics/laureates/1993/fogel-lecture.pdf.

22. Braudel, *Civilization and Capitalism*, 124; Overton, *Agricultural Revolution*, 90.

23. Braudel, *Civilization and Capitalism*, 132–33, 196.

24. Fogel, "New Findings," 7.

25. Hopkins, Eric, *Industrialisation and Society: A Social History, 1830–1951* (New York: Routledge, 2000), 22.

26. Fogel, Nobel Prize lecture.

27. Jack Kloppenburg, *First the Seed* (Cambridge: University of Cambridge, 1988), 46.

28. U.S. Department of Agriculture, "Growing a Nation: The Story of American Agriculture," chap. 1, online resource, http://www.agclassroom.org/gan/classroom/pdf/NarLesson1.pdf.

29. Dan Morgan, *Merchants of Grain* (Lincoln: iUniverse, Inc., 2000), 47.

30. J. L. Holechek, "Waste of the West: History of Public Lands Ranching," chap. 1, http://www.wasteofthewest.com/Chapter1.html.

31. Lowell Dyson, "American Cuisine in the Twentieth Century," *Food Review* 23, no. 1 (Jan.–April 2000): 2.

32. Harvey Levenstein, *Revolution at the Table: The Transformation of the American Diet* (Oxford: Oxford University Press, 1988), 7.

33. Gregory Clark, "The Agricultural Revolution and the Industrial Revolution England, 1500–1912," Ph.D. diss., University of California at Davis, 2002, http://www.econ.ucdavis.edu/faculty/gclark/papers/prod2002.pdf.

34. *Australian Sketcher,* April 23, 1881, 130.

35. Kloppenburg, *First the Seed,* 73; Marion Nestle, *Food Politics* (Berkeley: University of California Press, 2002), 32.

36. American Agricultural Economics Association, "Ray A. Goldberg: 2005 Fellow," http://www.aaea.org/fellows/f05goldberg.cfm.

37. J. Sawyer et al., "Concepts and Rationale for Regional Nitrogen Rate Guidelines for Corn," PM 2015 April 2006. Available at www.extension.iastate.edu/Publications/PM2015.pdf.

38. T. Crews et al., "Legume versus Fertilizer Source of Nitrogen: Ecological Tradeoffs and Human Needs," *Agriculture, Ecosystems, and Environment* 102 (2004), 279–97.

39. Ibid.

40. Fred Magdoff et al., eds., *Hungry for Profit* (New York: Monthly Review Press, 2000), 98.

41. Kloppenburg, *First the Seed,* 126.

42. Kloppenburg, 118.

43. U.S. Department of Agriculture, "Growing a Nation," 1880 and 1990–2000; A. Hallauer, December 2, 2005.

44. N. R. Kleinfield, "America Goes Chicken Crazy," *New York Times,* December 9, 1984.

45. National Chicken Council, "Per Capita Consumption of Poultry and Livestock," http://www.nationalchickencouncil.com/statistics/stat_detail .cfm ?id=8.

46. These statistics can be found in Jayachandran N. Variyam, "The Price Is

Right: Economics and the Rise in Obesity," *Amber Waves* (February 2005), http://www.ers.usda.gov/AmberWaves/February05/Features/ThePriceIs Right.htm; Fogel, "New Findings"; U.S. Department of Agriculture, "Food Consumption, Prices, and Expenditures, 1970–97," statistical bulletin no. 965, p. 7, http://www.ers.usda.gov/publications/sb965/sb965l.pdf.

47. David Johnson et al., "A Century of Family Budgets in the United States," *Monthly Labor Review*, U.S. Bureau of Labor Statistics, www.bls.gov/opub/mlr/2001/05/art3full.pdf.

48. Vernon W. Ruttan, "The Transition to Agricultural Sustainability," paper presented at the National Academy of Sciences colloquium "Plants and Population: Is There Time?" on December 5–6, 1998, at the Arnold and Mabel Beckman Center in Irvine, California, http://www.pnas.org/cgi/content/full/96/11/5960.

49. A. G. Tacon, "Contribution to Food Fish Supplies," United Nations Food and Agriculture Organization archives, http://www.fao.org/DOCREP/003/W7499E/w7499e17.htm; U.S. Department of Agriculture, "U.S. Meat Exports Continue Setting Records," http://www.fas.usda.gov/dlp2/highlights/2001/wmr_2002.pdf.

50. Alain Revel, *American Green Power* (Baltimore: Johns Hopkins University Press, 1981), 42.

51. Katy Mamen, "Current Issues and Trends Connected to the Vivid Picture Goals for a Sustainable Food System," Vivid Picture Project 2004, http://www.vividpicture.net/documents/4_Current%20Trends_and_Bkgd_Info.pdf.

52. Sophia Murphy, *Managing the Invisible Hand*, Institute for Agriculture and Trade Policy (2002), 24, http://www.tradeobservatory.org/library.cfm?RefID=25497.

53. Action Aid International, "Power Hungry," 23, http://www.actionaid.org.uk/_content/documents/power_hungry.pdf.

54. Murphy, *Invisible Hand*, 25.

55. Judy Putnam et al., "U.S. Per Capita Food Supply Trends," *Food Review* 25, no. 3 (Winter 2002), http://www.ers.usda.gov/publications/FoodReview/DEC2002/frvol25i3a.pdf.

56. National Research Council, *Alternative Agriculture* (Washington, DC: National Academy Press, 1989), 98 (115).

2. IT'S SO EASY NOW

1. AdBrands.net, "Nescafé: Brand Profile," http://www.mind-advertising.com/ch/nescafe_ch.htm.

2. Rachel Lauden, "A Plea for Culinary Modernism," *Gastronomica* 1, no. 1 (February 2001): 36–44.

3. Jean Heer, *Nestlé: 125 Years — 1866–1991* (Vevey: Nestlé, 1991), 30–33.

4. About.com Kraft, http://inventors.about.com/od/foodrelatedinventions/a/kraft_foods.htm.

5. Codex Alimentarius Commission, "Codex Alimentarius," Food and Agriculture Organization/World Health Organization, http://www.codexalimentarius.net/web/index_en.jsp; and Food and Agriculture Organization, "Understanding the Codex Alimentarius — Revised and Updated," http://www.fao.org/docrep/008/y7867e/y7867e03.htm#bm03.

6. Chris Bolling and Mark Gehlfar, "Global Food Manufacturing Reorients to Meet New Demands," in *New Directions in Global Food Markets*, http://www.ers.usda.gov/publications/aib794/aib794g.pdf.

7. Heer, *Nestlé*, 306, 340, 428.

8. Friedhelm Schwarz, *Nestlé: The Secrets of Food, Trust and Globalization* (Key Porter Books, 2002).

9. Heer, *Nestlé*, 261.

10. John Connor, personal communication with the author, January 19, 2006.

11. Harvey Levenstein, *Paradox of Plenty: A Social History of Modern Eating* (Oxford: Oxford University Press, 1993), 106.

12. John Connor et al., *The Food Manufacturing Industries: Structure, Strategies, Performance, and Policies* (Lexington, MA: Lexington Books, 1985), 66.

13. Ibid., 5.

14. David Barboza, "The Brazil Effect on Commodities," *New York Times*, January 27, 1999, http://select.nytimes.com/search/restricted/article?res=F30712F7385D0C748EDDA80894D1494D81.

15. John Connor, "Breakfast Cereals: The Extreme Food Industry," Dept. of Agriculture Economics, Purdue University, 1999.

16. Marion Nestle, *Food Politics* (Berkeley: University of California Press, 2002), 22.

17. Steve Silk, personal communication with the author, January 19 and April 6, 2006.

18. William Leach, personal communication with the author, April 26, 2006.

19. Anthony Gallo, "Food Advertising in the United States," *Agriculture Information Bulletin* 750: 173–80.

20. Charlene Price, "Food Service," in *The U.S. Food Marketing System, 2002* (Washington, DC: Economic Research Service, 2002), 34.

21. Dreyer's, "Light Takes Flight for Dreyer's Grand Ice Cream," press release, March 22, 2005, http://www.corporate-ir.net/ireye/ir_site.zhtml?ticker=DRYR&script=410&layout=-6&item_id=687750.

22. Steve Martinez, "The U.S. Food Marketing System," *Economic Research Report* 42 (Economic Research Service/USDA, May 2007): 8, http://www.ers.usda.gov/publications/err42/err42.pdf.

23. Tom Rourke, "Challenges in Meat Processing and Meal Solutions," presentation given at 2006 meeting of the American Meat Science Association, http://www.meatscience.org/pubs/rmcarchv/2006/presentations/16_3.pdf.

24. Ibid.

25. Malcolm Gladwell, "The Ketchup Conundrum," *The New Yorker,* September 6, 2004, http://www.gladwell.com/2004/2004_09_06_a_ketchup.html.

26. Kimberly Powell, "It's Easy Being Green: Ketchup Goes Green as Heinz EZ Squirt Hits Store Shelves," About.com, http://pittsburgh.about.com/library/weekly/aa101700a.htm.

27. *Just Food,* "UK: Birth of First Child Drives Different Food Demands," Just-food.com, June 20, 2002, http://www.just-food.com/article.aspx?id=70837&lk=s.

28. Allen Wysocki, "Major Trends Driving Change in the U.S. Food System," Institute of Food and Agricultural Sciences Extension, University of Florida, http://edis.ifas.ufl.edu/pdffiles/RM/RM00100.pdf.

29. Joel Ceausu, "Study finds Convenience, Not Health, Drives Consumption," Just-food.com, November 7, 2006, http://www.just-food.com/article.aspx?ID=96565&lk=dm; D. Bowers, "Cooking Trends Echo Changing Roles of Women," *Food Review* (January 2000, USDA): 28.

30. University of North Texas News Service, "Cooking Illiteracy Forcing Cookbook Publishers to Simplify Recipes," March 30, 2006, http://web2.unt.edu/news/story.cfm?story=9816.

31. Ceausu, "Study finds Convenience."

32. Just-drinks.com, "New Impulse Consumption Occasions," http://www.just-drinks.com/store/print_product.asp?art=26509.

33. Confectionerynews.com, "Fragmented Consumer Tastes Dictate Growth in Snack Market," May 7, 2004, http://www.confectionerynews.com/news/ng.asp?id=51958-fragmented-consumer-tastes.

34. Anita Awbi, "Handy Packaging Drives Snacking Mega-Trend," Foodproductiondaily.com, January 13, 2006, http://www.foodproduction daily.com/news/ng.asp?id=65087-walkers-kraft.

35. Foodproductiondaily.com, "Bilwinco Launches High-Speed Food Weigher," http://www.foodproductiondaily.com/news-by-product/news.asp?id=58831&idCat=0&k=bilwinco-launches-high; "Pot Filler Speeds Up Changeover Time, Claims Manufacturer," http://www.foodproductiondaily.com/news/ng.asp?n=66955-packaging-automation-pots-changeover; "Two Machines Aim to Speed Up Chicken Deboning, Cutting," http://www.foodproductiondaily.com/news-by-product/news.asp?id=60603&idCat=25&k=two-machines-aim; "Industry Briefs: Updated Wafer Machine; KitKat," http://www.foodproductiondaily.com/news-by-product/news.asp?id=60133&idCat=25&k=industry-briefs-updated; "Robotics: The Future of Food Processing?" http://www.foodproductiondaily.com/news/ng.asp?n=66874-k-robotix-robotics-anuga; and "Machine Makers Target Convenience Foods, Aseptic Trends," www.foodproductiondaily.com/news/ng.asp?n=66849-aseptic-convenience-zinetec.

36. Leffingwell and Associates, "2002–2006 Flavor and Fragrance Industry Leaders," http://www.leffingwell.com/top_10.htm.

37. Dr. Gary Reineccius, Department of Food Science and Nutrition, University of Minnesota, personal communication with the author, March 2006.

38. CPL Business Consultants, "Review of Bulking Agents," http://www.cplsis .com/index.php?reportid=175.

39. Duke University, "Vegetable Oil Industry," http://www.markets.duke.edu/ student_it/soc142_spring2002/team4/international.html; Alain Revel, *American Green Power* (Baltimore: Johns Hopkins University Press, 1981), 56.

40. Reineccius, interview with the author, March 2006.

41. Associated Press, "Popcorn maker to drop butter-flavor chemical," *Wenatchee World,* September 5, 2007.

42. Duke University, "Vegetable Oil Industry."

43. David Kiley, "SpongeBob: For Obesity or Health?" *BusinessWeek,* February 17, 2005, http://www.businessweek.com/bwdaily/dnflash/feb2005/nf 20050217_6978_db042.htm.

44. D. Bowers, "A Century of Change in America's Eating Patterns," *Food Review* 3, no. 1: 24.

45. Mark Jekanowski, "Causes and Consequences of Fast Food Sales Growth," *Food Review* (January–April 1999): 11, http://www.ers.usda.gov/publications/ foodreview/jan1999/frjan99b.pdf.

46. Jayachandran N. Variyam, "The Price Is Right: Economics and the Rise in Obesity," *Amber Waves* (Economic Research Services/USDA, February 2005), http://www.ers.usda.gov/AmberWaves/February05/Features/ThePrice IsRight.htm.

47. Institute of Food Technologists, 2006 Post-Show AM+FE newsletter, http:// www.ift.org/cms/?pid=1001460.

48. Price, "Food Service," 98.

49. Ronald Curhan, personal communication with the author, January 18, 2006.

50. Chris Mercer, "Nestlé Takes Global Ice Cream Lead," Dairyreporter.com, January 19, 2006, http://www.foodproductiondaily.com/news/ng.asp?id=65201.

51. Martinez, "The U.S. Food Marketing System," 8.

52. Reuters, "Slack Sales of Dressing and Drinks Slow Kraft," *New York Times,* October 24, 2006, http://www.nytimes.com/2006/10/24/business/24kraft .html.

53. Martinez, "The U.S. Food Marketing System."

54. Rourke, "Challenges in Meat Processing."

55. *CNN Money,* "Unilever Feasts on Deals," April 12, 2000, http://money .cnn .com/2000/04/12/europe/unilever.

56. John McMillin, Prudential Equity Group, *March–April 2005 Monthly Newsletter.*

57. Barboza, "The Brazil Effect."

58. *Food Business Review,* "Ready Meals: A Chinese Take-Away?" January 13, 2006, http://www.food-business-review.com/article_researchwire.asp?guid =C646CEA1-627E-46C7-8C79-C23A53C0977B&z=.

59. Mercer, "Nestlé Takes Global Ice Cream Lead."

60. *Food Production Daily,* "High Sugar Price May Boost Shift to Premium Products in China," March 28, 2006, http://www.foodproductiondaily.com/news/ng.asp?id=66685-Nestlé-pepsico-sugar.

61. AP-foodtechnology.com, "PepsiCo Plans to Up China Investment," January 31, 2006, http://www.ap-foodtechnology.com/news/ng.asp?id=65495.

62. Simon Montlake, "Milk Formula Goes on Trial in Asia," *Christian Science Monitor,* June 22, 2007, http://www.csmonitor.com/2007/0622/p05s01-woap.html?page=1.

63. Info America, "Latin America Market Report," http://www.infoamericas.com/expertise/industry-practices/fast-moving-consumer-goods.htm.

3. Buy One, Get One Free!

1. Barbara Kahn and Leigh M. McAlister, *Grocery Revolution: The New Focus on the Consumer* (Reading, MA: Addison-Wesley, 1997), 30.

2. Steve Martinez, "The U.S. Food Marketing System," *Economic Research Report* 42 (Economic Research Service/USDA, May 2007): 8, www.ers.usda.gov/publications/err42/err42.pdf.

3. David Robertson, "The State of the Union in Food Retail," Just-food.com, July 21, 2006, http://www.just-food.com/article.aspx?ID=95499&lk=dm.

4. Statistics in Marin Gjaja et al., "Dancing with the 800-Pound Gorilla," Boston Consulting Group, 2002, http://www.bcg.com/publications/publication_view.jsp?pubID=752&language=English, and Food Marketing Institute, "Key Facts, Median Average Store Size — Square Feet," June 2005, http://www.fmi.org/facts_figs/keyfacts/storesize.htm.

5. Gjaja, "Dancing with the 800-Pound Gorilla."

6. Wendy Pinkerton et al., "Trends Driving Consumer Meat Preferences," presentation given at 2006 meeting of the American Meat Science Association, http://www.meatscience.org/pubs/rmcarchv/2006/presentations/16_1_Pinkerton_Teigen.pdf.

7. Martinez, "The U.S. Food Marketing System," 8.

8. A. Krebs, "Wal-Mart Now Sells 15% of All Grocery Food in America," *Agribusiness Examiner* 246, http://www.organicconsumers.org/corp/walmart051403.cfm.

9. Statistics in Alan Barkema et al., "The New U.S. Meat Industry," Federal Reserve Bank of Kansas City, http://www.kc.frb.org/PUBLICAT/ECONREV/PDF/2q01bark.pdf, and Phil Kaufman, "Food Retailing," in *U.S. Food Marketing System 2002* (Economic Research Service/USDA): 27, http://www.ers.usda.gov/publications/aer811/aer811e.pdf.

10. Hao Wang, "Slotting Allowances and Retailer Market Power," Ph.D. diss., China Center for Economic Research, Beijing University, http://econwpa.wustl.edu/eps/io/papers/0411/0411009.pdf.

11. Federal Trade Commission, "Slotting Allowances in the Retail Grocery Industry" (November 2003): 11, http://www.ftc.gov/os/2003/11/slottingallow ancerpto31114.pdf.

12. Wang, "Slotting Allowances."

13. Interview with Hans Jorge Renk of Nestlé, May 2005.

14. Martinez, "The U.S. Food Marketing System," 8.

15. Global Insight, "Measuring the Impact of Wal-Mart on the U.S. Economy," Global Insight, Inc., November 4, 2005.

16. Ibid.

17. Roger Betancourt, personal communication with the author, March 15, 2006.

18. Food Marketing Institute, "Supermarket Facts Industry Overview," 2005, http://www.fmi.org/facts_figs/superfact.htm.

19. Kahn, *Grocery Revolution*, 109.

20. Linda Calvin et al., "Basic Determinants of Global Trade in Fruits and Vegetables," in *Global Trade Patterns in Fruits and Vegetables* (Economic Research Service/USDA, 2004), http://www.ers.usda.gov/publications/wrs0406/wrs 0406d.pdf.

21. Frank Greve, "Berry Boom," *McClatchy Newspapers*, in the *Wenatchee World*, April 9, 2007.

22. Sandra Cuellar, "Marketing Fresh Fruit and Vegetable Imports in the United States," Food Industry Management, Cornell University, April 2002, http:// aem.cornell.edu/research/researchpdf/rb0204.pdf.

23. Calvin et al., "Global Trade in Fruits and Vegetables."

24. Action Aid International, "Power Hungry," 13, http://www.actionaid.org.uk/ _content/documents/power_hungry.pdf.

25. Ibid.

26. McDonald's, "USA Nutrition Facts for Popular Menu Items," http://www .mcdonalds.com/app_controller.nutrition.index1.html.

27. See Michael Ollinger et al., "Structural Change in U.S. Chicken and Turkey Slaughter," *Agricultural Economic Report* 787 (November 2000), http://www .ers.usda.gov/publications/aer787, and David Leonhardt, "McDonald's: Can It Regain Its Golden Touch?" *BusinessWeek*, March 9, 1998, http://www .businessweek.com/1998/10/b3568001.htm.

28. Paul Aho, personal communication with the author, November 25, 2005.

29. N. R. Kleinfield, "America Goes Chicken Crazy," *New York Times*, December 9, 1984, http://query.nytimes.com/gst/fullpage.html?res=9405E7D71538 F93AA35751C1A962948260&sec=health&pagewanted=6.

30. Blake Lovette, personal communication with the author, December 5, 2005.

31. Ann Davis, "Tyson Foods Refines a Recipe by Energy Firms," *Wall Street Journal*, December 13, 2006, http://online.wsj.com/article/SB11649397447 0437634.html?mod=yahoo_hs&ru=yahoo.

32. Z. Ahmed et al., "Two Decades of Productivity Growth in Poultry Dressing

and Processing," *Monthly Labor Review* (April 1987), http://www.bls.gov/opub/mlr/1987/04/art5full.pdf.

33. Brewster Kneen, *Invisible Giant: Cargill and Its Transnational Strategies* (London: Pluto Press, 1995), 88.

34. Monte Mitchell, "Killing Floor Tales: Poultry Growers at Mercy of Industrialized Agriculture and Short, Tenuous Contracts Drawn Up by Food Giants," *Winston-Salem Journal,* June 20, 2004.

35. James MacDonald, "Overview of the U.S. Marketing Structure for Livestock," presentation given at workshops sponsored by the Pew Initiative on Food and Biotechnology, March 21–23, 2005, Rockville, MD, http://pewagbiotech.org/events/0321/proceedings.pdf.

36. Ollinger, "Structural Change in U.S. Chicken and Turkey Slaughter"; G. Morgan et al., "Economic Impact of the Mississippi Poultry Industry at the year 2002," Department of Poultry Science, Mississippi State University Extension, http://msucares.com/pubs/infobulletins/ib385.pdf.

37. Ahmed, "Two Decades of Productivity Growth."

38. Economic Research Service, "Briefing Rooms: Poultry and Eggs: Trade," http://www.ers.usda.gov/Briefing/Poultry/Trade.htm.

39. John Nalivka, personal communication with the author, May 9, 2006.

40. MacDonald, "Overview of the U.S. Marketing Structure."

41. Jerry Kelly, "National Meat Case Study," presentation given at 2006 meeting of the American Meat Science Association, http://www.meatscience.org/pubs/rmcarchv/2006/presentations/16_4_Kelly.pdf.

42. Doug Sutton, "Processors' Perspective on Current and Future Trends in Meat Quality," presentation given at 2006 meeting of the American Meat Science Association, http://www.meatscience.org/pubs/rmcarchv/2006/presentations/4_5_Sutton.pdf.

43. Dale Miller, "Straight Talk from Smithfield's Joe Luter," *National Hog Farmer,* May 1, 2000, http://nationalhogfarmer.com/mag/farming_straight_talk_smithfields.

44. Clint Peck, "Global Beef Production," course given at Montana State University, 2005, http://animalrange.montana.edu/courses/johnpaterson/global_beef_prod1.pdf.

45. Kristen Philipkoski, "Meat Stripper Gets Third Degree," *Wired,* January 19, 2004.

46. J. Brad Morgan, "Current Research in Case-Ready," in lecture given at Proceedings of the 55th Annual Reciprocal Meat Conference, July 28–31, 2002, Michigan State University, http://www.meatscience.org/Pubs/rmcarchv/2002/presentations/rmc_2002_055_1_0000_all.pdf, 69.

47. Clint Peck, "Beef Chat: The Wal-Mart Way," Beefmagazine.com, June 1, 2003, http://beef-mag.com/mag/beef_walmart.

48. John Nalivka, personal communication with the author, May 31, 2006.

49. MacDonald, "Overview of the U.S. Marketing Structure."

50. Kneen, *Invisible Giant,* 93.

51. MacDonald, "Overview of the U.S. Marketing Structure."

52. O. Peter Snyder, "Bloody Chicken," Hospitality Institute of Technology and Management, http://www.hi-tm.com/Documents/Bloody-chik.html.

53. Felicity Lawrence, "Special Report Supermarkets: Chicken," Theecologist.org, http://www.theecologist.org/archive_detail.asp?content_id=309.

54. Katy Mamen, "Current Issues and Trends Connected to the Vivid Picture Goals for a Sustainable Food System," 22, report by the Vivid Picture Project 2004, www.vividpicture.net/documents/4_Current%20Trends_and_Bkgd_Info.pdf.

55. Peg Luksik, "To Market, to Market, to Buy a Fat Pig," http://www.constitutional.net/Luksik/fatpig.html.

56. Miller, "Straight Talk From Smithfield's Joe Luter."

57. Ibid.

58. See Food and Agriculture Organization, "Farmers and Supermarkets in Asia," *Agriculture* 21 (May 2005), http://www.fao.org/AG/magazine/0505sp1.htm, and Just-food.com, "France: Carrefour Chairman Resigns as Retailer Eyes International Growth," March 8, 2007, http://www.just-food.com/article.aspx?ID=97746&lk=dm.

59. J. Kinsey, "Emerging Research and Policy Issues for a Sustainable Global Food Network," the Food Industry Center, University of Minnesota, July 2005, http://agecon.lib.umn.edu/cgi-bin/pdf_view.pl?paperid=17099&ftype=.pdf.

60. Just-food.com, "Mexico: Wal-Mart to Open 125 Stores This Year," February 14, 2007, http://www.just-food.com/article.aspx?ID=97504&lk=dm.

61. Just-food.com, "Wal-Mart Quiet on Chinese Acquisition Rumours," October 17, 2006, http://www.just-food.com/article.aspx?id=96339&lk=sd02.

62. Food and Agriculture Organization, "The Globalizing Livestock Sector: Impact of Changing Markets," provisional agenda from the nineteenth session, Rome, April 13–16, 2005, www.fao.org/docrep/meeting/009/j4196e.htm.

63. Ibid.

64. Ibid.

65. Food and Agriculture Organization, "Farmers and Supermarkets in Asia."

66. Just-food.com, "India: Wal-Mart Enters Market with Bharti Venture," November 27, 2006, http://www.just-food.com/article.aspx?id=96782&lk=s.

67. Food and Agriculture Organization, "Farmers and Supermarkets in Asia."

68. According to a transcript of the presentation, as reported by Ahmed ElAmin in *Food Production Daily,* "Smithfield Targets Romania for Expansion into Europe," September 8, 2006, http://www.foodproductiondaily.com/news/ng.asp?n=70416-smithfield-romania-pork.

4. TIPPING THE SCALES

1. HFCS accounts for 535 million bushels of corn a year, at $3.50 per bushel. From Corn Refiners Association. Statistics, 2004, via Answers.com Business & Finance, http://www.answers.com/topic/wet-corn-milling?cat=biz-fin.

2. Reuters, "Study ties hyperactivity in kids to food additives," September 5, 2007, at MSNBC, http://www.msnbc.msn.com/id/20612862/.

3. Robert Lustig, "Childhood Obesity: Behavioral Aberration or Biochemical Drive? Reinterpreting the First Law of Thermodynamics," *Nature Clinical Practice Endocrinology and Metabolism* 2 (2006): 447–58; doi:10.1038, http://www.nature.com/ncpendmet/journal/v2/n8/full/ncpendmeto220.html.

4. M. Lazar, "How Obesity Causes Diabetes," *Science* (January 21, 2005): 373.

5. Jeffrey Friedman, "A War on Obesity, Not the Obese," *Science* (February 7, 2003).

6. J. Hill and J. Peters, "Environmental Contributions to the Obesity Epidemic," *Science* 280, no. 5368 (May 29, 1998): 1371, http://www.sciencemag.org/cgi/content/abstract/280/5368/1371.

7. Marx Jean, "Cellular Warriors at the Battle of the Bulge," *Science* 299 (February 7, 2003).

8. Lowell Dyson, "American Cuisine in the Twentieth Century," *Food Review* (January–April 2000): 2.

9. Jayachandran Variyam, "The Price Is Right: Economics and the Rise in Obesity," *Amber Waves* (February 2005), http://www.ers.usda.gov/AmberWaves/February05/Features/ThePriceIsRight.htm.

10. Ibid.

11. Ibid.

12. U.S. Centers for Disease Control, "Prevalence of Overweight Among Children and Adolescents: United States, 1999–2002," U.S. Department of Health and Human Services, http://www.cdc.gov/nchs/products/pubs/pubd/hestats/overwght99.htm.

13. Cynthia Ogden et al., "Mean Body Weight, Height, and Body Mass Index, United States, 1960–2002," Centers for Disease Control, U.S. Department of Health and Human Services, http://www.cdc.gov/nchs/data/ad/ad347.pdf.

14. Martha Coventry, "Supersizing America: Obesity Becomes an Epidemic," *University of Minnesota Health News* (Winter 2004), http://www1.umn.edu/umnnews/Feature_Stories/Supersizing_America.html.

15. See A. Dannenberg et al., "Economic and Environmental Costs of Obesity — The Impact on Airlines," *American Journal of Preventative Medicine* 27 (2004): 3, and Associated Press, "Feds: Obesity Raising Airline Fuel Costs," *USA Today*, November 5, 2004.

16. Warren St. John, "One Size No Longer Fits All in America's Coffin Industry," *New York Times*, Sunday, September 28, 2003; Alison Hardie, "Super-size Me . . . Even After My Death," News.scotsman.com, http://news.scotsman.com/health.cfm?id=572772007; Caroline Innes, "Bodies Too Big for Crematorium," [Blackburn, UK] *Citizen*, http://www.blackburncitizen.co.uk/news/newsheadlines/display.var.859613.0.bodies_too_big_for_crematorium.php.

17. "Sick for Longer," *Endeavors* (Spring 2005), Internet press release summarizing research by the University of North Carolina–Chapel Hill, http://research.unc.edu/endeavors/spr2005/nutrition_news.php.

18. Fred Kuchler et al., "Societal Costs of Obesity: How Can We Assess When Federal Interventions Will Pay?" *Food Review* (Winter 2002): 33.

19. According to studies by, among others, Ken Ebihara, at Kyoto University.

20. David Ludwig, personal communication with the author, February 14, 2006.

21. U.S. Centers for Disease Control, "Overweight and Obesity: Economic Consequences," Division of Nutrition and Physical Activity, National Center for Chronic Disease Prevention and Health Promotion, CDC, National Institutes of Health, March 22, 2006, http://www.cdc.gov/nccdphp/dnpa/obesity/economic_consequences.htm.

22. Ibid.

23. "Major Trends in U.S. Food Supply, 1909–99," *Food Review* 23, no. 1, http://www.ers.usda.gov/publications/foodreview/jan2000/frjan2000b.pdf.

24. Eileen T. Kennedy et al., "Dietary-Fat Intake in the U.S. Population," *Journal of the American College of Nutrition* 18, no. 3 (1999): 207–12, http://www.jacn.org/cgi/content/full/18/3/207#T1.

25. Steven Milloy, "Food Police Indict SpongeBob," *Fox News Online,* January 19, 2006, http://www.foxnews.com/story/0,2933,182274,00.html.

26. Kate Zernike, "Lawyers Shift Focus from Big Tobacco to Big Food," *New York Times,* April 9, 2004.

27. Rick Berman interview with the author, January 7, 2007, in Berman's Washington, D.C., offices.

28. Randal O'Toole, "How Fat Are We?" *Liberty Unbound* 16, no. 11 (November 2002), http://libertyunbound.com/archive/2002_11/otoole-fat.html.

29. James Levine et al., "Inter-individual Variation in Posture Allocation: Possible Role in Human Obesity," *Science* 307, no. 5709 (January 28, 2005): 584–86; http://www.sciencemag.org/cgi/content/full/307/5709/584.

30. James Hill et al., "Obesity and the Environment: Where Do We Go from Here?" *Science* 299, no. 5608 (February 7, 2003): 853–55, doi:10.1126/science.1079857, http://www.sciencemag.org/cgi/content/abstract/299/5608/853.

31. Michael S. Rosenwald, "Why America Has to Be Fat: A Side Effect of Economic Expansion Shows Up in Front," *Washington Post,* January 22, 2006.

32. Todd Seavey, "Kristof Goes Berserk over High-Fructose Corn Syrup," HealthFactsAndFears.com, April 11, 2006.

33. Andrew Drewnowski et al., "Poverty and Obesity: The Role of Energy Density and Energy Costs," *American Journal of Clinical Nutrition* 79 (2004): 6–16.

34. "California Eating," *The Economist,* October 5, 2006; Katy Mamen, "Current Issues and Trends Connected to the Vivid Picture Goals for a Sustainable Food System," report by the Vivid Picture Project 2004, http://www.vividpicture.net/documents/4_Current%20Trends_and_Bkgd_Info.pdf; http://www.foodsecurity.org/PrimerCFSCUAC.pdf.

35. Behjat Hojjati and Stephanie J. Battles, "The Growth in Electricity Demand in U.S. Households, 1981–2001: Implications for Carbon Emissions," U.S. Energy Information Administration, Department of Energy, www.eia.doe.gov/

emeu/efficiency/2005_USAEE.pdf, 3; and Natural Resources Canada, "2003 Survey of Household Energy Use (SHEU) — Summary Report," March 22, 2006, http://www.oee.nrcan.gc.ca/Publications/statistics/sheu-summary/trends.cfm?attr=0.

36. B. Lin et al., "Nutrient Contribution of Food Away from Home," *Agriculture Information Bulletin* 750 (Economic Research Service/USDA, May 1999): 236, http://www.ers.usda.gov/publications/aib750/aib750l.pdf.

37. Kennedy et al., "Dietary-Fat Intake in the U.S. Population."

38. Sharon Elliott et al., "Fructose, Weight Gain, and the Insulin Resistance Syndrome," *American Journal of Clinical Nutrition* 76, no. 5 (November 2002): 911–22, http://www.ajcn.org/cgi/content/full/76/5/911?ijkey=1799c4e470262c457 d4cc28dc8f9514899ecba3c.

39. Walter Willett, personal communication with the author, July 30, 2006.

40. Steve Martinez, "The U.S. Food Marketing System," *Economic Research Report* 42 (Economic Research Service/USDA, May 2007): 8, http://www.ers.usda.gov/publications/err42/err42.pdf.

41. Assuming adult average intake of 2,250 calories per day and total U.S. sales of between $700 billion a year (Associated Press, "Ethanol boom, rising prices divide corn lobby," September 13, 2007, http://www.msnbc.msn.com/id/20760839) and $800 billion a year. See Marion Nestle, *Food Politics* (Berkeley: University of California Press, 2002).

42. Mark Jekanowski, "Causes and Consequences of Fast Food Sales Growth," *Food Review* (January–April 1999): 11, http://www.ers.usda.gov/publications/foodreview/jan1999/frjan99b.pdf.

43. Laura Miller, review of Greg Critser, *Fat Land*, in Salon.com, December 19, 2003, http://www.mcspotlight.org/media/press/mcds/saloncom191203.html.

44. Lisa Young and Marion Nestle, "The Contribution of Expanding Portion Sizes to the U.S. Obesity Epidemic," *American Journal of Public Health* (February 2002): 246.

45. Ibid.

46. Melanie Warner, "A Sweetener with a Bad Rap," *New York Times*, July 2, 2006, http://www.nytimes.com/2006/07/02/business/yourmoney/02syrup.html?ex=1153800000&en=27b87e1ed8aa3ac7&ei=5070.

47. Mark Pereira, personal communication with the author, February 11, 2006.

48. Tom Vanderbilt, "Self-Storage Nation: Americans are storing more stuff than ever," July 18, 2005, http://www.slate.com/id/2122832/.

49. Barbara Rolls et al., "Serving Portion Size Influences 5-Year-Old but Not 3-Year-Old Children's Food Intakes," *Journal of the American Dietetic Association* 100: 232–34.

50. Martinez, "The U.S. Food Marketing System."

51. Snack Food Association, "Convenience Is King," Snack Food Association newsletter 2006, http://www.sfa.org/pastnews.aspx.

52. Foodproductiondaily.com, "'Me-time' Drives Premium Foods Market," October 15, 2004, http://www.foodproductiondaily.com/news/ng.asp?id=55420.

53. Foodnavigator.com, "Fragmented Consumer Tastes Dictate Growth in Snack Market," May 7, 2004, http://www.foodproductiondaily.com/news/ng.asp?id=51958.

54. Foodproductiondaily.com, "Nestlé Takes a Break from Kit Kat Slogan," April 8, 2004, http://www.foodproductiondaily.com/news/ng.asp?id=53973.

55. Foodnavigator.com, "Fragmented Consumer Tastes."

56. Foodproductiondaily.com, "'Me-time' Drives Premium Foods Market."

57. Caroline E. Mayer, "TV Ads for Junk Food Tied to Obesity, Disease," Washington Post, December 7, 2005, http://sfgate.com/cgi-bin/article.cgi?f=/c/a/2005/12/07/MNGS1G44M71.DTL.

58. Mamen, "Current Issues and Trends."

59. Kristen Harrison et al., "Nutritional Content of Foods Advertised During the Television Programs Children Watch Most," American Journal of Public Health 95, no. 9 (September 2005): 1568–74, http://www.ajph.org/cgi/content/abstract/95/9/1568.

60. Public Health Institute, "2000 California High School Fast Food Survey," February 2000, http://www.phi.org/pdf-library/fastfoodsurvey2000.pdf.

61. Mamen, "Current Issues and Trends."

62. J. Tevlin, "General Mills Ad Campaign Turns Sour After Protest," Minneapolis Star Tribune, August 31, 2001.

63. Patti Valkenburg, "Identifying Determinants of Young Children's Brand Awareness: Television, Parents, and Peers," Amsterdam School of Communications Research, University of Amsterdam, 2005.

64. C. Cavadini et al., "U.S. Adolescent Food Intake Trends from 1965 to 1996," Archives of Disease in Childhood 83 (2000): 18–24.

65. "California Eating," The Economist, 40.

66. Martinez, "The U.S. Food Marketing System," 8.

67. Rosenwald, "Why America Has to Be Fat."

68. World Heath Organization, "Obesity in Europe," European Regional Office, http://www.euro.who.int/nutrition/obesity/20051216_3.

69. Genevieve Roberts, "Global Obesity Epidemic 'Out of Control,'" New Zealand Herald, November 1, 2004, http://www.globalpolicy.org/socecon/hunger/economy/2004/1101obesity.htm.

70. Amelia Gentleman, "India's Newly Rich Battle with Obesity," London Observer, December 4, 2005, http://www.guardian.co.uk/india/story/0,,1657330,00.html.

5. EATING FOR STRENGTH

1. Alain Revel, American Green Power (Baltimore: Johns Hopkins University Press, 1981), 88; Joy Harwood, "U.S. Flour Milling on the Rise — Effects of Increased Flour Consumption," Food Review (April–June 1991), http://findarticles.com/p/articles/mi_m3765/is_n2_v14/ai_11190348; Gary Vocke,

"The Changing Nature of World Agriculture: Feeding the World: the 1990s and Beyond," *National Food Review* (April–June 1990), http://findarticles .com/p/articles/mi_m3284/is_n2_v13/ai_9152949; Vocke, personal communication with the author, October 16, 2007.

2. Food and Agriculture Organization, "World Agriculture: Towards 2030/2050, Interim Report," 2006, http://www.fao.org/docrep/009/a0255e/a0255e04 .htm.

3. Ibid.

4. Quoted in Mark Ritchie et al., "Crisis by Design: A Brief Review of U.S. Farm Policy," League of Rural Voters Education Project, http://www.agobservatory .org/library.cfm?refID=97261.

5. W. Berry, "Nation's Destructive Farm Policy Is Everyone's Concern," Lexington *Herald-Leader,* July 11, 1999, http://www.agrenv.mcgill.ca/agrecon/ecoagr/ doc/berry.htm.

6. James Trager, *The Great Grain Robbery* (New York: Ballantine Books, 1975), 104.

7. Dan Morgan, *Merchants of Grain* (Lincoln: iUniverse, Inc., 2000), 156.

8. Trager, *Great Grain Robbery,* 85.

9. Charles Hanrahan et al., "U.S. Agricultural Trade: Trends, Composition, Direction, and Policy," Congressional Research Service Report for Congress, September 25, 2006, http://fpc.state.gov/documents/organization/74919.pdf.

10. Steven Blank, "Threats to American Agriculture," http://www.uga.edu/caes/ symposium01/sblank.html.

11. Information Please, "Number of Farms, Land in Farms, and Average-Size Farm: United States, 1990–2005," http://www.infoplease.com/ipa/A0883511.html.

12. Daryll Ray et al., "Rethinking U.S. Agricultural Policy: Changing Course to Secure Farmer Livelihoods Worldwide," Executive Summary (Agricultural Policy Analysis Center APAC, University of Tennessee, 2003), http://www .inmotionmagazine.com/ra03/rethinking.html.

13. K. Brown and A. Carter, "Urban Agriculture and Community Food Security in the United States," Community Food Security Coalition, October 2003, http://www.foodsecurity.org/PrimerCFSCUAC.pdf.

14. T. Crews et al., "Legume versus Fertilizer Sources of Nitrogen: Ecological Tradeoffs and Human Needs," *Agriculture, Ecosystems, and Environment* 102 (2004): 279–97.

15. Pew Initiative on Food and Biotechnology, panel discussion, February 13, 2003, http://www.connectlive.com/events/pewagbiotech021303/Pewagbio tech-021303.html.

16. Jasper Becker, *Hungry Ghosts: Mao's Secret Famine* (New York: Holt, 1996), 272; Jan Lahmeyer, "China: Historical Demographical Data of the Administrative Division" Werkgroep Seriële Publicaties, http://www.library.uu.nl/ wesp/populstat/Asia/chinap.htm.

17. Fred Gale et al., "China at a Glance: A Statistical Overview of China's Food and Agriculture," *China's Food and Agriculture: Issues for the Twenty-first Century,* 68, http://www.ers.usda.gov/publications/aib775/aib775e.pdf.

18. All in ibid. except hog production figures, which come from Shanghai's Nestlé Research and Development Center, "What China Consumes."

19. Iowa State University Extension Service, "World Feed Grain Stocks Approaching 1995–96 Levels," Educational Materials and Marketing Services, Iowa State University, January 23, 2004, http://www.extension.iastate.edu/ newsrel/2004/jan04/jan0413.html.

20. Edward Lotterman, "Broilers of the World, Unite!" *Fedgazette,* Federal Reserve of Minneapolis, April 1998, http://www.minneapolisfed.org/pubs/ fedgaz/98-04/broilers.cfm.

21. Martha Groves, "Asia's Woes Taking a Bite Out of U.S. Food Exports: Short-term Drop-off in Demand Could Be Offset by Reforms Stemming from Crisis," *Los Angeles Times,* March 7, 1998; Bonnie Setiawan, "The IMF Burden," *Inside Indonesia* (January–March 2004), http://www.serve.com/inside/ edit77/p12-13setiawan.html; P. McMichael, "Global Food Politics," in *Hungry for Profit,* Fred Magdoff et al., eds. (New York: Monthly Review Press, 2000), 125.

22. Nikos Alexandratos, "World Food and Agriculture: Outlook for the Medium and Longer Term," colloquium paper in *Proceedings of the National Academy of Sciences* 96, no. 11 (May 25, 1999): 5908–14, http://www.pnas.org/cgi/content/full/96/11/5908.

23. Christopher Delgado et al., "Country Case Studies of Structural Change in Markets and Policy Environment Impacting the Production, Processing and Marketing of Selected Livestock Products," *Agriculture* 21 (2002), http://www .fao.org/WAIRDOCS/LEAD/X6115E/x6115e06.htm#TopOfPage.

24. Knowledge@Wharton, "Are Emerging Markets Striking Back, or Out?" Wharton School of Business, June 22, 2006, http://knowledge.wharton .upenn.edu/article.cfm?articleid=1503.

25. Carolyn Dimitri et al., "The 20th Century Transformation of U.S. Agriculture and Farm Policy," *Electronic Information Bulletin* 3, June 2005, http:// www.ers.usda.gov/publications/EIB3/EIB3.htm.

26. Robert Schaeffer, "Free Trade Agreements: Their Impact on Agriculture and the Environment," cited in Philip McMichael, "Global Development and the Corporate Food Regime," presentation, Sustaining a Future for Agriculture conference, Geneva, November 2004, http://www.agribusiness accountability.org/pdfs//297_Global%20Development%20and%20the%20 Corporate%20Food%20Regime.pdf.

27. Equanet, "Tyson Foods-Product Development, International Growth and Renewable Energy Highlighted," November 14, 2006, http://equa-net.com/ news/ind_watch/?news=427.

28. United Nations, "Business Latin America," in UN Report on International Investment 2005, 93.

29. B. Babcock, "Farm Policy Amid High Prices," *Iowa Ag Review* (Fall 2006), http://www.card.iastate.edu/iowa_ag_review/fall_06/IAR.pdf.

30. Kevin Watkins et al., "Rigged Rules and Double Standards: Trade, Globaliza-

tion and the Fight Against Poverty," Oxfam Campaign Reports, November 2003, http://publications.oxfam.org.uk/oxfam/display.asp?isbn=0855985259.

31. Sophia Murphy, *Managing the Invisible Hand,* Institute for Agriculture and Trade Policy (2002), 24, http://www.tradeobservatory.org/library.cfm?RefID=25497.

32. Mindfully.org, "The Revolving Door: U.S. Government Workers and University Researchers Go Biotech. . . . and Back Again: A Question of Ethics," http://www.mindfully.org/GE/Revolving-Door.htm.

33. Cargill bulletin, April 1993, cited in Brewster Kneen, *Invisible Giant: Cargill and Its Transnational Strategies* (London: Pluto Press, 1995), 78.

34. *Cargill News,* November 1993, cited in ibid., 66.

35. Morgan, *Merchants,* 111.

36. Hanrahan, "U.S. Agricultural Trade."

37. Ibid.

38. Environmental Working Group, "Top Programs in the United States, 1995–2004," http://www.ewg.org:16080/farm/region.php?fips=00000#topprogs.

39. *Economist,* "Uncle Sam's Teat," September 9, 2006, 35.

40. Elanor Starmer et al., "Feeding the Factory Farm: Implicit Subsidies to the Broiler Chicken Industry," GDAE working paper no. 06-03, Global Development and Environment Institute, Tufts University, June 2006.

41. Paul Aho, personal communication with the author, November 12, 2005.

42. Blake Lovette, personal communication with the author, December 5, 2005.

43. Revel, *American Green Power,* 79.

44. *The Economist,* "Molecular Weight," November 2, 2006, http://www.economist.com/business/displaystory.cfm?story_id=E1_RTDPRGV.

45. A. Cha, "China Slow in Meeting Food Safety Standards," *Washington Post,* April 25, 2007.

46. Statistics in *People's Daily* [Beijing] Online, "Last Year Saw China's Soybean Import Hit a Record High in History," February 14, 2004, http://english.peopledaily.com.cn/200402/14/eng20040214_134838.shtml; Mark Ash et al., "Soybean Backgrounder," *Outlook,* April 2006, http://usda.mannlib.cornell.edu/usda/ers/OCS//2000s/2006/OCS-04-04-2006_Special_Report.pdf; and Mark Drabenstott et al., "Roar of the Dragon: The Asian Upside for U.S. Agriculture Commentary on the Rural Economy," *Main Street Economist,* May 2004, http://www.extension.iastate.edu/AGDM/articles/others/DraNovDec04.htm.

47. *Peoples' Daily,* "China Rivals World's Top Corn Exporters" Tuesday, April 15, 2003, http://english.peopledaily.com.cn/200304/15/eng20030415_115230.shtml; Daryll Ray, "Major U.S. export competitor in the distant future: EU? Brazil? It could be China!" http://apacweb.ag.utk.edu/weekdoc/153.doc.

48. Larry Rohter, "South America Seeks to Fill the World's Table," *New York Times,* December 12, 2004, http://select.nytimes.com/search/restricted/article?res=FA0B16FE34550C718DDDAB09.

49. Ibid.

50. Henning Steinfeld and Pius Chilonda, "Old Players, New Players: Food and

Agriculture Organization Livestock Report, 2006," UN FAO, http://www.fao.org/docrep/009/a0255e/a0255e04.htm.

51. OECD-FAO, *Agricultural Outlook 2007–2016*, 26, http://www.oecd.org/dataoecd/6/10/38893266.pdf.

52. Foreign Agricultural Service, "Situation and Outlook," *Oilseeds: World Market and Trade* (June 1998), http://www.fas.usda.gov/oilseeds/circular/1998/98-05/maytext.htm.

53. Foreign Agricultural Service, "Brazil Soybean Crush Declines as Soybean Exports Continue to Strengthen," *Oilseeds: World Market and Trade* (May 2006), http://www.fas.usda.gov/oilseeds/circular/2006/06-05/Maycov.pdf.

54. Statistics in Marcia Taylor, "Ready or Not," *Top Producer* (October 2006); Drabenstott et al., "Roar of the Dragon"; and Steinfeld et al., "Old Players, New Players."

55. D. Wood, "Food Safety Concerns Grow as Imports to U.S. Surge," *Christian Science Monitor,* May 8, 2007, http://www.csmonitor.com/2007/0508/p02s01-usgn.html?page=1.

56. Steve Martinez, "The U.S. Food Marketing System," *Economic Research Report* 42 (Economic Research Service/USDA, May 2007): 8, http://www.ers.usda.gov/publications/err42/err42.pdf.

57. OECD-FAO, *Agricultural Outlook 2007–2016*, 26.

6. The End of Hunger

1. Kofi Annan, "The Right to Food: Note by the Secretary-General," February 2002, http://www.unhchr.ch/Huridocda/Huridoca.nsf/0/990d43116ffe0be1c1256c5d00368067/$FILE/N0254654.doc.

2. Frances Moore Lappé, "Hunger Is Not a Place," *The Nation,* January 23, 2006, http://www.thenation.com/doc/20060123/lappe.

3. Food and Agriculture Organization, May 2007. Data provided to the author by Chris Barrett.

4. Segenet Kelemu et al., "Harmonizing the Agricultural Biotechnology Debate for the Benefit of African Farmers," *African Journal of Biotechnology* 2, no. 11 (November 2003): 394–416, http://www.ciat.cgiar.org/biblioteca/pdf/kelemu.pdf.

5. IPCC, "Climate Change 2007: Impacts, Adaptation and Vulnerability, Summary for Policymakers," April 7, 2007, http://www.ipcc.ch/SPM13apr07.pdf.

6. Melinda Smale et al., "Maize in Eastern and Southern Africa: 'Seeds' of Success in Retrospect," International Food Policy Research Institute, January 2003, http://www.ifpri.org/divs/eptd/dp/papers/eptdp97.pdf.

7. William Gaud, "The Green Revolution: Accomplishments and Apprehensions," address given to the Society for International Development in Washington, D.C., March 8, 1968, http://www.agbioworld.org/biotech-info/topics/borlaug/borlaug-green.html.

8. H. De Groote et al., "The Maize Green Revolution in Kenya Revisited," *Journal of Agricultural and Developmental Economics* 2, no. 1: 32–49, http://www .uneca.org/estnet/Ecadocuments/Towards_aGreen_Revolution_in_Africa .doc.

9. J. Enos, "In Pursuit of Science and Technology in Sub-Saharan Africa: The Impact of Structural Adjustment Programmes," United Nations University Institute for New Technologies, Maastricht, the Netherlands, 1995, accessed online, http://www.infoplease.com/year/1965.html; http://www.unu.edu/unupress/unupbooks/uu33pe/uu33pe00.htm#Contents.

10. Gaud, "The Green Revolution."

11. Marc Cohen of the International Food Policy Research Institute, personal communication with the author, August 16, 2006.

12. J. Enos, "In Pursuit of Science and Technology."

13. Ibid.

14. Gaud, "The Green Revolution."

15. Ibid.

16. Cited in Arne Bigsten, "Linkages from Agricultural Growth in Kenya," in John Mellor, ed., *Agriculture on the Road to Industrialization* (Baltimore: Johns Hopkins University Press, 1995), http://www.ifpri.org/pubs/books/mellor95.htm.

17. Donald Plucknett, "Saving Lives Through Agricultural Research," CGIAR, May 1991, http://www.worldbank.org/html/cgiar/publications/issues/issues1.pdf.

18. Vaclav Smil, *Feeding the World: A Challenge for the Twenty-first Century* (Cambridge, MA: MIT Press, 2000), 17.

19. Food and Agriculture Organization, "Anti-Hunger Programme: A Twin-Track Approach to Hunger Reduction," Rome, November 2003, http://www .fao.org/DOCREP/006/J0563E/j0563e06.htm#TopOfPage.

20. EurekaAlert, "Africa Fertilizer Summit Facts," American Association for the Advancement of Science, March 30, 2006, http://www.eurekaalert.org/staticrel.php?view+backgrounder0328.

21. References from the Food and Agriculture Organization's articles "Special Report: Crop and Food Supply Situation in Kenya," July 10, 2000, http:// www.fao.org/docrep/004/x7697e/x7697e00.htm; and "Agriculture, Trade and Food Security Issues and Options in the WT Negotiations," http://www .fao.org/DOCREP/003/X8731E/x8731e09.htm.

22. Holmén, "A Green Revolution for Africa."

23. Mark Rosegrant et al., "Global Water Outlook to 2025: Averting an Impending Crisis," International Food Policy Research Institute, September 2002, http://www.ifpri.org/pubs/fpr/fprwater2025.pdf.

24. Shaun Brunner, interview with the author, November 2, 2005.

25. Jan McGirk, "Growing Coffee: It's Black, No Sugar," MSNBC.com, August 2002, http://www.msnbc.msn.com/id/3072120/.

26. N. Stein, "Crisis in a Coffee Cup," *Fortune,* December 9, 2002, http://money .cnn.com/magazines/fortune/fortune_archive/2002/12/09/333463/index .htm.

27. Gerard Greenfield, "Vietnam and the World Coffee Crisis: Local Coffee Riots

in a Global Context," paper prepared for the Asia-Pacific Regional Land and Freedom Conference, March 1, 2004, http://www.greenbeanery.ca/bean/catalog/info_pages.php?pages_id=56?osCsid=bf11832ecf7e50291efcf155ca19fd6a.

28. Food and Agriculture Organization, "Agricultural Commodities: Profiles and Relevant WTO Negotiating Issues: Coffee," 2003, http://www.fao.org/DOCREP/006/Y4343E/y4343e05.htm.

29. Robert Collier, "World's Leading Java Companies Are Raking in High Profits but Growers Worldwide Face Ruin as Prices Sink to Historic Low," *San Francisco Chronicle*, May 20, 2001, http://www.globalexchange.org/campaigns/fairtrade/coffee/sfchron052001.html.

30. Celine Charveriat, "Bitter Coffee: How the Poor Are Paying for the Slump in Coffee Prices," Oxfam, May 16, 2001, http://www.oxfam.org.uk/what_we_do/issues/trade/downloads/bitter_coffee.pdf.

31. Miriam Wasserman, "Trouble in Coffee Lands," *Regional Review*, Federal Reserve Bank of Boston, Quarter 2, 2002, http://www.bos.frb.org/economic/nerr/rr2002/q2/coffee.htm.

32. Charveriat, "Bitter Coffee."

33. Collier, "World's Leading Java Companies."

34. Ibid.

35. Kenya Coffee Board, presentation, 2005, http://www.africacncl.org/(woxvojrstjkp1x45tfc33y55)/CCA_Summits/2005_Downloads/The%20Coffee%20Sector%20%20Presentation%20-%20Kenya.ppt.

36. Charveriat, "Bitter Coffee."

37. Ibid.

38. Stein, "Crisis in a Coffee Cup."

39. Donald G. McNeil Jr., "The Great Ape Massacre," *New York Times*, May 9, 1999, http://select.nytimes.com/search/restricted/article?res=FA0E11F63C580C7A8CDDAC0894D1494D81.

40. Kelemu et al., "Harmonizing the Agricultural Biotechnology Debate," 397; and Food and Agriculture Organization, "Anti-Hunger Programme."

41. Jeffrey Sachs et al., "The Millennium Villages Project," transcript of a panel discussion organized by Center for Global Development, March 14, 2006, http://www.cgdev.org/doc/event%20docs/3.14.06_Sachs/Sachs%20transcript.pdf.

42. IFPRI, "The Future of Small Farms, Workshop," June 26, 2005, http://www.ifpri.org/events/seminars/2005/smallfarms/sfproc/sfproc.pdf.

43. Food and Agriculture Organization, "In Praise of Family Poultry," *Agriculture* 21 (March 2002), http://www.fao.org/AG/magazine/0203sp1.htm.

44. Center for Global Development, "Jeffrey Sachs and the Millennium Villages," March 13, 2006, http://www.cgdev.org/content/article/detail/6660/; J. Hanlon, "Jeffrey Sachs Launches Millennium Village in Chibuto," *News Reports & Clippings* 99, Open University, July 1, 2006, http://www.gg.rhul.ac.uk/Simon/GG3072/Moz-Bull-99.pdf.

45. Hanlon, "Jeffrey Sachs Launches Millennium Village."

46. Sachs, "The Millennium Villages Project."

47. World Bank, "Kenya: Exporting out of Africa — Kenya's Horticultural Success Story, Thematic Case Studies," Reducing Poverty, 2003, http://info.worldbank.org/etools/reducingpoverty/case-Kenya-Horticultural.html.

48. Kamau Ngotho, "Kenya's Wealth in Foreign Hands," *Sunday Standard* (Nairobi), April 17, 2005, http://www.eastandard.net/archives/sunday/hm_news/news.php?articleid=18216.

49. *New Agriculturalist Online,* "Exacting Standards," May 1, 2003, http://www.new-agri.co.uk/03-3/focuson/focuson1.html.

50. Spensor Henson et al., "Case Study: Kenya's Horticultural Exports," Global Facilitation Partnership, presentation, April 17, 2004, http://www.gfptt.org/Entities/ReferenceReadingProfile.aspx?id=1f3b0d96-a006-4e0b-9abe-ea1216b0152a.

51. Fred Magdoff et al., eds., *Hungry for Profit* (New York: Monthly Review Press, 2000), 139.

52. George W. Bush, "Remarks by the President to Agriculture Leaders," the East Room, Office of the Press Secretary, June 18, 2001, http://www.whitehouse.gov/news/releases/2001/06/20010618-6.html.

53. S. Devereux, "State of Disaster: Causes, Consequences and Policy Lessons from Malawi," Action Aid International (Eldis), 2002, http://www.eldis.org/static/DOC9912.htm.

54. H. Steinfeld et al., "Livestock's Long Shadow: Environmental Issues and Options," Food and Agriculture Organization, Rome, 2006, 34.

55. World Bank, "Social Indicators: Health: Life Expectancy," http://www1.worldbank.org/prem/poverty/data/trends/mort.htm; World Bank, "World Development Indicators," http://devdata.worldbank.org/wdi2005/Section2.htm.

7. WE ARE WHAT WE EAT

1. Dr. Stewart J. Ritchie, personal communication with the author, December 21, 2006.

2. This account is taken from interviews with several of its participants, including Dr. Stewart Ritchie, DVM, Abbotsford, BC; Dr. Aleina Tweed, epidemiologist, Vaccine-Preventable Diseases, BC Centre for Disease Control; Dr. Victoria Bowes, DVM, MSc, ACPV avian pathologist, Animal Health Centre BC Ministry of Agriculture and Lands; Katherine Luke, researcher, and Dr. Tomy Joseph, both of the U.S. Centers for Disease Control; as well as from several documents, including "An Overview Of The 2004 Outbreak Of Highly Pathogenic Avian Influenza In British Columbia," presented at the 54th Western Poultry Disease Conference, Vancouver, BC, April 25, 2005, by Dr. Victoria A. Bowes.

3. Aleina Tweed, personal communication with the author, December 14, 2006.

4. World Bank, "Avian Flu: The Economic Costs," June 29, 2006, http://go .worldbank.org/9ZPSoUCAG0.

5. Michael Greger, *Bird Flu: A Virus of Our Own Hatching* (New York: Lantern Books, 2006). Excerpts on author's website: http://birdflubook.com/ a.php?id=99.

6. Dr. Victoria Bowes, personal communication with the author, December 14, 2006.

7. J. Rocourt et al., "The Present State of Foodborne Disease in OECD Countries," World Health Organization, Geneva, 2003, http://www.fao.org/ag/ againfo/subjects/en/health/diseases-cards/avian.html.

8. Statistics from U.S. Centers for Disease Control and Prevention, "Foodborne Illness," October 25, 2005, Division of Bacterial and Mycotic Diseases, http:// www.cdc.gov/ncidod/dbmd/diseaseinfo/foodborneinfections_g.htm#how manycases; Rocourt et al., "The Present State of Foodborne Disease," and B. Hargis et al., "Pre-Slaughter Control of Food-borne Bacterial Pathogens in Poultry Without Antibiotics," paper presented at 59th American Meat Science Association Reciprocal Meat Conference, June 2006, http://www .meatscience.org/pubs/rmcarchv/2006/presentations/3_2_Hargis.pdf.

9. Food and Agriculture Organization, "Bird Flu a 'Marker' for the Future: No Room for Neglect of Animal Disease, Says Top UN Official," press release, October 6, 2006, http://www.fao.org/docs/eims/upload//214351/ nabarro_6oct06.pdf.

10. J. Brooks et al., "The Spoilage Characteristics of Ground Beef Packaged in High-Oxygen and Low-Oxygen Modified Atmosphere Packages," paper presented at the 59th American Meat Science Association Reciprocal Meat Conference, June 2006, http://www.meatscience.org/pubs/rmcarchv/2006/presentations/16_7_Brookstry.pdf.

11. Food Facts for You!, http://www.wisc.edu/foodsafety/assets/foodfacts_2003/ foodfacts_july_2003.pdf.

12. G. Smith et al., "Traceback, Traceability and Source Verification in the U.S. Beef Industry," paper presented at the 21st Buiatrics Congress, 2000, Uruguay, http://ansci.colostate.edu/files/meat_science/gcs001.pdf.

13. U.S. Centers for Disease Control, "Addressing Emerging Infectious Disease Threats: A Prevention Strategy for the United States," Executive Summary, April 15, 1994, 43(RR-5), 1–18, http://www.cdc.gov/mmwr/preview/ mmwrhtml/00031393.htm; L. W. Riley et al., "Hemorrhagic Colitis Associated with a Rare *Escherichia coli* Serotype," *New England Journal of Medicine* 308, no. 12 (March 24, 1983): 681–85, http://content.nejm.org/cgi/content/ abstract/308/12/681?ijkey=cd9cabfe29461411c8b0427d753899a43c34ef0e&key type2=tf_ipsecsha.

14. T. R. Callaway, "Forage Feeding to Reduce Preharvest *Escherichia coli* Populations in Cattle, a Review," *Journal of Dairy Science* 86 (2003): 852–60, http:// jds.fass.org/cgi/content/full/86/3/852.

15. Robert L. Jamieson Jr., "Poster Child for a Safer Meat industry," *Seattle Post-*

Intelligencer, January 12, 2004, http://www.marlerclark.com/news/jackbox13.htm.

16. Elaine Porterfield, "Jack in the Box Ignored Safety Rules," Tacoma *News Tribune,* June 16, 1995.

17. Callaway, "Forage Feeding."

18. J. N. Sofos et al., "Salmonella Interventions for Beef," paper presented at the 59th Annual Reciprocal Meat Conference, http://www.meatscience.org/pubs/rmcarchv/2006/presentations/3_5_Sofostry.pdf.

19. Callaway, "Forage Feeding."

20. Gary Smith, "The Future of the Beef Industry," presentation at the Range Beef Cow Symposium, 2005, http://beef.unl.edu/beefreports/symp-2005-02-XIX.pdf.

21. Genevieve A. Barkocy-Gallagher, "Prevalence of *Escherichia coli* O157:H7 in Cattle and During Processing," presentation at the 55th Annual Reciprocal Meat Conference, http://www.meatscience.org/Pubs/rmcarchv/2002/presentations/rmc_2002_055_1_0015_barkocy-gallagher.pdf.

22. Stephen Hedges, "Durbin to USDA: Why delay in tainted-meat recall?" *Baltimore Sun,* October 17, 2007, http://weblogs.baltimoresun.com/news/politics/blog/2007/10/durbin_to_usda_why_delay_in_ta.html; Reuters, "USDA Urges Consumers to Be Safe During Massive Beef Recall," Foxnews.com, October 1, 2007, http://www.foxnews.com/story/0,2933,298780,00.html.

23. R. Nutting, "Food Safety Oversight Called a 'Sorry Mess,'" *Wenatchee World,* July 18, 2007.

24. Information in A. T. Kearney Research, "Fixing China's Food Safety Issues Will Require a $100 Billion Investment," company press release, June 26, 2007, http://www.atkearney.com/main.taf?p=1,5,1,190, and R. Weiss, "U.S. May Allow Chinese Chicken Imports," *Washington Post,* May 22, 2007, and the *Wall Street Journal,* "Commentary: Chinese Pet Food Tricks," May 15, 2007, at http://www.truthabouttrade.org/article.asp?id=7540.

25. A. Bridges, "Pet Food Recall Raises Questions about Safety of Imported Food," *Wenatchee World,* April 16, 2007.

26. Nutting, "Food Safety Oversight."

27. U.S. Department of Health and Human Services, "Expanded 'Mad Cow' Safeguards Announced to Strengthen Existing Firewalls Against BSE Transmission," Internet press release, January 26, 2004, http://www.hhs.gov/news/press/2004pres/20040126.html.

28. Food and Agriculture Organization, "The Changing Global Technological Context for Poultry and Pork Production," http://www.fao.org/WAIRDOCS/LEAD/X6115E/x6115e05.htm#TopOfPage, and Mary Gilchrist et al., "The Potential Role of Concentrated Animal Feeding Operations in Infectious Disease Epidemics and Antibiotic Resistance," *Environmental Health Perspectives,* February 2007, http://www.ehponline.org/members/2006/8837/8837.html.

29. Gilchrist, "The Potential Role of Concentrated Animal Feeding."

30. Keep Antibiotics Working, "Kennedy, Snowe & Slaughter Introduce AMA-

backed Bill," press release, February 12, 2007, http://www.keepantibiotics working.com/new/resources_library.cfm?RefID=97314.

31. Food and Drink Digital, "Tyson Chicken to Be Raised Antibiotics Free," June 28, 2007, http://www.fooddigital.com/NewsArticle.aspx?articleid=781.

32. D. E. Corrier et al., cited in James Dawe, "The Relationship between Poultry Health and Food Safety," *Poultry Informed Professional* (March–April 2004), http://www.avian.uga.edu/documents/pip/2004/0304.pdf.

33. See U.S. Centers for Disease Control, "Campylobacter Infections," fact sheet, Coordinating Center for Infectious Diseases, October 6, 2005, http://www.cdc.gov/ncidod/dbmd/diseaseinfo/campylobacter_g.htm, and Dawe, "The Relationship between Poultry Health and Food Safety."

34. Centers for Disease Control, "Disease Listing: Salmonellosis," Coordinating Center for Infectious Diseases, November 4, 2006, http://www.cdc.gov/ncidod/dbmd/diseaseinfo/salmonellosis_g.htm.

35. M. Brashears, "Multi-Drug Resistant Salmonella," paper presented at the 59th American Meat Science Association Reciprocal Meat Conference, June 2006, http://www.meatscience.org/pubs/rmcarchv/2006/presentations/3_6_Brashears.pdf.

36. Sofos et al., "Salmonella Interventions."

37. Dan Englejohn, "Performance Standards for Meat and Poultry," remarks at 59th American Meat Science Association Reciprocal Meat Conference, June 2006, 2, http://www.meatscience.org/pubs/rmcarchv/2006/presentations3_3_t.pdf.

38. Englejohn, a PowerPoint presentation accompanying remarks (ibid.), slide 7, http://www.meatscience.org/pubs/rmcarchv/2006/presentations/3_3_t.pdf.

39. Englejohn, "Performance Standards," 4.

40. Marion Nestle, *Safe Food: Bacteria, Biotechnology, and Bioterrorism* (Berkeley: University of California Press, 2003), 59.

41. The United States Court of Appeals for the Fifth Circuit, *Supreme Beef Processors, Inc., Versus United States Department of Agriculture*, no. 00-11008, December 6, 2001, http://caselaw.lp.findlaw.com/cgi-bin/getcase.pl?court=5th&navby=docket&no=0011008cv0.

42. Interview with Dan Glickman, *Frontline*, http://www.pbs.org/wgbh/pages/frontline/shows/meat/interviews/glickman.html.

43. Center for Infectious Disease Research and Policy, "California Debates Produce Safety Measures," news release, January 26, 2007, http://www.cidrap.umn.edu/cidrap/content/fs/food-disease/news/jan2607growers.html.

44. Dania Akkad, "California Ag Leaders Respond to *E. coli* Scare with Safety Plan," *Monterey County Herald*, viewed at http://archives.foodsafety.ksu.edu/fsnet/2007/1-2007/fsnet_jan_14.htm.

45. California Department of Health Services and U.S. Food and Drug Administration, "Investigation of an *E. coli* O157:H7 Outbreak Associated with Dole Pre-Packaged Spinach, Final," March 21, 2007, http://www.dhs.ca.gov/ps/fdb/local/PDF/2006%20Spinach%20Report%20Final%20redacted.PDF.

46. Ibid.

47. United States Department of Agriculture, "Scientists Seek to Sanitize Fruits and Vegetables," March 2002, http://www.ars.usda.gov/is/ar/archive/mar02/fruit0302.htm

48. Economic Research Service, "Commodity Highlight: Iceberg Lettuce," USDA June 22, 2006, http://www.ers.usda.gov/publications/vgs/2006/06jun/vgs315.pdf

49. Julie Schmit, "'Fresh Express Leads the Pack' in Produce Safety," USA Today, October 23, 2006, http://www.usatoday.com/money/industries/food/2006-10-22-fresh-express-usat_x.htm.

50. R. Alonso-Zaldivar, "Too Many Holes in Food Safety Net," Los Angeles Times, February 1, 2007.

51. World Bank, "Avian Flu: The Economic Costs."

52. Knowledge@Wharton, "Avian Flu: What to Expect and How Companies Can Prepare for It," Wharton School of Business, March 8, 2006, http://knowledge.wharton.upenn.edu/article.cfm?articleid=1402.

53. National Review, "Symposium: For the Birds or Sleeping with the Fishes?" March 14, 2006, http://www.nationalreview.com/script/printpage.p?ref=/symposium/symposium200603140821.asp.

54. Michael Leavitt, secretary of health and human services, remarks to health ministers of Central America, June 8, 2006, U.S. Department of Health and Human Services, http://www.hhs.gov/news/speech/2006/060608.html.

55. Brian Ross, "Ready or Not, Bird Flu Is Coming to America," ABC News, March 13, 2006, http://abcnews.go.com/GMA/print?id=1716820.

56. World Bank et al., "Enhancing Control of Highly Pathogenic Avian Influenza in Developing Countries through Compensation," http://www.fao.org/docs/eims/upload//217132/gui_hpai_compensation.pdf; Food and Agriculture Organization, "HPAI Risk, Bio-Security and Smallholder Adversity," http://www.fao.org/AG/AGAINFO/projects/en/pplpi/docarc/pb_hpaibiosecurity.html.

57. The Economist, "Under the Influence," February 10, 2007.

58. Ibid.

59. A. T. Kearney Research, "Fixing China's Food Safety Issues."

60. Neil King and Rebecca Blumenstein, "China Launches Public Response to Safety Outcry: Shift in Tactics Shows Greater Sophistication," Wall Street Journal, June 30, 2007, http://online.wsj.com/article/SB118316517029653742.html?mod=googlenews_wsj.

61. Ibid.

62. World Bank et al., "Enhancing Control of Highly Pathogenic Avian Influenza in Developing Countries through Compensation," page 8, http://www.fao.org/docs/eims/upload/217132/gui_hpai_compensation.pdf.

8. IN THE LONG RUN

1. Author's calculation, based on company's outstanding shares of 355.5 million and an analyst's estimate that each 10 cent increase in price of corn

translates into a 5 cent earnings loss per share. Analyst cited in Ann Davis, "Tyson Foods Refines a Recipe By Energy Firms," *Wall Street Journal,* December 13, 2007. http://online.wsj.com/article/SB116493974470437634 .html?mod=yahoo_hs& ru=yahoo (accessed March 29, 2007)

2. Lisa Haarlander, "U.S. Food Cost Up $47 per Person Due to Ethanol: Study," Reuters.com, May 17, 2007, http://www.reuters.com/article/ousiv/ idUSN1742753920070517.

3. Lester R. Brown, "Exploding U.S. Grain Demand for Automotive Fuel Threatens World Food Security and Political Stability," Earth Policy Institute, http://www.earth-policy.org/Updates/2006/Update60.htm; H. Steinfeld et al., "Livestock's Long Shadow: Environmental Issues and Options," Food and Agriculture Organization, Rome, 2006, 38.

4. Food and Agriculture Organization, "World Agriculture: Towards 2030/ 2050," interim report, 2006; Steinfeld, "Livestock's Long Shadow," 15.

5. Foreign Agricultural Service, "Tightening 2006/07 Global Grain Supplies to Boost Prices," May 2006, http://www.fas.usda.gov/grain/circular/2006/05-06/CGrains%2005-06.pdf.

6. Ibid.

7. "The OECD-FAO Agricultural Outlook 2007-2016," http://www.oecd.org/ dataoecd/6/10/38893266.pdf; Cees Bruggemans, "Evolving Global Inflation Shocks," *Weekly Comment,* First National Bank New Zealand, July 9, 2007, https://www.fnb.co.za/economics/servlet/Economics?ID=3152.

8. Food and Agriculture Organization, "World Agriculture 2030: Main Findings," press release, 2002, http://www.fao.org/english/newsroom/news/2002/ 7833-en.html.

9. Javier Blas and Jenny Wiggins, "UN Warns It Cannot Afford to Feed the World," *Financial Times,* July 15, 2007, http://www.ft.com/cms/s/7345310a-32fb-11dc-a9e8-0000779fd2ac.html.

10. Ibid.

11. Economic Research Service, "Briefing Rooms: Food Consumption," USDA, http://www.ers.usda.gov/Briefing/Consumption/; Vaclav Smil, *Feeding the World: A Challenge for the Twenty-first Century* (Cambridge: MIT Press, 2000), 230.

12. National Western Stock Show, http://www.nationalwestern.com/nwss/ home/index.asp?rpg=/nwss/international/roundtable.asp.

13. JSTOR: "Impact of Beef Imports on the U.S. Beef Industry," http:// links.jstor.org/sici?sici=10711031(196512)47%3A5%3C1594%3AIOBIOT%3E2 .0.CO%3B2-H (accessed May 4, 2007). No longer available online. For 1950 herd size, see Google reference: http://www.google.com/search?hl=en&q =United+States+beef+cattle+herd+1950. Also, for current herd size, see http://www.ag.ndsu.edu/news/columns/livestockma/marketadvisor-dry-weather-stalls-cattle-herd-buildup. And for decline in grazing lands, see http://www.ers.usda.gov/publications/EIB14/eib14e.pdf.

14. Based on National Chicken Council data showing beef prices in 1960 at fifty-

nine cents a pound; in 2006 prices would be $3.94, which is nearly twice the current price of two dollars a pound. National Chicken Council statistics, http://www.nationalchickencouncil.com/statistics/stat_detail.cfm?id=20.

15. Meat and Livestock Australia, "Global Liberalisation: Magellan Project, Phase 1," http://www.mla.com.au/NR/rdonlyres/6A127A33-DEF4-48CD-B1 D2-1510464963B1/0/Phase1Report.pdf; FAO/GIEWS, "Meat and Meat Products," *Food Outlook* 3 (June 2001): 12, http://www.fao.org/docrep/004/y08 49e/y0849e04.htm.

16. Smil, *Feeding the World*, 157.

17. Ibid.; also T. Crews et al., "Legume versus Fertilizer Sources of Nitrogen: Ecological Tradeoffs and Human Needs," *Agriculture, Ecosystems, and Environment* 102 (2004): 293.

18. Food and Agriculture Organization, "World Fisheries Production, by Capture and Aquaculture, by Country," *Yearbooks of Fishery Statistics 2004 Summary Tables,* ftp://ftp.fao.org/fi/STAT/summary/a-0a.pdf.

19. National Chicken Council, "Per Capita Consumption of Poultry and Livestock" (current as of March 26, 2007), http://www.nationalchickencouncil .com/statistics/stat_detail.cfm?id=8.

20. Norman Borlaug, "Feeding a World of 10 Billion People: The Miracle Ahead," lecture, http://www.nbipsr.org/nb_lect.html.

21. EarthTrends, "Energy and Resources, Country Profile: Italy," World Resources Institute, 2006, http://earthtrends.wri.org/text/energy-resources/ country-profile-91.html.

22. Food and Agriculture Organization, "Diet, Nutrition and the Prevention of Chronic Diseases: Report of a Joint WHO/FAO Expert Consultation," http:// www.fao.org/DOCREP/005/AC911E/ac911e05.htm.

23. Ibid.

24. Ibid.

25. Brad Knickerbocker, "Humans' Beef with Livestock: A Warmer Planet," *Christian Science Monitor,* February 20, 2007, http://www.csmonitor.com/ 2007/0220/p03s01-ussc.html.

26. Food and Agriculture Organization, "World Agriculture 2030: Main Findings."

27. Lester Brown, *Plan B 2.0: Rescuing a Planet Under Stress and a Civilization in Trouble* (New York: Norton, 2006), 177.

28. Katy Mamen, "Current Issues and Trends Connected to the Vivid Picture Goals for a Sustainable Food System," report by the Vivid Picture Project, December 2004, 11, http://www.foodsecurity.org/PrimerCFSCUAC.pdf.

29. Center for Environment & Population, U.S. National Report on Population and the Environment, http://www.cepnet.org/documents/USNatlReptFinal .pdf

30. Food and Agriculture Organization, "World agriculture: towards 2015/2030. Summary Report: Prospects by Major Sector: Crop production," http:// www.fao.org/docrep/004/y3557e/y3557e08.htm#TopofPage.

31. Mahendra Shah et al., *Food in the Twenty-first Century: From Science to Sustainable Agriculture,* World Bank 2000, http://www.worldbank.org/html/cgiar/publications/shahbook/shahbook.pdf.

32. Kenneth G. Cassman, "Ecological Intensification of Cereal Production Systems: Yield Potential, Soil Quality, and Precision Agriculture," *Proceedings of the National Academy of Sciences* 96, no. 11 (May 25, 1999): 5952–59, http://www.pnas.org/cgi/content/full/96/11/5952.

33. For statistics, see Food and Agriculture Organization, "Rice Crisis Looms in Asia," *Agriculture* 21 (1998), http://www.fao.org/AG/magazine/9809/spot1.htm, and Luis Pons, "Rice Yield Mystery Solved? Key Clues Revealed about Declines in Asia," *Agricultural Research* (September 2004), http://findarticles.com/p/articles/mi_m3741/is_9_52/ai_n6210362.

34. Vernon W. Ruttan, "The Transition to Agricultural Sustainability," *Proceedings of the National Academy of Sciences,* http://www.pnas.org/cgi/content/full/96/11/5960.

35. Food and Agriculture Organization, "Agricultural Biotechnology: Meeting the Needs of the Poor?" *The State of Food and Agriculture 2003-2004,* http://www.fao.org/docrep/006/Y5160E/y5160e07.htm#TopOfPage.

36. Shah et al., *Food in the Twenty-first Century,* 21.

37. J. M. Harris, *World Agriculture and the Environment* (New York: Garland, 1990), 115, cited in Ruttan, "The Transition to Agricultural Sustainability."

38. Chris Barrett, personal communication with the author, May 1, 2007.

39. Agriculture Energy Alliance, "Nitrogen Fertilizer Production," slide presentation, http://media.corporate-ir.net/media_files/irol/19/190537/Figures2_2006.pdf.

40. Vaclav Smil, "Phosphorus in the Environment: Natural Flows and Human Interferences," *Annual Reviews of Energy and Environment* 25 (2000): 53–88, http://home.cc.umanitoba.ca/~vsmil/pdf_pubs/aree2000-2.pdf.

41. Agriculture Energy Alliance, "Nitrogen Fertilizer Production."

42. *Citizen Gas, Natural Gas Conversion Factors,* company literature, http://www.citizensgas.com/pdf/EnergyConversion.pdf.

43. James Finch, "Natural Gas Investors to Benefit from Global Ethanol Boom," *Seeking Alpha,* April 30, 2007, http://energy.seekingalpha.com/article/33925; U.S. Department of Energy, "Energy Savers," http://www1.eere.energy.gov/consumer/tips/appliances.html.

44. Agriculture Energy Alliance, "Nitrogen Fertilizer Production."

45. Crews et al., "Legume versus Fertilizer."

46. Finch, "Natural Gas Investors to Benefit."

47. John Sawyer et al., "Concepts and Rationale for Regional Nitrogen Rate Guidelines for Corn," April 2006, available at www.extension.iastate.edu/Publications/PM2015.pdf.

48. *Environmental Valuation and Cost-Benefit News,* "Category: Nitrogen/Nitrates," May 13, 2007, http://envirovaluation.org/index.php?cat=169.

49. Crews et al., "Legume versus Fertilizer."

50. A. Mosier et al., "A New Approach to Estimate Emissions of Nitrous Oxide from Agriculture and Its Implications to the Global Nitrous Oxide Budget," *IGACtivities* newsletter, no. 12 (March 1998), http://www.igac.noaa.gov/newsletter/highlights/1998/n2o.php; and Crews et al., "Legume versus Fertilizer."

51. Robert Gilliom et al., "Testing Water Quality for Pesticide Pollution: U.S. Geological Survey Investigations Reveal Widespread Contamination of the Nation's Water Resources," *Environmental Science and Technology* (April 1, 1999): 164–69, http://pubs.acs.org/hotartcl/est/99/apr/test.html.

52. U.S. Environmental Protection Agency, "Consumer Factsheet on Atrazine," March 21, 2007, http://www.epa.gov/safewater/contaminants/dw_contamfs/atrazine.html.

53. M. Metcalfe et al., "The Economic Importance of Organophosphates," California Department of Food and Agriculture, October 2002, http://www.cdfa.ca.gov/exec/aep/aes/opca/docs/Organophosphates%20in%20CA%20Agriculture.pdf.

54. Mamen, "Current Issues and Trends."

55. William Freudenthal, "Toxicity, Organophosphates," Emedicine.com, June 2, 2006, http://www.emedicine.com/ped/topic1660.htm.

56. A. Lewis et al., "A Total System Approach to Sustainable Pest Management," *Proceedings of the National Academy of Sciences* 94 (November 1997): 12243–48.

57. Daryll E. Ray et al., "Potential Farm-Level Impacts of Proposed FQPA Implementation: The Tennessee Case," Agricultural Policy Analysis Center, Department of Agricultural Economics and Rural Sociology, University of Tennessee, http://ageconsearch.umn.edu/bitstream/123456789/14176/1/sp99ra02.pdf

58. Metcalfe, "The Economic Importance of Organophosphates."

59. Lewis, "A Total System Approach to Sustainable Pest Management."

60. Janet Raloff, "Wheat Warning — New Rust Could Spread Like Wildfire," *Science News Online* 168, no. 13 (September 24, 2005), http://www.sciencenews.org/articles/20050924/food.asp; "Plant Pathologists Fighting Global Threat to Wheat Supply," http://www.truthabouttrade.org/article.asp?id=7513.

61. Steve Suppan, personal communication with the author, March 13, 2006.

62. National Research Council, "Alternative Agriculture," National Academy of Sciences, Washington, DC, 1989.

63. Ibid.

64. G. Eshel and P. Martin, "Diet, Energy and Global Warming," Department of the Geophysical Sciences, University of Chicago, http://geosci.uchicago.edu/~gidon/papers/nutri/nutri3.pdf.

65. Cited in Brown, *Plan B 2.0*, 90.

66. John Ikerd, *A Return to Common Sense* (Flourtown, PA: R. T. Edwards, 2007), v–vii, http://edwardspub.com/books/171/preface.pdf.

67. Daniel Davis, "On the Precipice: Energy Security and Economic Stability on

the Edge," Association for the Study of Peak Oil and Gas, July 17, 2007, http://www.aspo-usa.com/assets/documents/Danny_Davis_On_the_Precipice.pdf.

68. Mark Shenk, Bloomberg.com, "$100-a-Barrel Oil May Be Only a Few Months Away," *International Herald Tribune*, July 24, 2007, http://www.iht.com/articles/2007/07/24/bloomberg/bxoil.php.

69. Ibid.

70. John Hendrickson, "Energy Use in the U.S. Food System," Center for Integrated Agricultural Systems, University of Wisconsin, Madison, January 1994, http://www.cias.wisc.edu/pdf/energyuse2.pdf.

71. Ibid.

72. Eshel and Martin, "Diet, Energy, and Global Warming."

73. Davis, "On the Precipice."

74. U.S. Global Change Research Program, "Climate Change Impacts on the United States: The Potential Consequences of Climate Variability and Change," U.S. Climate Change Science Program, 2000, http://www.usgcrp.gov/usgcrp/Library/nationalassessment/overviewagriculture.htm.

75. Robert Mendelsohn et al., "Climate Change Impacts of African Agriculture," http://www.ceepa.co.za/Climate_Change/pdf/(5-22-01)afrbckgrnd-impact.pdf.

76. M. C. Parry et al., "Climate Change and World Food security: A New Assessment," *Global Environmental Change* 9 (1999): 51–67.

77. Report of the Panel on Climate Variations, 1975. *Understanding Climate Change.* National Academy of Sciences. Washington, DC.

78. U.S. Global Change Research Program, "Climate Change Impacts on the United States: The Potential Consequences of Climate Variability and Change," U.S. Climate Change Science Program, 2000, http://www.usgcrp.gov/usgcrp/Library/nationalassessment/overviewagriculture.htm.

79. A. G. Tacon, "Contribution to Food Fish Supplies," United Nations Food and Agriculture Organization archives, 1997, http://www.fao.org/DOCREP/003/W7499E/w7499e17.htm; "U.S. Meat Exports Continue Setting Records," *International Agricultural Trade Report,* USDA Foreign Agricultural Service, May 14, 2001, http://www.fas.usda.gov/dlp2/highlights/2001/wmr_2002.pdf.

80. S. Postel, *Vital Signs 2002: The Trends That Are Shaping Our Future* (World Watch Institute, 2002), 134.

81. Food and Agriculture Organization, "Rice Faces the Future," *Agriculture* 21 (2005), http://www.fao.org/AG/magazine/0512sp2.htm.

82. Cited in Ruttan, "The Transition to Agricultural Sustainability."

83. Shah et al., *Food in the Twenty-first Century,* 21.

84. Jelle Bruinsma, ed., "World Agriculture: Towards 2015/2030, an FAO perspective," section 4.3.3, http://www.fao.org/DOCREP/005/Y4252E/y4252e06a.htm#TopOfPage.

85. M. Rosegrant et al., "Global Water Outlook to 2025: Averting an Impending Crisis," IFPRI, September 2002, http://www.ifpri.org/pubs/fpr/fprwater2025.pdf.

86. Mamen, "Current Issues and Trends," 21.

87. Food and Agriculture Organization, "Global Network on Integrated Soil Management for Sustainable Use of Salt-Affected Soils," http://www.fao.org/AG/AGL/agll/spush/intro.htm; and K. Mamen, "Current Issues and Trends," 21.

88. H. Yang et al., "Virtual water and the need for greater attention to rainfed agriculture," *Water* 21, Magazine of the International Water Association, April 2006, 14–15.

89. A. Kirby, "Dawn of a Thirsty Century," BBC online, June 2, 2000, http://news.bbc.co.uk/hi/english/static/in_depth/world/2000/world_water_crisis/default.stm.

90. World Bank, "China Agenda for Water Sector Strategy for North China," vol. 1, April 2, 2001, http://lnweb18.worldbank.org/eap/eap.nsf/Attachments/WaterSectorReport/$File/Vol1v13A4a1.pdf.

91. W. Jury and H. Vaux, "The Role of Science in Solving the World's Emerging Water Problems," Proceedings of the National Academy of Science, October 25, 2005, 10.1073/pnas.0506467102, http://www.pnas.org/cgi/reprint/0506467102v1.

92. World Bank, "China Agenda for Water," xi.

93. Ibid.

94. Based on a World Bank figure of 56 billion yuan, at a conversion rate of .129 yuan to the dollar.

95. Yang et al., "Virtual Water."

96. A. Zehnder, "Feeding a More Populous World," abstract from "The Role of Science in Solving the Earth's Emerging Water Problems," Arthur M. Sackler Colloquia, National Academy of Sciences, Irvine, California, October 8–10, 2004, http://www.nasonline.org/site/PageServer?pagename=SACKLER_water_zehnder.

97. Shah et al., *Food in the Twenty-first Century,* 21.

98. Food and Agriculture Organization, "World Agriculture 2030: Main Findings."

99. World Wildlife Fund, "Humanity's Ecological Footprint," *Living Planet Report,* 2006, http://www.panda.org/news_facts/publications/living_planet_report/footprint/index.cfm.

100. M. Rosegrant, "Water and Food to 2025: Policy Response to the Threat of Scarcity," International Food Policy Research Institute, 2002, http://www.ifpri.org/pubs/ib/ib13.pdf.

101. T. Crews et al., "Legume versus fertilizer."

102. Richard Brock, "Groups Seek Corn on CRP Land," *Corn and Soybean Digest,* February 21, 2007, http://cornandsoybeandigest.com/corn/corn-crp-land/.

103. Food and Agriculture Organization, "World Agriculture 2030: Main Findings."

104. World Bank, "At Loggerheads? Agricultural Expansion, Poverty Reduction, and Environment in the Tropical Forests," *Data and Research: Latin American and Caribbean,* 2007, http://go.worldbank.org/15945JIFX0.

105. William F. Laurance, Smithsonian Tropical Research Institute, Republic of Panama, personal communication with the author, August 26, 2007.

9. Magic Pills

1. *The Economist*, "Up from the Dead: Genetically Modified Foods Keep On Growing," May 4, 2006, http://www.economist.com/business/displaystory.cfm?story_id=E1_GRJDDVG.

2. Ian Sample, "Farmer quits GM trial after phone threats" *The Guardian*, December 16, 2006, http://www.guardian.co.uk/gmdebate/Story/0,,1973379,00.html.

3. Dick Taverne, "How Science Can Save the World's Poor: The Huge Benefits of GM Are Being Blocked by Blind Opposition," *Guardian*, March 3, 2004, http://www.guardian.co.uk/comment/story/0,,1160661,00.html.

4. Mike Phillips, interview with the author, May 17, 2007.

5. U.S. Supreme Court, *Diamond v. Chakrabarty*, 447 U.S. 303 (1980), http://caselaw.lp.findlaw.com/scripts/getcase.pl?navby=CASE&court=US&vol=447&page=303.

6. John Fetrow, "Economics of Recombinant Bovine Somatotropin on U.S. Dairy Farms," *AgBioForum* (1999): 103–10, http://www.agbioforum.org/v2n2/v2n2a07-fetrow.htm.

7. Carey Gillam, "Biotech Crops Mark First Decade with Wins, Losses," Food-news.com, http://list.web.net/archives/food-news/2006-January/000010.html.

8. G. Brookes and P. Barfoot, "GM Crops: The First Ten Years — Global Socio-Economic and Environmental Impacts," October 2006, pg. 11, http://www.pgeconomics.co.uk/pdf/global_impactstudy_2006_v1_finalPGEconomics.pdf.

9. Jay de Rocher, Targeted Growth's scientific director. Telephone interview with the author, March 23, 2007.

10. Theodore Crosbie, personal communication with the author, March 23, 2007.

11. Mark Arax and Jeanne Brokaw, "No Way Around Roundup: Monsanto's Bioengineered Seeds Are Designed to Require More of the Company's Herbicide," *Mother Jones* (January/February 1997), http://www.motherjones.com/news/feature/1997/01/brokaw.html.

12. Haley Stein, "Intellectual Property and Genetically Modified Seeds: The United States, Trade, and the Developing World," *Northwest Journal of Technology and Intellectual Property* (Spring 2005), http://www.law.northwestern.edu/journals/njtip/v3/n2/4/.

13. *Economist*, "The Green Gene Giant," April 24, 1997, http://www.economist.com/business/displaystory.cfm?story_id=E1_TQSJDG.

14. David Stipp, "Is Monsanto's Biotech Worth Less Than a Hill of Beans?" *Fortune* 141, no. 4 (February 21, 2000).

15. *The Economist*, "Up from the Dead."

16. John Ikerd, *A Return to Common Sense* (Flourtown, PA: R. T. Edwards, 2007).

17. David Holmgren, *Permaculture: Principles and Pathways Beyond Sustainability* (Holmgren Design Services, 2002).

18. Peter Hoffman, "Consumers Need to Know Truth About Organic Food," *Minneapolis Star Tribune,* March 26, 1998, http://www.hbci.com/~wenonah/new/organrul.htm.

19. Ibid.

20. Personal communication with K. Delate, May 30, 2007; Ray Hansen, "Organic Soy Profile," Iowa State University, February 2007, http://www.agmrc.org/agmrc/commodity/grainsoilseeds/soy/organicsoyprofile.htm.

21. David Swenson, Iowa State University, personal communication, August 30, 2007.

22. Testimony of Lynn Clarkson before the U.S. House of Representatives' Agriculture Committee's Subcommittee on Horticulture and Organic Agriculture, April 18, 2007, http://agriculture.house.gov/testimony/110/h70418/LClarkson.doc.

23. Richard Reynolds, "*Mother Jones* Releases USDA Memo Detailing Plans to Gut NOSB Recommendations on Organic Standards" *Mother Jones Magazine* press release, March 12, 1998, at http://www.purefood.org/Organic/usdaLeak.html.

24. Nancy Creamer, "CDC Has Never Compared *E. Coli* Risks of Organic, Traditional Food," press release, North Carolina State University Extension, Horticultural Science, February 1999, http://ipm.ncsu.edu/vegetables/veginews/veginw14.htm.

25. Susan Q. Stranahan, "Monsanto vs. the Milkman: A Maine dairy fights for the right to wear its hormone-free label," *Mother Jones* (January/February 2004), http://www.motherjones.com/news/outfront/2004/01/12_401.html?welcome=true.

26. Peter Aldhous, "More Heat than Light," *Nature* 420, December 19, 2002.

27. V. Prescott et al., "Transgenic Expression of Bean Amylase Inhibitor in Peas Results in Altered Structure and Immunogenicity," *Journal of Agricultural and Food Chemistry* 53 (2005): 9023–30.

28. The Institute of Medicine and the National Research Council of the National Academies, "Safety of Genetically Engineered Foods: Approaches to Assessing Unintended Health Effects," 2003, National Academies Press, Washington, in Innovest Strategic Advisors, "Monsanto and Genetic Engineering: Risks for Investors," Innovest Group, January 2005.

29. Laura M. Tarantino, "Biotechnology Consultation Agency Response Letter BNF No. 000090," CFSAN/Office of Food Additive Safety, Food and Drug Administration, http://www.cfsan.fda.gov/%7Erdb/bnfl090.html.

30. Innovest Strategic Advisors, "Monsanto and Genetic Engineering."

31. Ibid.

32. Pew et al., cited in ibid.

33. Brookes and Barfoot, "GM Crops: The First Ten Years."

34. Dan Charles, "Farmers Switch Course in Battle Against Weeds," Morning

Edition, National Public Radio, August 20, 2007, http://www.npr.org/templates/story/story.php?storyId=13746169; *Crop Protection Monthly,* "NEW Website Promotes Management of Resistance," *Crop Protection Monthly* 190 (September 30, 2005): 8.

35. Chuck Foresman, Syngenta pesticide blog, http://blog.syngenta-us.com/blogs/ask_the_expert/default.aspx?p=2; Mellon et al., "Environmental Effects of Genetically Modified Food Crops: Recent Experiences," Union of Concerned Scientists, http://www.ucsusa.org/food_and_environment/genetic_engineering/environmental-effects-of-genetically-modified-food-crops-recent-experiences.html.

36. Food and Agriculture Organization, "Agricultural Biotechnology: Meeting the Needs of the Poor?" http://www.fao.org/docrep/006/Y5160E/y5160e10.htm#TopOfPage.

37. Chris Webster, "Plant-derived Biologics Meeting — Q&A," Ames, Iowa, April 6, 2000, sponsored by Center for Biologics Evaluation and Research, FDA, http://www.fda.gov/cber/minutes/plnt2040600.pdf.

38. Brookes and Barfoot, "GM Crops: The First Ten Years," 11.

39. David Barboza, "The Power of Roundup; A Weed Killer Is a Block for Monsanto to Build On," *New York Times,* August 2, 2001, http://query.nytimes.com/gst/fullpage.html?sec=health&res=9C00EED8173CF931A3575BC0A9679C8B6; Anastasia L. Thatcher, "Continued Losses Put Pressure on Monsanto Product Launch," Senior Business Analyst, UHG New York, http://www.isb.vt.edu/articles/nov0405.htm.

40. Bob Hirschfeld, "Good Breeding," RegisteredRep.com, April 1, 2006, http://www.registeredrep.com/mag/finance_good_breeding/.

41. See ETC Group at http://gristmill.grist.org/story/2007/4/27/164811/882.

42. B. Lambrecht, *Dinner at the New Gene Café* (New York: Thomas Dunne, 2001), 113.

43. ETC Group at http://gristmill.grist.org/story/2007/4/27/164811/882.

44. Personal communication with the author, March 23, 2007.

45. Food and Agriculture Organization, "Agricultural Biotechnology."

46. Food and Agriculture Organization, "FAO Warns of 'Molecular Divide' between North and South," Rome, February 18, 2003, http://www.fao.org/english/newsroom/news/2003/13960-en.html.

47. Mark Tatge, "Piracy on the High Plains," *Forbes* 2000, April 12, 2004, http://www.forbes.com/forbes/2004/0412/135_print.html.

48. Food and Agriculture Organization, "FAO Warns of 'Molecular.'"

49. T. Crews et al., "Legume versus Fertilizer Sources of Nitrogen: Ecological Tradeoffs and Human Needs," *Agriculture, Ecosystems, and Environment* 102 (2004): 279–97.

50. S. Hill, "Taking Appropriate Next Steps to Progressive Change: Building on the Past and Risking Deep Transformation Towards More Sustainable Communities," revised keynote address from the APEN International Con-

ference, March 2006, http://www.regional.org.au/au/apen/2006/keynote/4003_hills.htm.

51. Vaclav Smil, *Feeding the World: A Challenge for the Twenty-first Century* (Cambridge, MA: MIT Press, 2000), 47.

10. FOOD FIGHT

1. For reports of the vandalism, see Valerie Elliott, "Animal activists free 15,000 farmed fish to their deaths," *London Times*, September 20, 2006, http://www.timesonline.co.uk/tol/news/uk/article644707.ece, and Raymond Hainey, "Fish Farms on Alert after Activists Release Thousands of Halibut," *The Scotsman*, September 18, 2006, http://business.scotsman.com/topics .cfm?tid=1080&id=1376852006.

2. *Archangel* magazine archives, 2006, http://www.arkangelweb.org/international/uk/20060915alfattackfishfarm.php.

3. Elliott, "Animal Activists."

4. Pisces, "Commercial Fishing, Fish Farming, and Fish Eating," http://www.pisces.demon.co.uk/factshe6.html.

5. Rebecca Goldburg, "Future Seascapes, Fishing, and Fish Farming," *Frontiers in Ecology*, 2005, http://www.frontiersinecology.org/specialissue/articles/Goldburg.pdf.

6. R. Goldburg et al., "Marine Aquaculture in the United States: Environmental Impacts and Policy," Pew Oceans Commission, 2001, http://www.pewtrusts.org/pdf/env_pew_oceans_aquaculture.pdf.

7. Shannon Keith, *Behind the Mask*, Uncaged Films, 2006, cited in Wikipedia, Animal Liberation Front (ALF), http://en.wikipedia.org/wiki/Animal_Liberation_Front.

8. T. Barrell, "Japan: The Last Big Paddy," ABC Radio National, December 31, 2006, http://www.abc.net.au/rn/streetstories/stories/2006/1798515.htm#transcript.

9. Ibid.

10. Scott Vlaun, interview with Bill Mollison, *Seeds of Change* Newsletter, 25, February 2002, http://www.seedsofchange.com/enewsletter/issue_25/issue_25.asp.

11. S. Setboonsarng and J. Gilman, "Alternative Agriculture in Thailand and Japan," Horizon Solutions site, April 28, 2003, http://www.solutions-site.org/artman/publish/article_15.shtml.

12. John Dyck, U.S. Economic Research Service, personal communication with the author, June 9, 2007.

13. G. Robertson et al., "Greenhouse Gases in Intensive Agriculture," *Science* 289, 1900–22.

14. Swenson, personal communication with the author.

15. U.S. Department of Agriculture, "American Farms," *Agriculture Fact Book,* 2001–2002, http://www.usda.gov/factbook/chapter3.htm.

16. John Ikerd, "Why Do Small Farmers Farm?" *Small Farm Today* magazine (September–October 2003), archived at http://web.missouri.edu/~ikerdj//papers/SFT-WhyFarm.htm.

17. USDA, *Agriculture Fact Book, American Farms, 2001–2002.*

18. Allison Linn, "Grocer merger raises unique legal questions: Will Whole Foods, Wild Oats merger hurt consumers, or help them?" *MSNBC,* June 12, 2007, http://www.msnbc.msn.com/id/19120095/.

19. *The Economist,* "Voting with your trolley," December 7, 2006, http://www.economist.com/business/displaystory.cfm?story_id=8380592.

20. International Society for Ecology and Culture, "Local Food Bringing the Food Economy Home," group website, http://www.isec.org.uk/pages/localfood.html.

21. Wal-Mart, "Maine's Ricker Hill Farm Teams with Wal-Mart to Celebrate America's Farmers," press release, November 6, 2006, http://www.walmartfacts.com/articles/4594.aspx.

22. The Rural Institute, "Ruralfacts: Update on the Demography of Rural Disability, Part One: Rural and Urban," http://rtc.ruralinstitute.umt.edu/RuDis/DescribeFigure2.htm.

23. James Randerson, "Focus on Distance Is Too Narrow, Say Researchers," *Guardian,* June 4, 2007, http://www.guardian.co.uk/uk_news/story/0,,2094651,00.html.

24. Ibid.

25. Caroline Saunders et al., "Food Miles: Comparative Energy/Emissions Performance of New Zealand's Agriculture Industry," Research Report No. 285, University in New Zealand, Christchurch, July 2006, http://www.lincoln.ac.nz/story_images/2328_RR285_s6508.pdf.

26. See http://www.truecostoffood.org/leaders.asp.

27. Ricardo Salvador, Kellogg Foundation, "Food Systems and Rural Development," press release, March 2007, http://ola.wkkf.org/fasupdate/2007/March/first.htm.

28. Pallavi Gogoi, "Organics: A Poor Harvest for Wal-Mart," *BusinessWeek,* April 12, 2007, http://www.businessweek.com/bwdaily/dnflash/content/apr2007/db20070412_005673.htm.

29. This quote, from an interview conducted by the author with Amy Schaefer, first appeared in an article by the author in *Seattle Metropolitan* magazine, April 2006.

30. T. Crews, "Legume versus fertilizer source of nitrogen."

31. Economic Research Service, "Food Consumption," USDA, May 25, 2007, http://www.ers.usda.gov/Briefing/Consumption/; Smil, *Feeding the World,* 230.

32. Economic Research Service, "Food Consumption"; Smil, *Feeding the World,* 230.

33. A. Collins and R. Fairchild, "Sustainable Food Consumption at a Sub-national Level: An Ecological Footprint, Nutritional and Economic Analysis," *Journal of Environmental Policy & Planning,* Vol. 9, No. 1, March 2007, 5–30.

34. George Lardner Jr. and Joby Warrick, "Pesticide Coalition's Text Enters House Bill; Industry, Farmers Trying to Blunt U.S. Regulation," *Washington Post.* May 13, 1990

35. See http://www.time.com/time/magazine/article/0,9171,951174,00.html.

36. See http://www.opensecrets.org/bush/cabinet/cabinet.gutierrez.asp.

37. Mindfully, *Rufus Yerxa Biography,* September 14, 2002, http://www.mindfully.org/WTO/Rufus-Yerxa-Trading-Places20aug02.htm.

38. U.S. Environmental Protection Agency, "U.S. Senate Confirms Linda J. Fisher as EPA Deputy Administrator," EPA press release, May 25, 2001, http://yosemite1.epa.gov/opa/admpress.nsf/b1ab9f485b098972852562e7004dc686/989209589ae67bc985256a570074ceb1?OpenDocument.

39. The Revolving Door, http://www.mindfully.org/GE/Revolving-Door.htm.

40. Oppenheimer World Bond Fund, *SEC Filing Information: Clayton K. Yeutter — Definitive Proxy Solicitation Material, Schedule 14A,* March 28, 1995, http://www.secinfo.com/dpwBe.a9.htm.

41. SourceWatch, "Charles F. Conner," Center for Media & Democracy, November 19, 2005, http://www.sourcewatch.org/index.php?title=Charles_ F._Conner. *New York Times,* "Question Is Raised on Hormone Maker's Ties to F.D.A. Aides," http://query.nytimes.com/gst/fullpage.html?res=9F02EED91731F93BA25757C0A962958260&scp=1&sq=Michael+Taylor+monsanto&st=nyt.

42. See http://weblog.greenpeace.org/ge/archives/US_ConsumerSurveyResults.pdf.

43. Andrew Martin, "Labels Lack Food's Origin Despite Law," *New York Times,* July 2, 2007, http://www.nytimes.com/2007/07/02/business/02label.html?scp=2&sq=country+of+origin+labeling&st=nyt.

44. *FoodNavigatorUSA,* "USDA Sets Date for Meat Labeling Changes," June 3, 2007, http://www.foodnavigator-usa.com/news-by-product/news.asp?id=74713&idCat=54&k=FSIS — meat-poultry-labeling.

45. Barry Lynn, "Breaking the Chain: The Antitrust Case against Wal-Mart," *Harper's Magazine,* July 2006, 35.

46. Ibid.

47. See Edward Alden, "U.S. Retaliation Against Egypt Hits Trade Plans," *Financial Times,* June 29, 2003, http://search.ft.com/ftArticle?queryText=US+retaliation+against+Egypt+hits+trade+plans&y=2&aje=true&x=11&id=030629001976&ct=0.

48. See R. Weiss, "U.S. May Allow Chinese Chicken Imports" and "China Has Sent U.S. Tainted Food for Years," *Washington Post,* May 22, 2007.

49. See http://www.energybulletin.net/31910.html.

50. See http://farm.ewg.org/sites/farmbill2007/cdlist.php.

51. Associated Press, "Side Deals Smooth Over Farm Bill Bumps," July 26, 2007,

New York Times, http://www.nytimes.com/aponline/us/AP-Farm-Bill.html?_
r=1&oref=slogin.

Epilogue: Nouvelle Cuisine

1. See http://www.brookings.edu/views/papers/mckibbin/200602.pdf.
2. *National Review,* "For the Birds or Sleeping with the Fishes? What to Make of the Avian-Flu Threat," March 14, 2006, http://www.nationalreview.com/script/printpage.p?ref=/symposium/symposium200603140821.asp.
3. See http://avianfluinvestor.blogspot.com/2006_02_01_archive.html.
4. Finfacts, "World Wheat Production in Peril from Stem Rust Outbreak in East Africa, Expert Panel Warns," September 8, 2005, http://www.finfacts.com/irelandbusinessnews/publish/article_10003193.shtml.
5. John Reilly et al., "Climate Change and Agriculture in the United States," *Potential Consequences of Climate Variability,* the National Assessment Synthesis Team, U.S. Global Change Research Program, 2000, http://www.usgcrp.gov/usgcrp/Library/nationalassessment/13Agri.pdf.
6. Vernon W. Ruttan, "The Transition to Agricultural Sustainability," paper presented at the National Academy of Sciences colloquium "Plants and Population: Is There Time?" on December 5–6, 1998, at the Arnold and Mabel Beckman Center in Irvine, California, and later published in the *Proceedings of the National Academy of Sciences* 96, no. 11 (May 25, 1999): 5960–67, http://www.pnas.org/cgi/content/full/96/11/5960.
7. See http://www.foodfirst.org/pubs/devreps/dr14.html.
8. See http://earthtrends.wri.org/pdf_library/country_profiles/agr_cou_332.pdf; http://jpe.library.arizona.edu/volume_9/baro02.pdf.
9. Peter Rosset, "The Greening of Cuba," *Earth Island Journal* 10, no. 1 (Winter 1994): 23, http://www.earthisland.org/journal/cuba.html.
10. See http://www.harpers.org/archive/2005/04/0080501.
11. Lester Brown, *Plan B 2.0* (New York: W. W. Norton, 2006).
12. M. Shpigel et al., "The Use of Effluent Water from Fish Ponds as a Food Source for the Pacific Oyster *Crassostrea gigas Tunberg,*" *Aquaculture and Fisheries Management* 24, 529–43, http://en.wikipedia.org/wiki/Integrated_Multi-Trophic_Aquaculture.
13. Carlos Duarte et al., "Rapid Domestication of Marine Species," *Science* 316 (5823): 382, http://www.sciencemag.org/cgi/content/full/316/5823/382.
14. See http://avianfluinvestor.blogspot.com/2006/03/avian-flu-its-your-fault.html.
15. See http://newstandardnews.net/content/index.cfm/items/3956/printmode/true.
16. See http://www.hhs.gov/news/speech/2006/060608.html.
17. John Thackara, *In the Bubble: Designing in a Complex World* (Cambridge, MA: MIT Press, 2005), 5.

Bibliography

Barkema, Alan, et al. "The New U.S. Meat Industry." Federal Reserve Bank of Kansas City, 2001.

Becker, Jasper. *Hungry Ghosts: Mao's Secret Famine.* New York: Holt, 1996.

Bergsten, C. Fred, et al. *China: The Balance Sheet.* New York: PublicAffairs, 2006.

Brown, Lester. *Plan B 2.0.* New York: W. W. Norton, 2006.

——— . *Who Will Feed China?* New York: W. W. Norton, 1995.

Burnett, John. *Plenty and Want.* Middlesex, UK: Penguin, 1966.

Braudel, Fernand. *Civilization and Capitalism, 15th–18th Century,* Vol. I: *The Structure of Everyday Life.* Berkeley: University of California Press, 1992.

California Food Emergency Response Team. "Investigation of an *E. coli* O157:H7 Outbreak Associated with Dole Pre-Packaged Spinach," Final. March 21, 2007. Department of Health Services, U.S. Food and Drug Administration, Sacramento, CA. http://www.dhs.ca.gov/ps/fdb/local/PDF/2006%20Spinach%20Report%20Final%20redacted.PDF.

Cohen, Mark. *The Food Crisis in Prehistory.* New Haven, CT: Yale University Press, 1977.

——— . *Health and the Rise of Civilization.* New Haven, CT: Yale University Press, 1991.

Connor, John, et al. *The Food Manufacturing Industries: Structure, Strategies, Performance, and Policies.* Lexington, MA: Lexington Books, 1985.

——— . *Food Processing: An Industrial Powerhouse in Transition.* Lexington, MA: Lexington Books, 1988.

Cordain, Loren. *The Paleo Diet.* Hoboken, NJ: John Wiley and Sons, 2002.

Critser, Greg. *Fat Land: How Americans Became the Fattest People in the World.* Boston: Houghton Mifflin, 2003.

Diamond, Jared. *Collapse.* New York: Viking, 2005.

Dyson, Lowell. "American Cuisine in the Twentieth Century." *Food Review* 23, no. 1 (January–April 2000): 2–7.

Fogel, Robert. "New Findings about Trends in Life Expectation and Chronic Disease." Graduate School of Business Selected Paper Series, no. 76. Chicago: University of Chicago, 1996.

Gibbons, Ann. "Solving the Brain's Energy Crisis." *Science* 280 (May 1998): 1345–47.

Greig, W. Smith, et al. *Economics and Management of Food Processing.* Westport, CT: AVI Publishing, 1984.

Gustafson, R. H., and R. E. Bowen. "Antibiotic Use in Animal Agriculture." Agricultural Research Division, American Cyanamid Co., Princeton, NJ.

Guthman, Julie. *Agrarian Dreams: The Paradox of Organic Farming in California.* Berkeley: University of California Press, 2004.

Heer, Jean. *Nestlé: 125 Years — 1866–1991.* Vevey, Switzerland: Nestlé, 1991.

Hallberg, Milton. *Economic Trends in U.S. Agriculture and Food Systems Since World War II.* Ames: Iowa State University Press, 2001.

Jukes, T. "Antibiotics in Animal Feeds and Animal Production." *Journal of BioScience* 22: 526–34.

Kahn, Barbara E., and Leigh M. McAlister. *Grocery Revolution: The New Focus on the Consumer.* Reading, MA: Addison-Wesley, 1997.

Kiple, Kenneth, and Kriemhild Ornelas, eds. *The Cambridge World History of Food.* Cambridge, UK: Cambridge University Press, 2000.

Kloppenburg, Jack. *First the Seed.* Cambridge, UK: University of Cambridge, 1988.

Kneen, Brewster. *Invisible Giant: Cargill and Its Transnational Strategies.* London: Pluto Press, 1995.

Lauden, Rachel. "A Plea for Culinary Modernism." *Gastronomica* 1, no. 1 (February 2001): 36–44.

Levenstein, Harvey. *Paradox of Plenty: A Social History of Modern Eating.* Oxford, UK: Oxford University Press, 1993.

——— . *Revolution at the Table: The Transformation of the American Diet.* Oxford, UK: Oxford University Press, 1988.

Magdoff, Fred, et al., eds. *Hungry for Profit.* New York: Monthly Review Press, 2000.

Mamen, Katy. "Current Issues and Trends Connected to the Vivid Picture Goals for a Sustainable Food System." Report by the Vivid Picture Project 2004.

Morgan, Dan. *Merchants of Grain.* Lincoln: iUniverse, 2000.

National Research Council. *Alternative Agriculture.* A report for the National Academy of Sciences by the Committee on the Role of Alternative Farming Methods in Modern Production Agriculture. Washington, DC: National Academies Press, 1989.

Nestle, Marion. *Food Politics.* Berkeley: University of California Press, 2002.

——— . *Safe Food: Bacteria, Biotechnology, and Bioterrorism.* Berkeley: University of California Press, 2003.

Orden, David, Robert Paarlberg, and Terry Roe. *Policy Reform in American Agriculture: Analysis and Prognosis.* Chicago: University of Chicago Press, 1999.

Overton, Mark. *Agricultural Revolution in England.* Cambridge, UK: University of Cambridge Press, 1996.

Perkins, John H. *Geopolitics and the Green Revolution: Wheat, Genes, and the Cold War.* New York: Oxford University press, 1997.

Pinstrup-Andersen, Per, and Ebbe Schiøler. *Seeds of Contention: World Hunger and the Global Controversy over GM Crops.* Baltimore: Johns Hopkins University Press, 2000.

Pollan, Michael. *The Omnivore's Dilemma.* New York: Penguin Press, 2006.

Revel, Alain. *American Green Power.* Baltimore: Johns Hopkins University Press, 1981.

Rosegrant, Mark, et al. "Global Water Outlook to 2025: Averting an Impending Crisis." International Food Policy Research Institute. Washington, DC, September 2002.

Shouying, Lui, and Luo Dan, eds. *Can China Feed Itself? Chinese Scholars on China's Food Issue.* Beijing: Foreign Languages Press, 2004.

Smil, Vaclav. *Feeding the World: A Challenge for the Twenty-first Century.* Cambridge, MA: MIT Press, 2000.

Tannahill, Reay. *Food in History.* New York: Three Rivers Press, 1995 (revised edition).

Thackara, John. *In the Bubble: Designing in a Complex World.* Cambridge, MA: MIT Press, 2005.

Trager, James. *The Great Grain Robbery.* New York: Ballantine Books, 1975.

Watson, James, ed. *Golden Arches East: McDonald's in East Asia.* Stanford, CA: Stanford University Press, 1997.

Index